ADVANCES IN CHEMICAL ENGINEERING
Volume 21

Intelligent Systems in Process Engineering

Part I: Paradigms from Product and Process Design

ADVANCES IN
CHEMICAL ENGINEERING

Series Editors

JAMES WEI
School of Engineering and Applied Science
Princeton University
Princeton, New Jersey

JOHN L. ANDERSON
Department of Chemical Engineering
Carnegie Mellon University
Pittsburgh, Pennsylvania

KENNETH B. BISCHOFF
Department of Chemical Engineering
University of Delaware
Newark, Delaware

MORTON M. DENN
College of Chemistry
University of California at Berkeley
Berkeley, California

JOHN H. SEINFELD
Department of Chemical Engineering
California Institute of Technology
Pasadena, California

GEORGE STEPHANOPOULOS
Department of Chemical Engineering
Massachusetts Institute of Technology
Cambridge, Massachusetts

ADVANCES IN CHEMICAL ENGINEERING
Volume 21

Intelligent Systems in Process Engineering
Part I: Paradigms from Product and Process Design

Edited by

GEORGE STEPHANOPOULOS
CHONGHUN HAN

*Laboratory for Intelligent Systems in Process Engineering
Department of Chemical Engineering
Massachusetts Institute of Technology
Cambridge, Massachusetts*

ACADEMIC PRESS
San Diego   New York   Boston   London   Sydney   Tokyo   Toronto

This book is printed on acid-free paper. ∞

Copyright © 1995 by ACADEMIC PRESS, INC.

All Rights Reserved.
No part of this publication may be reproduced or transmitted in any form or by any means, electronic or mechanical, including photocopy, recording, or any information storage and retrieval system, without permission in writing from the publisher.

Academic Press, Inc.
A Division of Harcourt Brace & Company
525 B Street, Suite 1900, San Diego, California 92101-4495

United Kingdom Edition published by
Academic Press Limited
24-28 Oval Road, London NW1 7DX

International Standard Serial Number: 0065-2377

International Standard Book Number: 0-12-008521-6

PRINTED IN THE UNITED STATES OF AMERICA
95 96 97 98 99 00 QW 9 8 7 6 5 4 3 2 1

"All virtue is one thing: Knowledge"
        PLATO

However,

"The chance of the quantum theoretician is not the ethical freedom of the Augustinian"
               NORBERT WIENER

George Stephanopoulos dedicates
this editorial work to
Eleni-Nikos-Elvie
with love and gratitude

Chonghun Han dedicates
this editorial work to
Jisook and *Albert*

CONTENTS VOLUME 21

CONTRIBUTORS TO VOLUME 21 xi
PROLOGUE xix

Modeling Languages: Declarative and Imperative Descriptions of Chemical Reactions and Processing Systems
CHRISTOPHER J. NAGEL, CHONGHUN HAN, AND GEORGE STEPHANOPOULOS

I. Introduction 2
   A. The Five Premises of a Modeling System 3
   B. Review of Modeling Systems for Process Simulation 7
   C. Modeling Systems in Chemistry 10
II. LCR: A Language for Chemical Reactivity 13
   A. Modeling Elements of LCR 13
   B. Semantic Relations among Modeling Elements in LCR 26
   C. Syntax of LCR 33
III. Formal Construction of Representations for Chemicals and Reactions . . . 36
   A. Extension of LCR's Modeling Objects 36
   B. The "Model-Class Decomposition Digraph" (MCDD) 50
   C. Generation and Representation of Reaction Pathways 53
   D. Creation of Contextual Reaction Models 58
   E. Case Study: Ethane Pyrolysis 64
IV. MODEL.LA.: A Modeling Language for Process Engineering . . . 73
   A. Basic Modeling Elements 73
   B. Semantic Relationships 75
   C. Hierarchies of Modeling Subclasses 76
   D. Syntax 78
V. Phenomena-Based Modeling of Processing Systems 78
   A. The "Chemical Engineering Science" Hierarchies of Modeling Elements . . . 79
   B. Formal Construction of Models 82
   C. Multifaceted Modeling of Processing Systems 82
   D. Computer-Aided Implementation of MODEL.LA. 87
   References 90

Automation in Design: The Conceptual Synthesis of Chemical Processing Schemes
CHONGHUN HAN, GEORGE STEPHANOPOULOS, AND JAMES M. DOUGLAS

I. Introduction 94
   A. Conceptual Design of Chemical Processing Schemes 96
   B. Issues in the Automation of Conceptual Process Design . . . 98

II. Hierarchical Approach to the Synthesis of Chemical Processing Schemes:
   A Computational Model of the Engineering Methodology 103
   A. Hierarchical Planning of the Process Design Evolution 104
   B. Goal Structures: Bridging the Gap between Design Milestones 107
   C. Design Principles of the Computational Model 117
III. HDL: The Hierarchical Design Language 122
   A. Multifaceted Modeling of the Process Design State 123
   B. Modeling the Design Tasks 129
   C. Elements for Human–Machine Interaction 134
   D. Object-Oriented Failure Handling 138
   E. Management of Design Alternatives 138
IV. Concept Designer: The Software Implementation 139
   A. Overall Architecture 139
   B. Implementation Details 143
V. Summary 144
   References 145

Symbolic and Quantitative Reasoning: Design of Reaction Pathways through Recursive Satisfaction of Constraints

MICHAEL L. MAVROVOUNIOTIS

I. Reaction Systems and Pathways 148
II. Catalytic Reaction Systems 151
   A. Basic Concepts, Terminology, and Notation 151
   B. Previous Work on the Construction of Mechanisms 154
   C. Structure of the Algorithm 155
   D. Features of the Algorithm 159
   E. Examples 160
III. Biochemical Pathways 169
   A. Features of the Pathway Synthesis Problem 173
   B. Formulation of Constraints 175
   C. Algorithm 176
   D. Examples 179
IV. Properties and Extensions of the Synthesis Algorithm 183
V. Summary 185
   References 185

Inductive and Deductive Reasoning: The Case of Identifying Potential Hazards in Chemical Processes

CHRISTOPHER NAGEL AND GEORGE STEPHANOPOULOS

I. Introduction 188
   A. Predictive Hazard Analysis 190
   B. Incompleteness of Conventional Hazard Analysis Methodologies 192
   C. Premises of Traditional Approaches 193
   D. Overview of Proposed Methodology 194
II. Reaction-Based Hazards Identification 195
   A. System Foundations 196
   B. Modeling Languages and Their Role in Hazards Identification 198

	C. Generations of Reactions and Evaluation of Thermodynamic States	205
III.	Inductive Identification of Reaction-Based Hazards	209
	A. Hazards Identification Algorithm	211
	B. Properties of Reaction-Based Hazards Identification	214
	C. An Example in Reaction-Based Hazard Identification: Aniline Production	217
IV.	Deductive Determination of the Causes of Hazards	221
	A. Methodological Framework	222
	B. Variables as "Causes" or "Effects"	225
	C. Construction of Variable-Influence Diagrams	227
	D. Characterization of Variable-Influence Pathways	232
	E. Assessment of Hazards-Preventive Mechanisms	235
	F. Fault-Tree Construction	238
	G. An Example of Reaction-Based Hazard Identification: Reaction Quench	241
V.	Conclusion	253
	References	254

Searching Spaces of Discrete Solutions:
The Design of Molecules Possessing Desired Physical Properties

KEVIN G. JOBACK AND GEORGE STEPHANOPOULOS

I.	Introduction	258
	A. Brief Review of Previous Work	260
	B. General Framework for the Design of Molecules	264
II.	Automatic Synthesis of New Molecules	267
	A. The Generate-and-Test Paradigm	267
	B. The Search Algorithm	271
	C. Case Study: Automatic Design of Refrigerants	283
	D. Case Study: Automatic Design of Polymers as Packaging Materials	284
III.	Interactive Synthesis of New Molecules	290
	A. Illustration of Interactive Design	291
	B. Case Study: Interactive Design of Refrigerants	296
	C. Case Study: Interactive Design of an Extraction Solvent	299
	D. Case Study: Interactive Design of a Pharmaceutical	301
IV.	The *Molecule-Designer* Software System	304
	A. General Description	304
	B. Interactive-Design-Relevant Sections	305
V.	Concluding Remarks	307
	References	309

COMBINED INDEX APPEARS AT THE END OF VOLUME 22 611
CONTENTS OF VOLUMES IN THIS SERIAL 621

CONTRIBUTORS VOLUME 21

Numbers in parentheses indicate the pages on which the authors' contributions begin.

JAMES M. DOUGLAS, *Department of Chemical Engineering, University of Massachusetts, Amherst, Massachusetts 01003* (43)

CHONGHUN HAN, *Laboratory for Intelligent Systems in Process Engineering, Department of Chemical Engineering, Massachusetts Institute of Technology, Cambridge, Massachusetts 02139* (1, 43)

KEVIN G. JOBACK, , *Molecular Knowledge Systems, Inc., Nashua, New Hampshire 03063* (257)

MICHAEL L. MAVROVOUNIOTIS, *Department of Chemical Engineering, Northwestern University, Evanston, Illinois 60208* (147)

CHRISTOPHER J. NAGEL, *Molten Metal Technology, Inc., Waltham, Massachusetts 02154* (1, 187)

GEORGE STEPHANOPOULOS, *Laboratory for Intelligent Systems in Process Engineering, Department of Chemical Engineering, Massachusetts Institute of Technology, Cambridge, Massachusetts 02139* (1, 43, 187, 257)

CONTENTS VOLUME 22

CONTRIBUTORS TO VOLUME 22 xvii
PROLOGUE xix

Nonmonotonic Reasoning:
The Synthesis of Operating Procedures in Chemical Plants
CHONGHUN HAN, RAMACHANDRAN LAKSHMANAN, BHAVIK BAKSHI,
AND GEORGE STEPHANOPOULOS

I. Introduction 314
   A. Previous Approaches to the Synthesis of Operating Procedures 316
   B. The Components of a Planning Methodology 318
   C. Overview of the Chapter's Structure 324
II. Hierarchical Modeling of Processes and Operations 324
   A. Modeling of Operations 325
   B. Modeling of Process Behavior 329
III. Nonmonotonic Planning 334
   A. Operator Models and Complexity of Nonmonotonic Planning 336
   B. Handling Constraints on the Temporal Ordering of Operational Goals 337
   C. Handling Constraints on the Mixing of Chemicals 339
   D. Handling Quantitative Constraints 343
   E. Summary of Approach for Synthesis of Operating Procedures 348
IV. Illustrations of Modeling and Nonmonotonic Operations Planning 351
   A. Construction of Hierarchical Models and Definition of Operating States . . . 351
   B. Nonmonotonic Synthesis of a Switchover Procedure 359
V. Revamping Process Designs to Ensure Feasibility of Operating Procedures . . . 368
   A. Algorithms for Generating Design Modifications 369
VI. Summary and Conclusions 374
   References 375

Inductive and Analogic Learning:
Data-Driven Improvement of Process Operations
PEDRO M. SARAIVA

I. Introduction 378
II. General Problem Statement and Scope of the Learning Task 381
III. A Generic Framework to Describe Learning Procedures 384
   A. A Generic Formalism 385
   B. Major Departures from Previous Approaches 385
IV. Learning with Categorical Performance Metrics 389
   A. Problem Statement 389
   B. Search Procedure, S 391

	C. Case Study: Operating Strategies for Desired Octane Number	394
V.	Continuous Performance Metrics	396
	A. Problem Statement	396
	B. Alternative Problem Statements and Solutions	398
	C. Taguchi Loss Functions as Continuous Quality Cost Models	401
	D. Learning Methodology and Search Procedure, S	403
	E. Case Study: Pulp Digester	405
VI.	Systems with Multiple Operational Objectives	408
	A. Continuous Performance Variables	409
	B. Categorical Performance Variables	409
	C. Case Study: Operational Analysis of a Plasma Etching Unit	413
VII.	Complex Systems with Internal Structure	417
	A. Problem Statement and Key Features	417
	B. Search Procedures	424
	C. Case Study: Operational Analysis of a Pulp Plant	426
VIII.	Summary and Conclusions	431
	References	432

Empirical Learning through Neural Networks: The Wave-Net Solution

ALEXANDROS KOULOURIS, BHAVIK R. BAKSHI, AND GEORGE STEPHANOPOULOS

I.	Introduction	438
II.	Formulation of the Functional Estimation Problem	441
	A. Mathematical Description	444
	B. Neural Network Solution to the Functional Estimation Problem	449
III.	Solution to the Functional Estimation Problem	451
	A. Formulation of the Learning Problem	451
	B. Learning Algorithm	465
IV.	Applications of the Learning Algorithm	471
	A. Example 1	471
	B. Example 2	474
	C. Example 3	477
V.	Conclusions	479
VI.	Appendices	480
	A. Appendix 1	480
	B. Appendix 2	481
	C. Appendix 3	482
	References	483

Reasoning in Time: Modeling, Analysis, and Pattern Recognition of Temporal Process Trends

BHAVIK R. BAKSHI AND GEORGE STEPHANOPOULOS

I.	Introduction	487
	A. The Content of Process Trends: Local in Time and Multiscale	488
	B. The Ad Hoc Treatment of Process Trends	490

C. Recognition of Temporal Patterns in Process Trends 492
D. Compression of Process Data 493
E. Overview of the Chapter's Structure 494
II. Formal Representation of Process Trends 495
A. The Definition of a Trend 496
B. Trends and Scale-Space Filtering 500
III. Wavelet Decomposition: Extraction of Trends at Multiple Scales 507
A. The Theory of Wavelet Decomposition 508
B. Extraction of Multiscale Temporal Trends 516
IV. Compression of Process Data through Feature Extraction
and Functional Approximation 527
A. Data Compression through Orthonormal Wavelets 527
B. Compression through Feature Extraction 530
C. Practical Issues in Data Compression 530
D. An Illustrative Example 532
V. Recognition of Temporal Patterns for Diagnosis and Control 535
A. Generating Generalized Descriptions of Process Trends 538
B. Inductive Learning through Decision Trees 541
C. Pattern Recognition with Single Input Variable 543
D. Pattern Recognition with Multiple Input Variables 544
VI. Summary and Conclusions 545
References 546

Intelligence in Numerical Computing: Improving Batch Scheduling Algorithms through Explanation-Based Learning

Matthew J. Realff

I. Introduction 550
A. Flowshop Problem 552
B. Characteristics of Solution Methodology 553
II. Formal Description of Branch-and-Bound Framework 555
A. Solution Space Representation—Discrete Decision Process 555
B. The Branch-and-Bound Strategy 557
C. Specification of Branch-and-Bound Algorithm 563
D. Relative Efficiency of Branch-and-Bound Algorithms 564
E. Branching as State Updating 566
F. Flowshop Lower-Bounding Scheme 568
III. The Use of Problem-Solving Experience in Synthesizing
New Control Knowledge 570
A. An Instance of a Flowshop Scheduling Problem 570
B. Definition and Analysis of Problem-Solving Experience 573
C. Logical Analysis of Problem-Solving Experience 578
D. Sufficient Theories for State-Space Formulation 579
IV. Representation 581
A. Representation for Problem Solving 583
B. Representation for Problem Analysis 588
V. Learning 593
A. Explanation-Based Learning 594

       B. Explanation 598
       C. Generalization of Explanations 601
  VI. Conclusions 607
      References 608

INDEX 611
CONTENTS OF VOLUMES IN THIS SERIAL 621

CONTRIBUTORS VOLUME 22

Numbers in parentheses indicate the pages on which the authors' contributions begin.

BHAVIK R. BAKSHI, *Department of Chemical Engineering, Ohio State University, Columbus, Ohio 43210* (313, 347, 485)

CHONGHUN HAN, *Laboratory for Intelligent Systems in Process Engineering, Department of Chemical Engineering, Massachusetts Institute of Technology, Cambridge, Massachusetts 02139* (313)

ALEXANDROS KOULOURIS, *Laboratory for Intelligent Systems in Process Engineering, Department of Chemical Engineering, Massachusetts Institute of Technology, Cambridge, Massachusetts 02139* (437)

RAMACHANDRAN LAKSHAMANAN, *Department of Chemical Engineering, University of Edinburgh, Edinburgh, Scotland, United Kingdom* (313)

MATTHEW J. REALFF, *School of Chemical Engineering, Georgia Institute of Technology, Atlanta, Georgia 30332* (549)

PEDRO M. SARAIVA, *Department of Chemical Engineering, University of Coimbra, 3000 Coimbra, Portugal* (377)

GEORGE STEPHANOPOULOS, *Laboratory for Intelligent Systems in Process Engineering, Department of Chemical Engineering, Massachusetts Institute of Technology, Cambridge, Massachusetts 02139* (437, 485)

PROLOGUE

The adjective "intelligent" in the term "intelligent systems" is a misnomer. No one has ever claimed that an intelligent system in an engineering application possesses the kind of intelligence that allows it to *induce* new knowledge, (1) or "to contemplate its creator, or how it evolved to be the system that it is". (2) Åström and McAvoy (3) have suggested terms such as "knowledgeable" and "informed" to accentuate the fact that these software systems depend on large amounts of (possibly) fragmented and unstructured knowledge. For the purposes of this book, the term "intelligent system" always implies a computer program, and although the quotation marks around the adjective intelligent may be dropped occasionally, no one should perceive it as a computer program with attributes of human-like intelligence. Instead, the reader should interpret the adjective as characterizing a software artifact that possesses a computational procedure, an algorithm, which attempts to "model and emulate," and thus automate an engineering task that used to be carried out *informally* by a human. Whether or not this models the actual cognitive process in a human is beyond the scope of this book.

In the wide spectrum of engineering activities, collectively known as *process engineering* and encompassing tasks from product and process development through process design and optimization to process operations and control, so-called intelligent systems have played an important role. Ten years ago the broad introduction of knowledge-based expert systems created a pop culture that started affecting many facets of process engineering work. Expert systems were followed by their cousins, fuzzy systems, and the explosion in the use of neural networks. During the same period, the object-oriented programming (OOP) paradigm, one of the most successful "products" of artificial intelligence, has led to a revolutionary rethinking of programming practices, so that today OOP is the paradigm of choice in software engineering. After 10 years of work, 15 books/monographs/edited volumes, over 700 identified papers in archival research and professional journals, 65 reviews/tutorial/industrial survey papers, about 150 Ph.D. theses, and several thousand industrial applications worldwide, (4) the area of what is known as "intelligent systems" has turned from fringe to mainstream in a large number of process engineering activities. These include monitoring and analysis of process operations, fault diagnosis, supervisory control, feedback control, scheduling and planning of process operations, simulation, and process and product design. The early emphasis on tools and methodologies, originated by research in artificial intelligence, has given place to more integrative approaches, which focus more on the engineering problem and its characteristics. So, today, one does not encounter as frequently as 10 years ago conference sessions with titles including terms such as "expert systems," "knowledge-based systems," or "artificial intelligence." Instead one sees many more mature contributions, from both the academic and industrial worlds, in mainstream engineering sessions, with significant components of what one would have earlier termed "intelligent systems." The evolving complementarity in the use of approaches from artificial intelligence, systems and control theory, mathematical programming, and statistics is a strong indication of the maturity that the area of intelligent systems is reaching.

A. THE CURRENT SETTING

The explosive growth of academic research and industrial practice in the synthesis, analysis, development, and deployment of intelligent systems is a natural phase in the saga of the Second Industrial Revolution. If the First Industrial Revolution in 18th century England ushered the world into an era characterized by machines that extended, multiplied, and leveraged human *physical capabilities*, the Second, currently in progress, is based on machines that extend, multiply, and leverage human *mental abilities*. (5) The thinking man, *Homo sapiens*, has returned to its Platonic roots where "all virtue is one thing, knowledge." Using the power and versatility of modern computer science and technology, software systems are continuously developed to preserve knowledge for it is perishable, clone it for it is scarce, make it precise for it is often vague, centralize it for it is dispersed, and make it portable for it is difficult to distribute. The implications are staggering and have already manifested themselves, reaching the most remote corners of the earth and the inner sancta of our private lives. In this expanding pervasiveness of computers, intelligent systems can affect and are affecting the way we educate, entertain, and govern ourselves, communicate with each other, overcome physical and mental disabilities, and produce material wealth. Computer-based deployment of "knowledge" has been thrust by modern sociologists into the center of our culture as the force most effective in resolving inequities in the distribution of biological, historical and material inheritance. But what is the tangible evidence? Software systems have been composed to do the following: (5–7) (i) harmonize chorales in the style of Johann Sebastian Bach and automate musical compositions into new territories; (ii) write original stanzas and poems with thematic uniformity, which could pass as human creations for about half of the polled readers; (iii) compose original drawings and "photographs" of nonexistent worlds; (iv) "author" complete books.

Equally impressive are the results in engineering and science. Characterized as "knowledgeable," "informed," "expert," "intelligent," or any other denotation, software systems have expanded tremendously the scope of automation in scientific and engineering activities. (4, 8–13)

B. THE THEORETICAL SCOPE AND LIMITATIONS OF INTELLIGENT SYSTEMS

So what? a skeptic may ask. Are the above examples manifestations of the computer's long-awaited, human-like intelligence? No one familiar with Gödel's theorem of incompleteness would ask such a question, (14,15) for this theorem states that it is not possible to create a formal system that is both consistent and complete. As such, you cannot create a software system based on some sort of a formal system, i.e., a consistent set of axioms, which can reflect upon itself and discover (not invent) a new dimension of knowledge. (1)

Indeed, whenever you focus your attention on any of the so-called intelligent systems, and you take the time to learn the mechanisms they use to generate their marvelous and wondrous behavior, you come up with the anti-climactic realization that everything is quite ordinary and perfectly expectable with no surprises or mystical insights. Such reaction reminds us of how Sherlock Holmes reacted when a man questioned the brilliance of his deductive reasoning in solving one of his cases:

> Mr. Jabez Wilson laughed heavily. "Well, I never!" said he. "I thought at first that you had done something clever, but I see that there was nothing in it, after all." "Begin to think, Watson," said Holmes, "that I made a mistake in explaining. 'Omne ignotum promagnifico,' you know, and my poor little reputation, such as it is, will suffer shipwreck if I am so candid."

Similarly, Alan Turing, the father of the digital computer and creator of the Turing Test for checking the "intelligence" of a machine, put it this way:

> The extent to which we regard something as behaving in an intelligent manner is determined as much by our own state of mind and training as by the properties of the object under consideration. If we are able to explain and predict its behavior or if there seems to be little underlying plan, we have little temptation to imagine intelligence. With the same object, therefore, it is possible that one man would consider it as intelligent and another would not; the second man would have found out the rules of its behavior.

C. The Character of the Ten Paradigms

All the paradigms of intelligent systems in this volume have plans and assume extensive amounts of knowledge. As such they are ordinary computer programs and they emulate a precise computational procedure, which uses a predefined set of data. In the Aristotelian form, "all instruction given or received (by the intelligent systems) by way of argument, proceeds from preexistent knowledge." Consequently, one should see all cases put forward by the individual chapters as nothing more than paradigms for new uses of the computer. Every one of them carries out deduction from a predefined set of knowledge, using explicit reasoning strategies. The reader should not search for inductive generation of new knowledge, even when the terms "induction" and "inductive reasoning" have been loosely employed in some chapters. Instead, the reader should see each chapter as a computer-based paradigm in *capturing, articulating,* and *utilizing* various forms of knowledge. As a result, the reader will notice that the ten chapters of this volume serve as a paradigm of an integrative attitude to the modeling and processing of knowledge. Nowhere in this volume will the reader find artificial debates on the superiority of a numerical over a symbolic approach or vice versa. On the contrary, the engineering methodologies advanced by the individual chapters indicate that *all available knowledge should be acquired, modeled, and used* within a framework that requires interaction and/or integration of processing methodologies from artificial intelligence, systems and control theory, operations research, statistics, and others.

It is this integrative attitude that today characterizes most of the work in the area of "intelligent systems for process engineering," as the editors of this volume have indicated in a recent review article. (4) It is this need for integrative approaches that has moved the applications of artificial intelligence into the mainstream of engineering activities. This is certainly the pivotal feature that characterizes the ten paradigms discussed in the subsequent chapters.

The ten chapters of this volume advance ten distinct paradigms for the use of ideas and methodologies from artificial intelligence in conjunction with techniques from various other areas. They represent the culmination of research efforts which started in 1986 at the *Laboratory for Intelligent Systems in Process Engineering* (LISPE) of the Chemical Engineering Department at MIT, and currently are spread over a half a dozen academic institutions. Each chapter, as the corresponding title indicates, is centered around two themes. The first theme (represented by the first part of a title) is drawn from the artificial intelligence techniques discussed in the specific chapter, while the second theme (represented by the second part of the title) focuses on a process engineering problem. It should be noted, though, that it is the process engineering problem, its formulation and characteristics, that sets the tone for every chapter. The various components of the corresponding intelligent systems serve specific needs. Nowhere will the reader find the "a technique in search for a problem to solve" attitude, which has led to the distortion of several engineering problems and the malignant proliferation of techniques. As a result, even if future developments suggest a change in the techniques used, the formulation of the engineering problems may retain the bulk of the essential features proposed by each of the ten chapters.

D. THEMES COVERED BY THE TEN CHAPTERS

Let us now give a brief synopsis of the themes advanced by each of the ten chapters. The five chapters of Volume 21 (Part I) advance paradigms which are related to product and process design, while the five chapters of Volume 22 (Part II) focus on aspects of process operations.

Volume 21: Product and Process Design

Chapter 1. *MODELING LANGUAGES:*
Declarative and Imperative Descriptions of Chemical Reactions and Processing Systems

To model is to represent reality, and modeling as an essential task of any engineering activity is always *contextual*. Within the scope of differing engineering contexts, the same physical entity, e.g. molecule, chemical reaction, or process flowsheet, is represented with a broad variety of models. An enormous amount of effort is expended in the development and maintenance of a *Babel of models*, sporting different languages and being at cross purposes with each other, although like their biblical counterpart they share a common progenitor—in this case, the fundamentals of chemistry/physics and the principles of chemical engineering science. Creating a language that supports the expeditious generation of consistent models has become the key to unlocking the power of computer-aided tools, and unleashing the explosive synergism between human and computer. However, a modeling language is of little use if it only creates representations of physical entities as "things unto themselves" without meaningful semantic designation to what it purports to represent. Furthermore, the model of an entity should contain all knowledge that has some bearing on the representation of that entity, be that *declarative* or *imperative* (procedural) in character. Chapter 1 describes two modeling languages; LCR (*L*anguage for *C*hemical *R*easoning) to represent molecules and chemically reactive systems, and MODEL.LA. (*MODEL*ing *LA*nguage) for the representation of processing systems. Both are based on the same principles and have, to a large extent, a common structure. Both have been based on ideas and techniques which originated in artificial intelligence, and both have been implemented in a similar object-oriented programming environment.

Chapter 2. *AUTOMATION IN DESIGN:*
The Conceptual Synthesis of Chemical Processing Schemes

If you really know how to carry out an engineering task, then you can instruct a computer to do it automatically. This self-evident truism can be used as the litmus test of whether a human "really" knows how to, say, design an engineering artifact. Experience has shown that engineers have been able to automate the process of design in very few instances, thus demonstrating the presence of serious flaws in (a) their understanding of how to do design and/or (b) their ability to clearly articulate the design methodology, both of which can be traced to the inherent difficulty of making the "best" design decisions. The pivotal element in automating the design process is *modeling the design process itself*, which includes the following modeling tasks: (1) modeling the *structure of design tasks* that can take you from the initial design specifications to the final engineering artifact; (2) representing the *design decisions* involved in each task, along with the assumptions, simplifications, and methodologies needed to frame and make the design decisions; (3) modeling the *state of the evolving design*, along with the underlying rationale. Chapter 2 shows how one can use ideas and techniques from artificial intelligence, e.g., symbolic modeling, knowledge-based systems, and logic, to construct a computer-implemented model of the design process itself. Using Douglas' hierarchical approach as the conceptual model of the design process, this chapter shows how to generate models of the design tasks' structure, design decisions, and the state of design, thus leading to automation of large segments of the synthesis of chemical processing schemes. The result is a *human-aided, machine-based* design paradigm, with the computer "knowing" how the design is done, what the scope of design is, and how to provide explanations and the rationale for the design decisions and the resulting final design. Such a paradigm is in sharp contrast with the traditional *computer-aided, human-based* prototype, where the computer carries out numerical calculations and data fetching from files and databases, but has no notion of how the design is done, knowledge resting exclusively in the province of the individual human designer.

Chapter 3. *SYMBOLIC AND QUANTITATIVE REASONING:*
Design of Reaction Pathways through Recursive Satisfaction of Constraints

Given a fixed, predetermined set of elementary reactions, to compose reaction pathways (mechanisms) which satisfy given specifications in the transformation of available raw materials to desired products is a problem encountered quite frequently during research and development of chemical and biochemical processes. As in the assembly of a puzzle, the pieces (available reaction steps) must fit with each other (i.e., satisfy a set of constraints imposed by the precursor and successor reactions) and conform with the size and shape of the board (i.e., the specifications on the overall transformation of raw materials to products). Chapter 3 draws from *symbolic and quantitative reasoning* ideas of AI which allow the systematic synthesis of artifacts through a *recursive satisfaction of constraints* imposed on the artifact as a whole and on its components. The artifacts in this chapter are mechanisms of catalytic reactions and pathways of biochemical transformations. The former require the construction of *direct* mechanisms, without cycles or redundancies, to determine the basic legitimate chemical transformations in a reacting system. The latter are the chemical engines of living cells, and they represent legitimate routes for the biochemical conversion of substrates to products either desired from a bioprocess or essential for cell survival. The algorithms discussed in this chapter could be used in one of the following two settings: (a) Synthesize alternative pathways of chemical/biochemical reactions as a means to interpret overall transformations which are experimentally observed. (b) Synthesize reaction pathways in the course of exploring new, alternative production route. This chapter discusses examples in both directions. Although it is concerned only with constraints on the directionality and stoichiometry of elementary reactions, the ideas can be extended to include other types of constraints arising, for example, from kinetics or thermodynamics.

Chapter 4. *INDUCTIVE AND DEDUCTIVE REASONING:*
The Case of Identifying Potential Hazards in Chemical Processes

All reasoning carried out by computers is *deductive*; i.e., any software has all the necessary data, stored in various forms in a database, and possesses all the necessary algorithms to operate on the set of data and *deduce* some results. Many researchers in the area of cognitive psychology make similar claims on the reasoning mechanisms of human beings. The fact remains, though, that both humans and machines can use very simple "algorithms" on small sets of data and produce results which could not have been visible to the "naked eye" of direct reasoning. In such cases, we tend to talk about the *inductive* capabilities of either of the two. These ideas are nowhere more prominent than in the area of *hazards identification and analysis*. One often hears, "if I knew that the conversion of A to B could be catalyzed by the presence of C then I would have foreseen the last disaster, and have done something about it," with the speaker converting a problem of *inductive* identification (i.e., induce the possibility of a hazard from the list of chemicals) into an issue of deductive reasoning. Chapter 4 demonstrates that the identification of hazards is essentially an interplay between inductive and deductive reasoning. Through inductive reasoning one attempts to generate all potential hazardous top-level events which can be justified by the presence of a set of chemicals. The reasoning is called inductive because it has the potential to generate specific knowledge that was not "visible" ahead of time. Once the potentially harmful top-level events have been identified, deductive reasoning attempts to "walk" through the processing scheme, its unit operations, and their design or operating characteristics (assumptions, or decisions), and generate the preconditions which would enable the occurrence of a specific top-level event. The inductive reasoning procedures operate on a set of chemicals and create in an *exhaustive, bottom-up* manner many alternative reaction pathways, some of which could lead to a hazard, e.g., release of large amounts of energy over a short period of time. On the other hand, the deductive reasoning procedures are *goal-directed* and operate in a *top-down* manner. Chapter 4 develops the detailed framework for the implementation of these ideas which, among other benefits, offers the following advantages: (a) Formalizing the hazards identification problem and unifying the methodological approaches at any stage of the design activities and (b) systematizing the generation and evaluation of mechanisms for the prevention of hazards, or containment of their effects.

Chapter 5. *SEARCHING SPACES OF DISCRETE SOLUTIONS:*
The Design of Molecules Possessing Desired Physical Properties

Strings of letters make words. From words to verses and stanzas, a poet composes a work with its own dynamic behavior, e.g., emotional impact on the reader, which transcends the character of its components. In an analogous manner, atoms form functional groups and these in turn yield molecules with distinct behavior, e.g., physical properties. It takes a Homeric or Shakespearean genius to convert letters to an epic with a predefined desired impact. It suffices to efficiently search a space of combinatorial alternatives in order to identify the molecules which satisfy the desired constraints on a set of physical properties. Often the requisite scientific knowledge is fragmented, dispersed, and nonformalized, making the deductive search for the desired molecules inefficient or impossible. The inductive "genius" of a scientist or engineer is needed to break the impasse in such cases. By evolution or revolution one needs to respond to tighter and shifting product specifications and identify new solvents, pharmaceuticals, imaging chemicals, herbicides and pesticides, refrigerants, polymeric materials, and many others. Chapter 5 sketches the characteristics of an intelligent, computer-aided tool to support the synthetic search for the desired molecules. With functional groups as the "letters" of an alphabet, automatic and interactive procedures compose and screen classes of potential molecules. The automatic synthesis algorithm defines and searches the space of discrete solutions (molecules) through a hierarchical sequence of the space's representations. However, one should never overestimate the effectiveness of search algorithms in locating the desired solutions. Quite frequently one needs to resort to human-driven, abductive jumps. Chapter 5 also describes how automatic search can become interwoven with effective man–machine interaction. Thus, the resulting computer-aided tool, the *Molecule Designer*, constitutes a paradigm of an intelligent system with two distinct but integrated and complementary capabilities. Examples of the synthesis of refrigerants, solvents, polymers, and pharmaceuticals illustrate the logic and features of the design procedures in the *Molecule Designer*.

Volume 22: Process Operations

Chapter 6: NONMONOTONIC REASONING:
The Synthesis of Operating Procedures in Chemical Plants

The inherent difficulty of planning a sequence of actions to take you from one point to another usually increases as more obstacles are placed in your way. The number of these obstacles (constraints) that you must circumvent determines the complexity of the task, because any time you run into one of them you must *backtrack* and try an alternative step or path of steps. Such *serial* (or *linear* or *monotonic*) construction of a plan is fraught with pitfalls and repeated backtracking. The more the constraints, the more inefficient the monotonic planning. If, on the other hand, an action-step (a Clobberer) leads to the violation of a constraint, then *do not backtrack*. Take another action-step (a White Knight) which, when it precedes a Clobberer, negates the impact of the Clobberer, and you never need to backtrack. So the more constraints the more efficient your planning process. Such *nonserial* (or *nonlinear*, or *nonmonotonic*) reasoning has become the essence of all modern and efficient planners, whether they are *logic-based* and *explicit*, or *implicit* enumerators of alternative plans. The purpose of Chapter 6 is twofold: (i) To introduce the ideas of *nonmonotonic reasoning* in the planning of process operations. (ii) To demonstrate how nonmonotonic planning can be used to synthesize operating procedures for chemical processes, either off-line for standard tasks (e.g., routine start-up or shut-down), or on-line for real-time response to large departures from desired conditions. It is shown that hierarchical modeling of process operations and operators is essential for the efficient deployment of nonmonotonic planning, and that the tractability of the resulting algorithms is strictly dependent on the form of the operators. In this regard, the modeling needs in this chapter draw heavily from the material of Chapter 1. Nonmonotonic planners handle with superb efficiency constraints on (a) the temporal ordering of operations, (b) avoidable mixtures of chemical species, and (c) bounding quantitative conditions on the state of a process. Consequently, they could be used to generate explicitly all feasible operating procedures, leaving a far smaller search space for the selection of the optimum procedure by a numerical optimizer.

Chapter 7. *INDUCTIVE AND ANALOGICAL LEARNING:*
Data-Driven Improvement of Process Operations

Informed and systematic observation of naturally generated data can lead to the formulation of interesting and effective generalizations. While some statisticians believe that experimentation is the only safe and reliable way to "learn" and achieve operational improvements in a manufacturing system, other statisticians and all the empirical machine learning researchers contend that by looking at past historical records and sets of examples, it is possible to extract and generate important new knowledge. Chapter 7 draws from *inductive and analogical learning* ideas in an effort to develop systematic methodologies for the extraction of structured new knowledge from operational data of manufacturing systems. These methodologies do not require any a priori decisions/assumptions either on the character of the operating data (e.g., probability density distributions) or on the behavior of the manufacturing operations (e.g., linear or nonlinear structured quantitative models), and they make use of *instance-based learning* and *inductive symbolic learning* techniques developed in artificial intelligence. They are aimed to be complementary to the usual set of statistical tools that have been employed to solve analogous problems. Thus, one can see the material of Chapter 7 as an attempt to fuse statistics and machine learning in solving specific engineering problems. The framework developed in this chapter is quite generic and can be used to generate operational improvement opportunities for manufacturing systems (a) which are simple or complex (with internal structure), (b) whose performance is characterized by one or multiple objectives, and (c) whose performance metrics are categorical (qualitative) or continuous (real numbers). A series of industrial case studies illustrates the learning ideas and methodologies.

Chapter 8. *EMPIRICAL LEARNING THROUGH NEURAL NETWORKS:*
The Wave-Net Solution

Empirical learning is an ever-lasting and ever-improving procedure. Although *neural networks* (NN) captured the imagination of many researchers as an outgrowth of activities in artificial intelligence, most of the progress was accomplished when empirical learning through NNs was cast within the rigorous analytical framework of the *functional estimation problem*, or *regression*, or *model realization*. Independently of the name, it has been long recognized that, due to the inductive nature of the learning problem, to achieve the desired accuracy and generalization (with respect to the available data) in a dynamic sense (as more data become available) one needs to seek the unknown approximating function(s) in functional spaces of varying structure. Consequently, a recursive construction of the approximating functions at multiple resolutions emerges as a central requirement and leads to the utilization of wavelets as the basis functions for the recursively expanding functional spaces. Chapter 8 fuses the most attractive features of a NN, i.e., representational simplicity, capacity for universal approximation, and ease in dynamic adaptation, with the theoretical soundness of a recursive functional estimation problem, using wavelets as basis functions. The result is the *Wave-Net (Wavelet Network)*, a multiresolution hierarchical NN with localized learning. Within the framework of a Wave-Net where adaptation of the approximating function is allowed, we have explored the use of the L^∞ error measure as the design criterion. One may cast any form of data-driven empirical learning within the framework of a Wave-Net to address a variety of modeling situations encountered in engineering problems, such as design of process controllers, diagnosis of process faults, and planning and scheduling of process operations. Chapter 8 discusses the properties of a Wave-Net and illustrates its use on a series of examples.

Chapter 9. REASONING IN TIME:
    Modeling, Analysis, and Pattern Recognition of Temporal
    Process Trends

The plain record of a variable's numerical values over time does not invoke appreciable levels of cognitive activity in a human. Although it can cause a fervor of numerical computations by a computer, the levels of cognitive appreciation of the variable's temporal behavior remain low. On the other hand, if one presents the human with a graphical depiction of the variable's temporal behavior, the level of cognition increases and a wave of reasoning activities is unleashed. Nevertheless, when the human is presented with scores of graphs depicting the temporal behavior of interacting variables, his/her reasoning abilities are severely tested. In such a case, the computer will happily continue crunching numbers without ever rising above the fray and thus developing a "mental" model, interpreting correctly the temporal interactions among the many variables. Reasoning in time is very demanding, because time introduces a new dimension with significant levels of additional freedom and complexity. While the real-valued representation of variables in time is completely satisfactory for many engineering tasks (e.g., control, dynamic simulation, planning and scheduling of operations), it is very unsatisfactory for all those tasks which require decision-making via logical reasoning (e.g., diagnosis of process faults, recovery of operations from large unsolicited deviations, "supervised" execution of start-up or shut-down operating procedures). To improve the computer's ability to reason efficiently in time, we must first establish new forms for the representation of temporal behavior. It is the purpose of Chapter 9 to examine the engineering needs for temporal decision-making and to propose specific models which encapsulate the requisite temporal characteristics of individual variables and composite processes. Through a combination of analytical techniques, such as *scale-space filtering, wavelet-based, multiresolution decomposition of functions,* and modeling paradigms from artificial intelligence, Chapter 9 develops a concise framework that can be used to model, analyze, and synthesize the temporal trends of process operations. Within this framework, the modeling needs for logical reasoning in time can be fully satisfied, while maintaining consistency with the numerical tasks carried out at the same time. Thus, through the modeling paradigms of this chapter, one may put together intelligent systems which use consistent representations for their logical-reasoning and numerical tasks.

Chapter 10. *INTELLIGENCE IN NUMERICAL COMPUTING:*
Improving Batch Scheduling Algorithms through Explanation-Based Learning

Learning comes from reflection upon accumulated experience and the identification of patterns found among the elements of past experience. All numerical algorithms used in scientific and engineering computing are based on the same paradigm: *execute a predetermined sequence of calculation tasks and produce a numerical answer.* The implementation of the specific numerical algorithm is oblivious to the experience gained during the solution of a specific problem and, in the next encounter, a different, or even the same problem is solved through the execution of exactly the same sequence of calculation steps. The numerical algorithm makes no attempt to reflect upon the structure and patterns of the results it produced, or to reason about the structure of the calculations it has performed. Chapter 10 shows that this need not be the case. By allowing an algorithm to reflect upon and reason with aspects of the problems it solves and its *own structure of computational tasks, the algorithm can learn* how to carry out its tasks more efficiently. Such *intelligent numerical computing* represents a new paradigm, which will dominate the future of scientific and engineering computing. But, in order to unlock the computer's potential for the implementation of truly intelligent numerical algorithms, the *procedural* depiction of a numerical algorithm must be replaced by a *declarative* representation of the algorithmic logic. Such a requirement upsets an established tradition and imposes new educational challenges, which most educators and educational curricula have not, as yet, even recognized. This chapter shows how one can take a branch and bound algorithm, used to identify optimal schedules of batch operations, and endow it with the ability to learn to improve its own effectiveness in locating the optimal scheduling policies for flowshop problems. Given that most batch scheduling problems are NP-hard, it becomes clear how important it is to improve the effectiveness of algorithms for their solution. Using the Ibaraki framework, a branch and bound algorithm is declaratively modeled as a *discrete decision process*. Then explanation-based machine learning strategies can be employed to uncover patterns of generic value in the experience gained by the branch and bound algorithm from solving specific instances of scheduling problems. The logic of the uncovered patterns (i.e., new knowledge) can be incorporated into the control strategy of the branch and bound algorithm when the next problem is to be solved.

GEORGE STEPHANOPOULOS AND CHONGHUN HAN

REFERENCES

1. Stephanopoulos, G., Computers, Systems, Languages and Other Fragments. *CAST Newsletter*, Spring (1994).
2. Antsaklis, P. J. and Passino, K. M., eds., "An Introduction to Intelligent and Autonomous Control." Kluwer, Norwell, MA, 1993.
3. Åström, K. J. and McAvoy, T. J., Intelligent Control, *J. Proc. Cont.* **2**(3), 115 (1992).
4. Stephanopoulos, G. and Han, C., "Intelligent Systems in Process Engineering: A Review," *Proc. PSE*, Kyongju, Korea, 1994.
5. Kurtzweil, R., "The Age of Intelligent Machines." MIT Press, Cambridge, MA, 1990.
6. Mandelbrot, B. B., "The Fractal Geometry of Nature." W. H. Freeman, New York, NY, 1983.
7. Davis, P. J. and Hersh, R., "Descartes' Dream: The World According to Mathematics." Harcourt Brace Jovanovich, San Diego, CA, 1986.
8. Stephanopoulos, G. and Mavrovouniotis, M.L., eds., Artificial Intelligence in Chemical Engineering—Research and Development, *Comp. Chem. Eng.* **12**(9/10), (1988).
9. Stephanopoulos, G., Artificial Intelligence and Symbolic Computing in Process Engineering Design. *In* "Foundations of Computer-Aided Process Design" (J. J. Siirola, I. E. Grossmann, and G. Stephanopoulos, eds.), p. 21, Elsevier, New York, 1989.
10. Stephanopoulos, G., Artificial Intelligence: What Will Its Contributions Be to Process Control? *In* "The Second Shell Process Control Workshop" (D. M. Prett, C. E. Garcîa, and B. L. Ramaker, eds.), p. 591, Butterworths, Stoneham, MA, 1990.
11. Stephanopoulos, G., Brief Overview of AI and Its Role in Process Systems Engineering. *In* "Process Systems Engineering, Vol. I," CACHE, 1992.
12. Mavrovouniotis, M., ed., "Artificial Intelligence in Process Engineering." Academic Press, San Diego, CA, 1990.
13. Quantrille, T. E. and Liu, Y. A., "Artificial Intelligence in Chemical Engineering." Academic Press, San Diego, CA, 1991.
14. Hofstadter, D. R., "Gödel, Escher, Bach: An Eternal Golden Braid." Vintage Books, New York, 1980.
15. Penrose, R., "The Emperor's New Mind." Oxford University Press, Oxford, UK, 1989.

MODELING LANGUAGES: DECLARATIVE AND IMPERATIVE DESCRIPTIONS OF CHEMICAL REACTIONS AND PROCESSING SYSTEMS

Christopher J. Nagel, Chonghun Han, and George Stephanopoulos

Laboratory for Intelligent Systems in Process Engineering
Department of Chemical Engineering
Massachusetts Institute of Technology
Cambridge, Massachusetts 02139

I. Introduction	2
A. The Five Premises of a Modeling System	3
B. Review of Modeling Systems for Process Simulation	7
C. Modeling Systems in Chemistry	10
II. LCR: A Language for Chemical Reasoning	13
A. Modeling Elements of LCR	13
B. Semantic Relations among Modeling Elements in LCR	26
C. Syntax of LCR	33
III. Formal Construction of Representations for Chemicals and Reactions	36
A. Extension of LCR's Modeling Objects	36
B. The "Model Class Decomposition Digraph" (MCDD)	50
C. Generation and Representation of Reaction Pathways	53
D. Creation of Contextual Reaction Models	58
E. Case Study: Ethane Pyrolysis	64
IV. MODEL.LA.: A Modeling Language for Process Engineering	73
A. Basic Modeling Elements	73
B. Semantic Relationships	75
C. Hierarchies of Modeling Subclasses	76
D. Syntax	78
V. Phenomena-Based Modeling of Processing Systems	78
A. The "Chemical Engineering Science" Hierarchies of Modeling Elements	79
B. Formal Construction of Models	82
C. Multifaceted Modeling of Processing Systems	82
D. Computer-Aided Implementation of MODEL.LA.	87
References	90

To model is to represent reality, and modeling as an essential task of any engineering activity is always *contextual*. Within the scope of differing engineering contexts, the same physical entity, such as a molecule, chemical reaction, or process flowsheet, is represented with a broad variety of models. An enormous amount of effort is expended in the development and maintenance of a *Babel of models*, sporting different languages and being at cross-purposes with each other, although like their biblical counterparts share a common progenitor: in this case, the fundamentals of chemistry and physics and the principles of chemical engineering science. Creating a language that supports the expeditious generation of consistent models has become the key to unlocking the power of many computer-aided tools, and unleashing the explosive synergism between human and computer. However, a modeling language is of little use if it only creates representations of physical entities as "things onto themselves" without meaningful semantic designation to what it purports to represent. Furthermore, the model of an entity *should contain all knowledge* that has some bearing to the representation of that entity, whether that is *declarative* or *imperative* (procedural) in character. In this chapter we will describe two modeling languages: LCR (Language for Chemical Reasoning) to represent molecules and chemically reactive systems, and MODEL.LA. (MODELing LAnguage) for the representation of processing systems. Both are based on the same principles and have, to a large extent, a common structure. Both have been based on ideas and techniques, which originated in artificial intelligence, and both have been implemented in a similar object-oriented programming environment.

I. Introduction

Since the early efforts, computer-aided modeling of physical systems has been generally organized around unilateral computations that perform predetermined operations on fixed inputs to produce the desired outputs. This is still a common practice since it helps engineers and scientists to manage tedious calculations and coordinate and control diverse numerical tasks, by producing models that use very efficiently computer resources. However, a series of inherent weaknesses have pointed out the disadvantages of the traditional approach: (1) the time and cost associated with computer model development are high; (2) the resulting models are difficult to document and maintain adequately; (3) the reuse of computer-aided models is minimal, as these models tend to be task-specific and are often intrinsically linked to solution procedures; (4) the models cannot be synthesized automatically by the computer in the course of automatic execution of an engineering task; and (5) for interactive modeling the modeler is required to be highly skilled in programming. As a result of these weaknesses the duplication of modeling efforts has been enormous. Accumulated modeling knowledge is almost impossible to use, since the underlying modeling context (purpose, assumptions, simplifications) has never been documented and rationalized. So, why must every new modeling effort start from scratch (Meyer, 1987). Furthermore, automatic generation of models at higher abstractions, cannot be done. Finally, the fragmentation of modeling efforts and the ad hoc character of their computer implementation have led to internal inconsistencies among the various models used in different process engineering tasks. A typical example of this phenomenon is found in the diversity of models used to support process-control-related engineering tasks (Stephanopoulos, 1990) such as design of process controllers, controller adaptation mechanisms, optimization of process operations, and diagnosis of process faults.

A. THE FIVE PREMISES OF A MODELING SYSTEM

To overcome the weaknesses discussed above, any modeling system should be built on the following premises.

1. Articulate All Declarative Knowledge

The model of an entity, whether a processing unit, a molecule, or a chemical reaction, should contain explicitly all relevant information, including the following (Stephanopoulos *et al.*, 1990a):

(a) *Underlying assumptions.* The form of the relationships that describe the behavior of a given entity depends on a series of assumptions, such as the operational mode; the assumed mechanisms for mass, energy, or momentum transfer; the mechanistic pathway for a chemical reaction; and the constitutive models for the estimation of physical properties.

(b) *Simplifications* made by the modeler (human, or another computer program) in order to limit the model's validity over a given range of conditions, or to underscore the relative importance of various physicochemical phenomena.

(c) *Scope of engineering task*, i.e., what the model is intended for. Typical examples are (1) the variety of models required by the different levels of the hierarchical synthesis of process flowsheets (Stephanopoulos, 1990a, 1990b); (2) the variety of models needed for feedback and adaptive control, diagnosis, or planning of process operations (see the fifth chapter in this volume).

(d) *Relationships among the various variables and parameters.* These relationships express the fundamental principles of physics and chemistry as well as the assumed mechanisms for rate and equilibrium phenomena, and the correlations for the estimation of physical properties. These relationships could be quantitative, qualitative, or semiquantitative (e.g., order-of-magnitude, or ordinal), depending on the level of the available knowledge, and could express static or dynamic behavior.

2. Separate Declarative from Procedural Knowledge

The declarative knowledge that is articulated in a given model represents the "what is" knowledge about the modeled entity. Since the same declarative model could be used for a variety of engineering tasks, it is essential that it be separated from the procedural knowledge, i.e., the "how to" knowledge of a particular engineering methodology. Previous modeling approaches relied on a tight integration of declarative and procedural knowledge with the sequential modular simulators (e.g., ASPEN; see Evans *et al.*, 1979) exemplifying the tightest integration, and the equation-oriented simulators (e.g., SPEEDUP; Perkins and Sargent, 1982; Pantelides, 1988) offering a weaker but still dominant integration.

3. Hierarchical and Multiview Representation of Entities

A modeling system should allow the representation of the various entities at any level of detail, using multiple, coexisting abstractions, which can communicate with each other. This is a critical requirement as it sets the desired modeling system apart from many other modeling environments. For example, a processing system could be represented in any of the following abstractions (Fig. 1): (a) an overall plant, (b) a network with generalized reaction and separation sections, (c) a network of reactors and abstract separation sections, or (d) a network of processing units. All models, depicting the four abstraction views in Fig. 1, should be consistent with each other and should allow the transfer of information from more abstract to more detailed representations and vice versa. In addition, two distinct and different versions of the same abstraction (e.g., Fig. 2) have many modeling components in common. The modeling system should allow common handling of the common modeling elements. For example, the modeling relations for the generalized reaction and separation sections

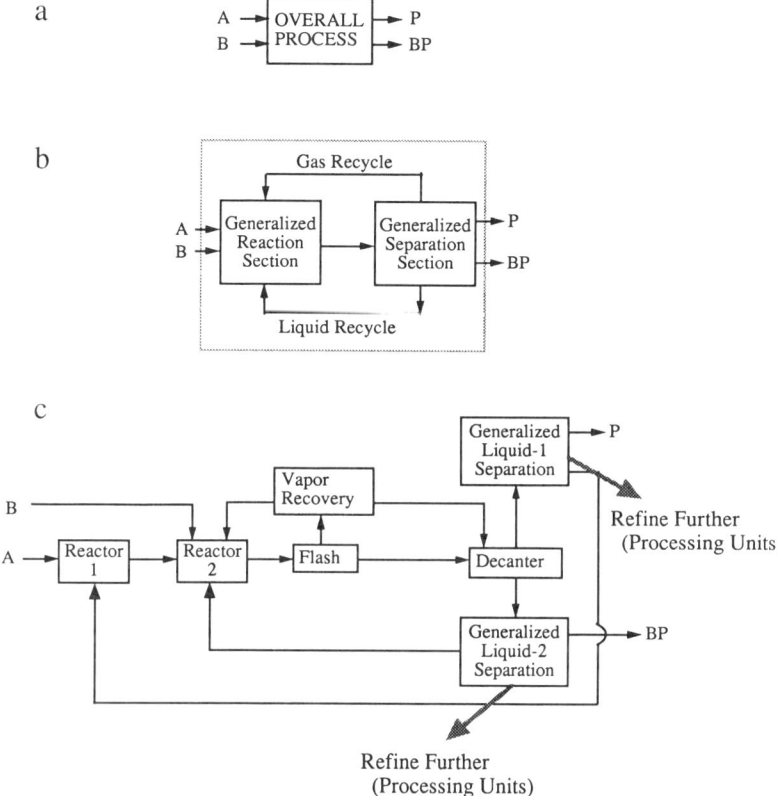

FIG. 1. Three abstraction levels for the representation of a processing scheme.

of the two versions in Fig. 2 are the same. If we change the models in one of the versions, the representation of the other version should be updated automatically.

4. Automatic Generation and Modification of Models

Once the modeler has described the elements of the modeling premise 1—i.e., made the underlying assumptions and simplifications and defined the scope of the engineering task—the modeling system should automatically generate the modeling relationships and their elements. In other words, the modeling language should possess declarative representations of relationships, of the terms that made up such relationships, of the variables that make up the terms, and of the semantic connections that

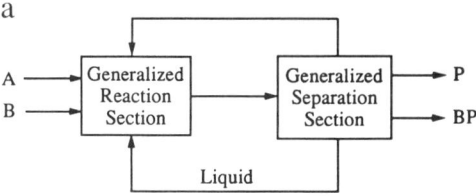

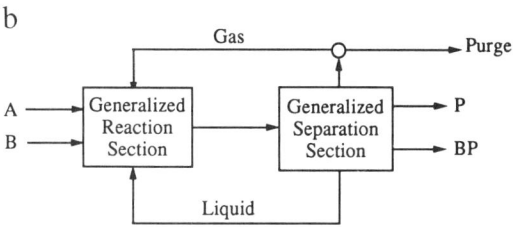

Fig. 2. Two distinct versions of a plant at the same abstraction.

assign physical meaning to the variables, terms, and relationships. Furthermore, once the modeling system has developed the model of an entity, it should be capable of updating the model automatically, if the modeling assumptions and simplifications have changed.

5. Simulation and Reasoning

The modeling relationships of the various entities should possess enough granularity to allow structural reorganization for efficient numerical simulation and logical reasoning. For example, given the static quantitative model of a process, the modeling system should possess the necessary modeling information to allow the construction of the corresponding cause and effect, directed graph, or automatically analyze the degrees of freedom and generate the simplest numerical procedures for simulation.

The preceding five premises imply requirements that go beyond the scope of traditional computer-aided modeling systems. The modeling language should allow interaction with the modeler at a high level of abstraction, while the generated models should be explicit, transparent, fully documented, and usable without the need to consult manuals for assumptions or input/output language conventions.

B. Review of Modeling Systems for Process Simulation

In recent years several modeling systems have appeared in the literature, all of them aimed at eliminating the perceived modeling bottleneck in generating static or/ and dynamic simulations of processing systems. Consequently, all the modeling systems to be discussed in the subsequent paragraphs do not conform to all of the five premises, discussed in the previous section, and consequently they cannot be used outside the simulation tasks that they were designed to serve.

1. ASCEND (Advanced System for Computations in ENgineering Design)

As its name implies, ASCEND is a general computer-aided environment with a modeling language that was developed to support the expeditious generation of mathematical models that are needed for the simulation and design of processing systems (Piela, 1989; Piela *et al.*, 1991). It follows an object-oriented paradigm, but it is strongly typed. Its main features can be summarized as follows:

a. Building Blocks. ASCEND uses three types of generic modeling elements: `model`, `atom`, and `type`. Models are structured entities that are built hierarchically from (1) instances of other models, atoms, or/ and types; and (2) relationships between these instances. Atoms are primitive variables, which are used to represent physical quantities, design variables, etc. Types are elementary declarations of real, integer, string, Boolean, etc. They are predefined in ASCEND.

b. Relationships. Several operators define relationships among the various elements of ASCEND. Three of them are used to define inheritance relationships: (1) the ***refines*** operator is used to link a subclass to a class in an inheritance hierarchy; (2) the ***is-a*** operator is used to declare an object as the instance of a class (or, a type); and (3) the ***is-refined-to*** operator changes the type associated with a previously declared instance. Two relationships are used to group instances together: (1) the ***are-alike*** operator coerces all members of a group of instances to take the type associated with the most refined instance; (2) the ***are-the-same*** operators is similar to the previous one, but merges the operands into a single structure.

c. General Features. ASCEND uses the inherited mechanism of the object-oriented paradigm in order to provide expeditious extension of its "vocabulary." It contains both instances and classes, but instances are not allowed to have local attributes. ASCEND generates mathematical relationships that are expressed as implicit equations or solved assignments. It uses a fairly concise syntax to express a fairly broad set of modeling statements.

2. OMOLA (*O*bject-oriented *MO*deling *LA*nguage)

OMOLA (Andersson, 1989; Nilsson, 1993) was designed to support the model generation and simulation needs during the computer-aided synthesis and analysis of control systems. It is oriented toward the creation of structured dynamic systems. Its main features can be summarized as follows:

a. Modeling Elements. All models are created as subclasses of the mother class, `Model`, thus inheriting all the attributes of the mother class. It is possible to create specialized versions of an existing model, M1, by creating a subclass from the class, M1. The description of each model is organized around four modeling elements; `Terminal`, `Parameter`, `Variable`, and `Realization`. The first is used as a point of communication between the models of the various entities. `Parameter` and `Variable` are used to describe model attributes that are time-independent and time-varying, respectively. `Realization` is the element that is used to define the mathematical relationships describing the behavior of an entity. Each of the four modeling elements is described by a structured class with a series of attributes. Specialized subclasses emanate from `Terminal` and `Realization`, thus allowing an expansion in the vocabulary of the language itself.

b. Functionality. OMOLA allows the representation of complex structured systems through the aggregation of individual models. The language offers no semantic relationships to link the generated models or their attributes to any physical concepts. Consequently, it cannot be used to automate the generation of models from a set of physical modeling assumptions. Also, it does not allow for the articulation of qualitative or semiquantitative knowledge.

3. MODASS (*MOD*eling *ASS*istant)

MODASS (Sørlie, 1990) is more of a toolbox for the construction of models. Although it allows declarative description of models and their elements, it also permits the incorporation of subroutines as model components.

a. Modeling Elements. The modeling elements of MODASS are assembled in a hierarchy of classes. At the top is the class, `model`, containing very few common attributes. At the next level we encounter the subclass, `Model-Element`, which contains the basic building blocks of the actual model. Further specializations (i.e., subclasses) provide narrow specifications of the sets of equations used for modeling specific entities. To represent the definitional boundary of an entity and the gateways for the information exchange between different entities, MODASS uses the classes, `Process-Terminal` and `Boundary-Process`, respectively. Specializations of the latter class allow the tailoring of the modeling element to mass, energy, and information transfer.

b. Functionality. MODASS has very limited network of semantic relationships among the different modeling elements, but it offers a satisfactory set of tools to the user for (1) the navigation through the tree of library models, (2) the symbolic manipulation of the modeling equations, and (3) the examination of models for inconsistencies and errors. Its lack of meaningful semantic relationships among the various modeling elements and the physical entities they represent makes MODASS a modeling system with limited scope, and largely outside the framework of the five premises, discussed in Section I.A.

4. Hybrid Phenomena Theory

Hybrid phenomena theory (HPT) (Woods, 1993) is not a modeling language in strict terms, but a theoretical framework for the generation of models that capture the topological, phenomenological, and behavioral aspects of the entity being modeled. The *topological* model of an entity includes a set of objects, (e.g., pipes, vessels, valves) and a set of logical relationships among the objects. The *phenomenological* model of the entity contains a set of objects, describing instances of physical phenomena that occur in the system being modeled. These phenomena are active processes such as heat transfer, material accumulation, and convective flow. The *behavioral* model is a state-space model providing a quantitative description of the entity's dynamic (or, static) behavior. The topological, phenomenological, and behavioral modeling components of an entity are tightly integrated and information flows among them as needed.

a. Modeling Elements. The object-oriented character of HPT places the `quantity` as the mother-class of the subclasses `constant`, `parameter`, and `variable`, whose further specialization has produced the subclass `state-variable`. A second hierarchy of classes, emanating from the mother-class `object`, captures the information available for various physical objects, such as materials (represented by class `stuff` and its specializations), processing units (from class `process-equipment`), devices (from class `device`). A hierarchy of `conditions` provides the logical mechanism for monitoring the values of data and taking action when specific conditions are met.

b. Functionality. The modeling elements, described above, are used as the building blocks for the construction of higher-level modeling objects, such as `views` and `phenomena`, which describe physical interactions. The topological model of HPT provides a Boolean representation of the model's components, whereas the phenomenological model supplies a qualitative description of the entity's behavior. It is the behavioral, state–space model that carries the full quantitative description.

5. Critique of Modeling Systems for Simulation

All the systems, described in the previous paragraphs, were created with a very specific purpose in mind: to expedite the generation and modification of quantitative models to support the needs of static or dynamic simulation or design. As a result, they are very difficult (if possible at all) to be used for other engineering tasks, which may have a different use for the generated models (e.g., diagnosis). Their common weakness results from two essential deficiencies: first, they cannot capture all forms of knowledge (e.g., qualitative or semiquantitative relationships); second, they do not possess a rich set of semantic relations, which allow intelligent response to queries about (a) the assumptions that produced a given model, (b) the interrelationship of physical phenomena captured in the modeling relations, (c) the propagation of information flow among different views of the same entity, etc. In Section IV we will see how the modeling language MODEL.LA. can overcome these deficiencies.

C. MODELING SYSTEMS IN CHEMISTRY

With the exception of computational chemistry, computer-aided chemical reasoning has not enjoyed the success experienced in other scientific disciplines. This has often been attributed to the conceptual nature of chemistry and the inexactness of known relationships (Dugundji and Ugi,

1973). However, the synthesis of new chemical reaction paths and processing alternatives that support their production remains an attractive objective within the general field of process synthesis. The availability of raw materials and energy, changing ecological and health considerations, and shifting requirements of the market, reinforce each other: to create the need to identify new routes or new processes to obtain existing chemicals or develop new chemicals to meet perceived requirements.

Chemists were the first to look into computer-aided organic synthesis (CAOS) in an attempt to answer Woodward's question (Woodward, 1956, 1963) and furthered by Corey (1967): "How does a chemist choose a pathway for the synthesis of a large organic molecule, given either the great diversity of organic structures and reactions, or, in contrast, the critical importance of each step to ultimate success?" The works of Corey, Wipke, Ugi, Hendrickson, Jorgensen, Moreau, Gelernter, and others support the CAOS effort (see Table I). Excellent reviews can be found in the works of Vernin and Chanon (1986) and Wipke *et al.* (1974).

For any problem of computer-aided organic synthesis, the quality and quantity of required data will vary, the most appropriate strategies may be different, but the questions central to the advancement of computer-aided chemical reasoning remain the same:

(a) How should the molecules and reactions be represented?
(b) What is the necessary degree of descriptional completeness, and at what cost can it be achieved?

Just as Lavoisier stated 200 years ago, "It is time to rid chemistry of obstacles of every kind... this reform must be brought about by perfecting the language," we postulate that a high-level (computer) language must be developed for meaningful advancement to occur in computer-aided chemical reasoning.

Domain-specific modeling languages are "very high level" and of "special purpose." They have a *theme*, that is, the class of ideas that is optimized to communicate. They allow the user to employ terms and constructs that lie closer to the informal terminology and modes of speech customary in the discussion of domain-specific problems; however, in constructing such a language, one should strike for domain-specific generality of the language. Thus, the same modeling language should satisfy all the needs: chemical synthesis, reaction analysis, pathway optimization, and innovative design of reaction or processing networks. The implication of this requirement is clear: *A modeling language in chemical reasoning should be fully declarative, and in no way should its generality be compromised by the specificity of the methodologies of the chemical tasks themselves.*

TABLE I
COMPUTER AIDS IN CHEMISTRY

Number	Authors	Program name	First published
1	Corey-Wipke	OCSS	1969
2	Corey	LHASA	1971
3	Ugi-Gasteiger	—	1971
4	Hendrickson	—	1971
5	Bersohn	—	1971
6	Weise	AHMOS	1973
7	Gelernter	SYNCHEM	1973
8	Barone-Chanon	SOS	1973
9	Powers	DINASYN	1973
10	Brownscombe	EXTRUS	1973
11	Brownscombe	HEXARR	1973
12	Wipke	SECS	1974
13	Ugi-Gasteiger	CICLOPS	1974
14	Benedek	SIMUL	1974
15	Dubois	SYNOPSYS	1975
16	Whitlock	—	1976
17	Donova	HEDOS	1976
18	Powers	REACT	1977
19	Govind	REPAS	1977
20	Djerassi	REACT	1977
21	Pensak	LHASA (Du Pont)	1977
22	Kaufmann	PASCOP	1978
23	Moreau	MASSO	1978
24	Gelernter	SYNCHEM2	1978
25	Ugi-Gasteiger	EROS	1978
26	Yoneda	GRACE	1978
27	Gasteiger	PSYCHE	1979
28	Weise	GSS	1979
29	Barone-Chanon	SAS	1979
30	Stolow-Joncas	LHASA (Educ)	1980
31	Agnihotri	CHIRP	1980
32	Jorgensen	CAMEO	1980
33	Gund	SECS (Merck)	1980
34	Hendrickson	SYNGEN	1981
35	Hippe	SCANSYNTH	1981
36	Hippe	SCANPHARM	
37	Hippe	SCANMAT	
38	Zefirov	FLAMINGOES	1981
39	Ghose	—	1981
40	Schubert	ASSOR	1981
41	Zin	—	1982
42	Erdos	ASR	1983
43	Kaufmann	PSYCHO	1984
44	Seidel	—	1984
45	Hara	PFP	1984
46	Wipke	SST	1984
47	Cense	MICROSYNTHESE	1985
48	Barone-Chanon	TAMREAC	1985

In the absence of any formal modeling language for the representation of chemicals and chemical reactions, it is important to state a concise set of requirements that the desired language should satisfy. These requirements, consistent with the five premises of Section I.A are

1. Represent molecules, reactions, and reaction pathways at various levels of detail.
2. Allow easy extension to new classes of molecules and reactions.
3. Allow the contextual representation of chemicals' reactivity, i.e., reactivity influenced by the relative position of atoms, the surrounding conditions and presence or absence of specific molecules.
4. Support all scientific or empirical knowledge brought together during the synthesis of reaction pathways.

In Section III we will see how LCR (Language for Chemical Reasoning) meets these requirements.

II. LCR: A Language for Chemical Reasoning

In this section we will describe the components of a modeling language, called LCR, that was developed to represent the knowledge about chemically reacting systems. First, we will describe the *basic modeling elements*, which constitute the building blocks for the representation of the declarative knowledge about molecules, reactions, and pathways. Second, we will discuss the semantic relationships among the basic modeling elements. These semantic relationships establish the "meaning" behind the linguistic expressions, defining knowledge about molecules and reactions. Third, we will present the syntax used by the language for the description of chemically reacting systems. LCR was implemented on a Symbolics 3650. It consists of approximately 50,000 lines of LISP (excluding code for the interfaces). Extensive discussion on the use of LCR for pathway generation can be found in Nagel (1991). The fourth chapter in this volume also discusses how LCR has been used to generate the reaction-based potential hazardous events in a chemical plant.

A. MODELING ELEMENTS OF LCR

A single chemical structure may exhibit multiple chemical behaviors, depending on the conditions in the surrounding environment. As a result, LCR uses two groups of modeling elements; the first provides those elements that are needed to describe the structure of the molecule (e.g., atoms, bonds, connectivity), and the second class contains those elements that characterize its chemical behavior and are related to the atom's electronic configuration.

1. Modeling Elements Defining Chemical Structures

LCR uses a graph theoretic representation of molecular structures, wherein nodes are atoms and edges are bonds. Three modeling elements are employed to create and describe chemical structures. Each modeling element has been implemented as a *class*, in an object-oriented programming environment, and is described by a set of attributes and a set of procedures.

a. Modeling Element 1: `atom`. An atom is the building block of chemical structures? Each `atom` encapsulates its own structural information, modeling relationships, and modeling assumptions. The attributes describing the object class, `atom`, are shown in Table II. Additional attributes may be associated with the `atom` class to refine the class description. For example, the attributes, formal-charge, oxidation-state, steric-hindrance, chirality, may be included into an atom's description to refine the class description. Several of the attributes given in Table II warrant further explanation: Attributes "identifier" and "old-identifier" are pointers used to trace an atom's lineage; attribute "database-atom" contains an object `db-atom` that manages invariant information indigenous to a particular atomic species (e.g., atomic-symbol, atomic-number, atomic-weight). The attribute "parent-abc" associates an atom to a given chemical structure; its value provides the vehicle for moving between the atom representation and the "parent-abc" representation. "Parent-group" is used similarly. An `atom` may have an association with a functional group or set of groups, as well as a parent-abc. As in the attribute case of "parent-abc," the value of "parent-group" provides a conduit for accessing information residing at the `group` representational level. As will be shown later, all representational levels can be accessed from any individual level. The freedom of information flow between levels enhances representational expressiveness and facilitates efficient reasoning.

A partial listing of the methods operating on the class `atom` is also shown in Table II. Methods with the "compute" prefix (e.g., *compute-parent-groups*, *compute-hybridization*, and *compute-formal-charge*) are used to evaluate attribute values. Predicate methods (i.e., methods returning a Boolean value) are represented by the "-p" suffix. These methods are

TABLE II
Select Attributes and Methods for atom, bond, and atom-bond-configuration Classes

atom	bond	atom-bond-configuration
atom attributes	**bond attributes**	**atom-bond-configuration attributes**
identifier	identifier	identifier
old-identifier	character	atoms
name	strength	bonds
type	length	empirical-formula
chiral-p	type	molecular-weight
formal-charge	atoms	charge
electronegativity	parent-abc	equivalent-atoms
electron-withdrawing-substituent-pacidity	progenitors	equivalent-bonds
	db-bond	weakest-bond
connectivity	**bond-methods**	heat-of-formation
hybridization	methods	entropy-of-formation
conjugated-p	find-bond-chains	free-energy-of-formation
p-orbitals	cleave-bond	homo
open-approach-p	compute-bond-strength	lumo
radical-orbital	remove-bond	progenitors
bond-angle	add-bond	environment
parent-abc	modify-bond	...
progenitors	create-bond	**atom-bond-configuration-methods**
parent-groups	**bond-selectors**	setup-atom-bond-graph-descriptor
neighbor-atoms	same-bond-type-p	atom-bond-graph-descriptor
neighbor-groups	bond-equivalence-p	setup-atom-bond-graph-from-old-graph
bonds	alpha-bonds	setup-atom-bond-configuration
database-atom	beta-bonds	make-graph-from-symmetric-connectivity-matrix
atom methods	gamma-bonds	
compute-connectivity-number	equivalent-bonds-p	make-bonds-from-connectivity-list
compute-parent-groups	terminal-bond-p	make-atom-bond-configuration
compute-neighbor-groups	internal-neighbor bonds	equivalent-atom-bond-configuration-p
compute-hybridization	terminal-bond-for-additions-p	make-old-arc-to-new-arc
compute-p-orbitals		identify-bond-printed-representation
compute-formal-charge		correct-bond-number-p
compute-radical-orbital		atom-symmetry-identification
identify-electron-withdrawing-substituent		bond-symmetry-identification
		enumerate-all-atom-chains
find-atom-chains		identify-atom-bond-configuration-environment
atom-backbone		
atom-degree		identify-atom-bond-configuration
higher-degree-atom-p		compute-equivalent-bonds
compute-open-approach-p		compute-weakest-bond
identify-conjugated-p		compute-equivalent-atoms
atom-specific-selections		...
create-atom		
abstract-grouping		
1,2-mobile-atom-p		
1,5-mobile-atom-p		
mobile-univalent-atom-p		

helpful when a characteristic of the atom is used by several different procedures, whether they are internal or external to the object class that contains them. For example, various operations may require knowledge about an atom's mobility as in the case of radical rearrangements; rearrangement alternatives are accessed using methods illustrated by *1,2-mobile-atom-p* and *1,5-mobile-atom-p*.

The remaining methods evaluate properties of an atom or properties of the parent structure. For example, *atom-backbone* identifies the skeleton of interest; *find-atom-chains* enumerates all the paths emanating from the specified atom; *abstract-atom-grouping* produces metagroups around reaction centers. Partitioning of the parent structure into active and inactive sites enables efficient manipulation during pathway construction, as we will see in Section III. New atoms are constructed using the method, *create-atom*. Selector functions, implemented as methods, are also built into the class atom. These methods provide an efficient means of collecting atom, which contain some specified desired property. Illustrative examples of selector methods include *collect-sp2-atoms*, *collect-oxygen-atoms*, *collect-beta-neighbors*, and *collect-terminal-atoms*.

b. Modeling Element 2: bond. Atoms are connected by bonds. Like atoms, each bond has associated with it structural information and models that manipulate this information to deduce new features that describe it. Bonds know the atoms that describe it. It is through these elements (bonds) that information is transferred from entity to entity, and new connectivity information is deduced.

A partial listing of the describing attributes and methods operating on object class bond is presented in Table II. Notice that the atom's attribute "value" allows information flow from the bond representation to an atom representation, whereas the "parent-abc" attribute value allows movement between various levels of abstraction within the representation (e.g., groups and structures). In addition to evaluating attribute values, bond methods are used to facilitate the construction of chemical structures. To achieve this purpose, *cleave-bond*, *add-bond*, *remove-bond*, *modify-bond*, and *create-bond* are provided. Like atom selectors, bond selectors are used to collect bonds exhibiting a designated property to facilitate processing.

c. Modeling Element 3: atom-bond-configuration. Atoms and bonds containing connectivity information and spatial relations represent an instance of the atom-bond-configuration class. All chemical structures can be defined by an instance of the atom-bond-configuration class. This structure, although in the strictest sense is not a primitive, has been elevated to a primitive to reduce complexity in subsequent reasoning. Associated with the `atom-bond-configuration` class are all methods and properties indigenous to an abstract chemical structure, such as physical and spatial properties. Thus, each specialized chemical structure is constructed from an atom-bond-configuration. An atom-bond-configuration in conjunction with the constituent atoms and bonds allows us to capture and isolate the structural features of a configuration from the feature that characterizes its behavior.

Generic chemical structures are represented by the object class atom-bond-configuration. These attributes describing this class include "name," "identifier," "atoms," "bonds," "empirical formula," and "frontier molecular orbital" (FMO). Since atoms and bonds are objects, the values of the attributes describing these entities are the set of objects constituting the atom list and the bond list, respectively. As a consequence, the values of the attributes describing these objects are easily accessible to procedures invoked by the atom-bond-configuration class. For example, the evaluation of an atom-bond-configuration's highest occupied molecular orbital (HOMO) requires evaluation of the atomic orbitals (AOs) that compose it. This information, which is resident in the atoms composing the atom-bond-configuration, is accessed through selector functions that are applied by the *compute-HOMO* procedure of the class atom-bond-configuration.

Attributes common to an `atom-bond-configuration` as well as methods operating on the instances of this class, are shown in Table II. Methods of particular interest include general setup methods, such as *map-old-abc-to-new-abc*, *equivalent-atom-bond-configuration-p*, and *identify-atom-bond-configuration-environment*. Setup methods, as expected, provide a means for instantiation. *Map-old-abc-to-new-abc* is a utility method that maintains the system pointers, which is an important feature when competing pathways are simultaneously analyzed. *Equivalent-atom-bond-configuration-p* determines when two instances of the atom-bond-configuration are equivalent, whereas *identify-atom-bond-configuration-environment* accesses information on the reaction environment. Although the reaction environment is specified for most pathways of synthetic interest, in the broader domain of computer-aided chemical reasoning many systems exist in which the environment is not known a priori.

2. Modeling Elements Defining Reactive Behavior of Chemicals

The reactive behavior of a chemical structure is determined by the interaction between low-level transformations of an instance of atom-bond-configuration (abc), such as bond cleavage, bond formation, electron distribution, and the conditions in the surrounding environment. To capture all the necessary information, LCR used the following classes of modeling elements.

a. Modeling Element 4: chemical-behavior. The electronic state that characterizes chemical reactivity is captured in chemical behavior. This may result from internal or external influences or both. Assessment of these electronic states characterize radical, nucleophilic, and electrophilic behavior or combinations thereof.

The set of behaviors that defines the spectrum of a species' reactivity is determined by its electronic state, which is dictated by internal and external influences. These influences may be established by the ground state, the excited state, or the effect of an external environment on the state. The chemical-behavior class uses a set of operations that define the chemical behavior of a chemical structure and proceed as follows:

Step 1. A set of operations, S, is used to deduce and assess the structural character, s, of an abc instance: abc $\xrightarrow{S} s$.

Step 2. Given s, a set of operations, I evaluates the electronic character i of the molecular structure: $s \xrightarrow{I} i$.

Step 3. Given a set of k species with structure $s_1, s_2, \ldots, s_k$, and electronic characters, $i_1, i_2, \ldots$, a set of operations, E, assess the electronic character, e, of the surrounding environment, $\{s_1, s_2, \ldots, s_k; i_1, i_2, \ldots, i_k\} \xrightarrow{E} e$.

Step 4. Given s, i, and e of a specific chemical structure, a set of operations B establishes the set of potential instances of chemical-behavior (or cb) for the specific chemical structure:
$\{s, i, e\} \xrightarrow{B}$ cb.

The operations S, I, E, and B are methods invoked by the instances of the class chemical-behavior and form the basis for the assignment of specific reactivity of a given abc. During their execution call on other methods, they are encapsulated by the structural classes, atom, bond, and abc. An important feature of cb is that detail can be managed and persued on demand to assist in an evaluation. The structure permits the association of a set of potential behaviors with a species. These tasks can be performed at run time and the assignment made *dynamically*. For example, whether an alcohol acts as a bulk solvent, weak acid, weak nucleophile, or form an alcoholate anion, is dynamically assigned by chemical-behavior (cb) once the reaction environment is known.

b. Modeling Element 5: reaction-environment. The set of properties characterizing the environment in which a reaction is occurring is contained in reaction-environment. Attributes typical of this element include; "temperature," "pressure," "wavelength," "surface type," "species concentration," "pH," "species present," etc. Methods built into the modeling element evaluate attribute values and access information in other modeling elements.

c. Modeling Element 6: ab-initio-operator. Low-lying transformations that are characterized by bond cleavage, bond formation, and electron distribution are captured in the modeling element ab-initio-operator. These transformations may be grouped as mass transfer operations and energy transfer operations for abstraction purposes. Axioms, physical laws, rules, or constraints, obtained from the physical sciences are encoded to prevent the generation of an atom-bond-configuration obtained by applying infeasible transformations. Such knowledge may include conservation of mass, conservation of energy, charge support, Pauli exclusion principle, or valence constraints.

A specific chemical transformation or rearrangement is accomplished by calling on a set of base operations, called ab-initio-operators (K_{ai}'s), to perform the necessary elementary structural changes and electron redistribution. The base operations selected are evaluated on the reaction sites of the specified reactant(s). These low-lying operations have embedded models or constraints, such as conservation of mass and energy, that prevent the generation of infeasible structures as specified by the scope of knowledge that defines them. For example, if we have not specified operators capable of constructing delocalized bonds, then we should not expect structures that emanate from this knowledge. We have found that the knowledge embedded in these models is nearly always structural in nature reflecting for example, knowledge about orbital symmetry, atom coordination, or charge. As will be discussed later, K_{ai} operators can be used directly to generate the upper bound of theoretically feasible transformations.

d. Modeling Element 7: composite-operator. The class composite-operator contains knowledge about user specifications and mechanistic operations. Generic rate information, such as relative magnitude of rate constants (e.g., rate constants of photochemical reactions lie between

10^{10} and 10^{12} s^{-1}) and electronic state information (e.g., excited, radical, charged) are also contained in composite-operator. This information is used to limit an operator's range of applicability.

The class composite-operator contains procedures that transform a set of chemical species of predisposed behaviors into a set of products, provided that an optional set of prespecified conditions is satisfied. These conditions can be specified by the user or imposed by the system. They may encompass virtually any symbolically encodable concept: structural character, chemical behavior, spatial orientation, reaction conditions, free-energy requirements, enthalpy considerations, toxicity, etc. An instance of the composite-operator (e.g., K) calls on the following procedures to carry out the corresponding tasks:

$K_{get\text{-}sites}$, to identify the potential sites for reaction, by assessing the chemical-behavior of each atom-bond-configuration.

K_t, to transform a set of potential reaction sites, utilizing instances (e.g., K_{ai}) of ab-initio-operator. Successful applications of K_{ai} operators generate new instances of atom-bond-configuration.

K_f, to return a Boolean value of "true," when the encoded set of prespecified conditions are achieved, and "false," otherwise.

LCR uses a set of predefined instances of K to represent a set of known transformations utilizing specific ab-initio-operators. New instances K of composite operators can be easily built by the user through the aggregation of different sets of ab-initio-operators, using the facilities of LCR.

3. Modeling Elements for Reactions and Pathways

The successful application of a composite operator on a chemical structure, or a set of structures, constitutes a reaction:

$$R: \{abc_1\} \xrightarrow{K} \{abc_2\}.$$

a. Modeling Element 8: reaction. This modeling class captures the information that is specific to a given transformation. An instance of reaction contains information about the species determining which are the reactants and which are the products. It also contains a reference to the instance of the composite operator K, which is responsible for the transformation, and the instance of reaction-environment, which supplies the reaction conditions.

Instances of reaction constitute the building blocks for the construction of a reaction pathway P, which is modeled as an ordered list of reactions:

$$P \triangleq \{R_1, R_2, \ldots, R_n\}, \text{ where } R_1: \{abc_{i-a}\} \xrightarrow{K_i} \{abc_i\}.$$

The transitivity of semantic relations in LCR (see Section I.C) allows the system to automatically construct an instance of reaction as the abstract representation of a pathway. For example, if R_1: A → C and R_2: B → C, then LCR constructs automatically R_3: A → C and includes it in the library of reactions for future pathway contraction.

b. Modeling Element 9: context. A context is a consistent set of assumptions that characterize a species behavior or structural character. If one wants to change the description of some modeling elements (i.e., associate a new set of assumptions with them), but also preserve the former version, then one assigns a new context to the new version, keeping all other information the same. The use of context is particularly helpful when a reactant can exist in several forms, such as resonance structures, or keto enol forms. If a context is associated with a pathway, assumptions leading to the selection of a particular behavior can be modified and the pathway adjusted accordingly without reinitializing the entire system.

4. Modeling Elements to Describe Quantitative Relationships

In all the previous modeling elements, the need arises for establishing equations, inequalities, Boolean, ordinal or order-of-magnitude relationships among the attributes (variables and parameters) of the modeling elements. To capture these quantitative relationships, LCR uses the modeling classes constraint and generic-variable as well as their various subclasses. Since these modeling elements have been borrowed from the modeling language MODEL.LA., we will defer their description to Section IV. For the time being it suffices to say that generic-variable provides a structured representation of any variable or parameter, and constraint does the same for a set of different types of relationships.

5. Subclasses of Modeling Elements

Each of the nine modeling elements described above possesses a basic structure of attributes and methods and are inherited by any of their instances. Nevertheless, we have found that the construction of specialized subclasses, emanating from a specific class, enriches the vocabulary of

LCR and brings it closer to the linguistic constructs used by chemists and engineers. For example, a functional group can be described by an instance of the class `atom-bond-configuration`, but having a distinct modeling element, called `group`, with its own specialized attributes and methods, it facilitates the identification of such groups in molecules and streamlines the reasoning around such specific structures. Nevertheless, in order to preserve the fact that a group is also in atom-bond-configuration, we create the class `group`, as a subclass of the class `atom-bond-configuration`. Inheritance mechanisms of object-oriented programming allow the modeling class, `group`, to inherit all attributes and methods of the mother-class, `atom-bond-configuration`. Using this inheritance mechanism, we have developed expanded trees of modeling subclasses for four of the basic modeling elements: `atom-bond-configuration, chemical-behavior, ab-initio-operator,` and `composite-operator`.

a. The `atom-bond-configuration` *Classes*. Four main subclasses emanate from the abc class: (1) `group`, representing a substructure of the atom-bond-configuration—Additional groups may be added to or built from the list of conventional functional groups (e.g., acid from oxo and alcohol); (2) `ion`, representing any atom-bond-configuration that has a net electrical charge; (3) `radical`, representing any atom-bond-configuration that has an open shell; and (4) `molecule`, representing any atom-bond-configuration that is not an ion or a radical. Further specialization of these primary subclasses, either through the mixing of the primary subclasses (e.g., the class radical-ion may be formed by mixing radical and ion) or by specializing the class description (e.g., creating a class organic under molecule), allows extension of the generic abc-class.

Consider, for example, specialization of the class `molecule` (Fig. 3) into `organic-molecule` and `inorganic-molecule` and suppose `inorganic-molecule` contains a method for electron counting. This method is inherited to `organometallic-molecule`; how the method is used and combined with other methods is controlled by `organometallic-molecule`. Likewise, if methods and attributes of `organic-molecule` facilitate reasoning about the organometallic system, they can be included as well.

b. The `chemical behavior` *Classes*. Chemical behavior is classified into two main subclasses: `external-influences` and `internal-influences` (Fig. 4). Unlike the abc-class, the cb-class structure provides a means for communicating attribute values between classes, independent

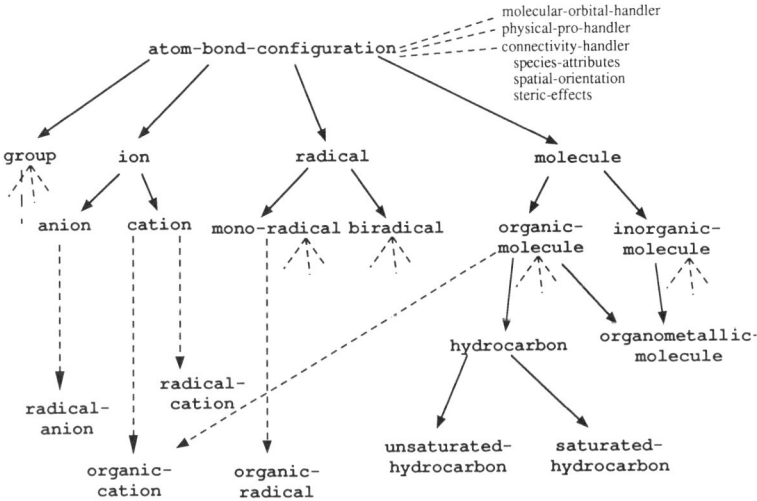

FIG. 3. Hierarchy of atom-bond-configuration subclasses.

of their position within the class structure. This is necessary because chemical behavior is often a combination of effects (e.g., the ground state of a species is influenced by its inherent electronic structure and its external environment). LCR provides a means of combining these effects. These operations constitute the basis for reactivity assignment. They are distributed throughout the hierarchy and are used to classify electronic states. The classes composing this hierarchy categorize electronic state and not the derived properties associated with an electronic state. Nucleophilicity, for example, is derived from the class pertinent-occupied-molecular-orbital (POMO). Radical behavior is derived from the class singly-occupied-molecular-orbital (SOMO).

As in the abc hierarchy, each class composing the chemical behavior hierarchy contains only those methods and attributes that pertain to it. Concepts derived at a particular class are used in classes of higher specialization so that more sophisticated concepts can be deduced and discriminating properties elucidated at the proper level. For example, the class POMO contains methods for evaluating the POMO, whether it is the HOMO, the n-HOMO, or any other occupied molecular orbital. These attributes may then be used at more specialized level to expand on the features of an electronic state that gives rise to a particular conceptual behavior. Nucleophilicity illustrates this point, because it may arise from either a negatively charged ion or a neutral atom (i.e., nucleophilicity can

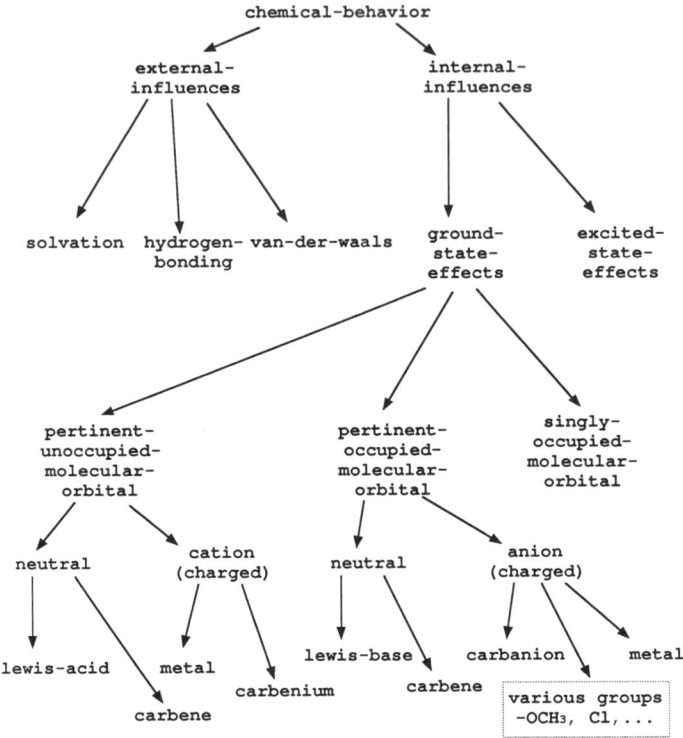

FIG. 4. Hierarchy of chemical-behavior subclasses.

result from lone pairs that are bonded, as in the case of the lone pairs on nitrogen; or nonbonded, as in the case of the electrons comprising a π bond).

An important distinction between the abc and cb hierarchy is that the cb hierarchy of classes is not treated as task specialization by operators calling for the assignment of chemical behavior. Since cb is a set of potential behaviors, {b}, it is necessary to associate behaviors to a species that lies on the same class level [e.g., pertinent unoccupied molecular orbital (PUMO) and POMO] as well as on different, more specialized, levels of the same class (e.g., Lewis-base and POMO). This is necessary because the species dynamically assumes the requisite behavior—(weak) nucleophile or Lewis base—when it is required by the procedure $K_{get\text{-}sites}$.

c. *The* ab-initio-operator *Classes*. Five primary subclasses were developed to model elementary transformations (Fig. 5) (1) bond-

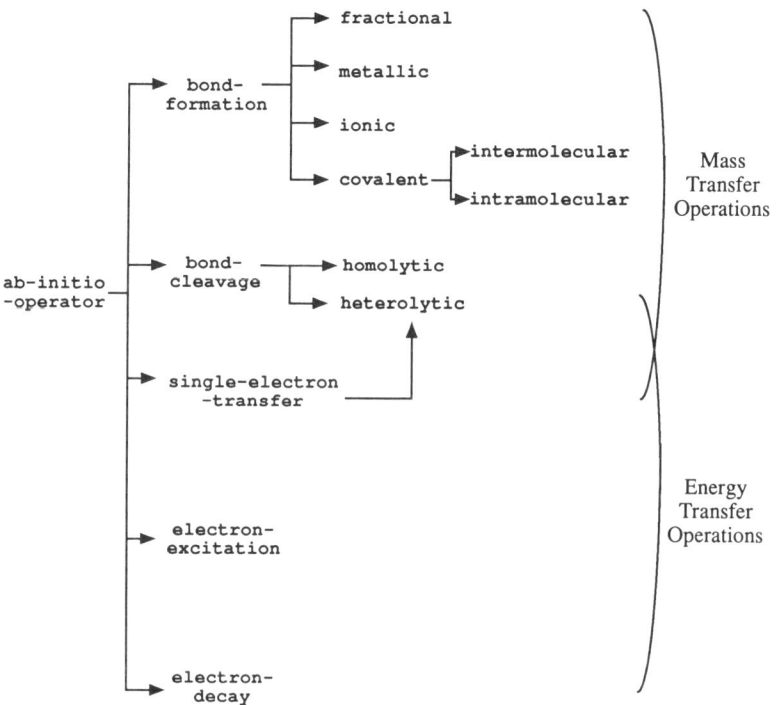

FIG. 5. Hierarchy of ab-initio-operator subclasses.

formation, (2) bond-cleavage, (3) single-electron-transfer, (4) electron-excitation, and (5) electron-decay. The operations stem from two types of operations: mass transfer and energy transfer. Each subclass contains methods that prevent the generation of theoretically infeasible structures. As in the abc-class, the subclasses of ab-initio-operator can be mixed, together with their methods, to form various new classes. For example, the subclass heterolytic-bond-cleavage is formed by combining the parent classes of single-electron-transfer and bond-cleavage. Together, these operations afford the generation of any elementary reaction mechanism.

d. The composite-operator *Classes.* The hierarchy of composite-operator classes is shown in Fig. 6. As described earlier, composite operators are built from sets of ab-initio-operators. They utilize spatial

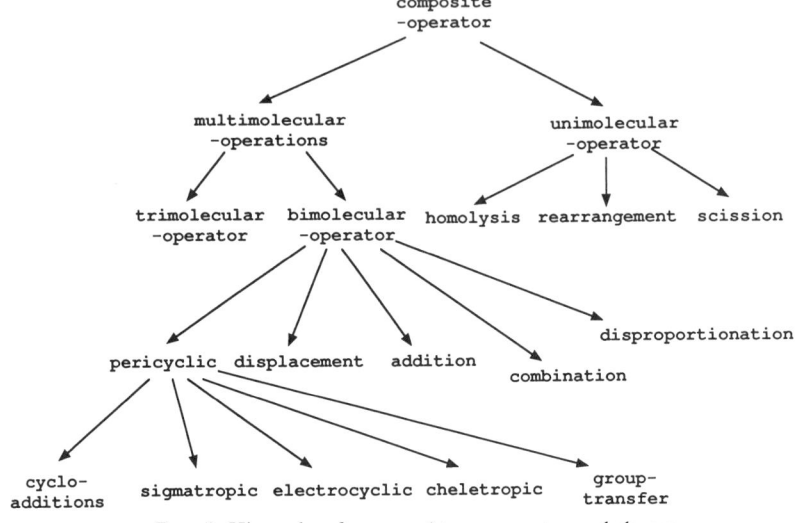

FIG. 6. Hierarchy of composite-operator subclasses.

operators, which modify the orientation of an abc. Although many hierarchies can be developed utilizing various classification schemes, for efficiency reasons, we have chosen a two-staged classification built on particle reaction requirements and specialized by structural reaction requirements. Because of this formulation, there are only two primary subclasses extending from composite-operator: multimolecular-operator and unimolecular-operator. Each subclass is then further specialized according to the structural requirements of the composite operator. For example, the bimolecular operation of radical coupling requires two radical substrates.

This classification limits the combinatorial explosion encountered in conventional computer-aided synthesis approaches, because composite operators know a priori when they apply. For example, the application of bimolecular operators to a list of chemical species results in the selective application of radical-operator to radicals contained in the list and molecule-operator to species that are molecules.

B. Semantic Relations among Modeling Elements in LCR

If the modeling elements of LCR represent the modes of a network, the semantic relationships constitute the links (edges) of the network. There exist different types of links, with each one establishing a different interpretation on the relationship of the connected nodes. Thus, whereas the modeling objects carry by themselves no assertional importance, the semantic relationships establish assertions and thus impose interpretation, i.e., meaning, to the representational objects.

1. Entity-Attribute Semantics

The first two semantic relationships establish the structure of a modeling element; they allow the declaration of the attributes and methods of a modeling element. Although all object-oriented systems possess, by definition, the first two semantic relationships and no special computer-aided provisions are needed, they have been included here for completeness.

*a. Semantic Relationship 1: "**is-attribute-of**."* This relationship enables the association of a modeling object as an attribute of an other object. It is useful when the present scope of LCR's modeling elements needs to be extended. Without this relationship a modeling would have remained a conventional low-level data structure. For example, the attributes of the object representing hydrocarbons are declared as follows:

```
hydrogen-atoms      is-attribute-of      hydrocarbon
aromatic-p          is-attribute-of      hydrocarbon
```

*b. Semantic Relationship 2: "**is-method-of**."* This relationship declares that a given computational procedure is "owned" by a particular modeling object, and that the procedure describes part of the object's behavior or computational ability. The following are typical examples of the use of this semantic relationship:

```
compute-aliphatic-p          is-method-of      hydrocarbon
collect-longest-carbon-chain is method-of      hydrocarbon
```

2. Specialization Semantics

The following two semantic relationships establish the links between a basic modeling element and its derivative subclasses of modeling objects and instances. Both of them are isomorphic mappings.

*a. Semantic Relationship 3: "**is-a**."* This is needed to establish *sub/superset* links and is used to define various new subclasses of modeling objects, emanating from a specific class. Thus, starting with a small number of basic modeling elements, this semantic relationship allows a modeling language to be expanded and remain open-ended. Each new class of modeling objects is a structure isomorphic to the modeling element from which it was derived, but with more specializations. Typical examples of its use are

`molecule`	*is-a*	`abc`
`hydrocarbon`	*is-a*	`organic-molecule`
`addition`	*is-a*	`bimolecular operator`

b. Semantic Relationship 4: "is-a-member-of." This relationship is needed to relate isomorphic structures, emanating from the same class of modeling objects, with identical specialization. For example, assume that specific assumptions on the modeling of free radicals lead to the class of modeling objects called `radical`. Then if specific free radicals, `radical-1` and `radical-2`, are modeled under the same set of assumptions, the resulting models will have the same internal structure as `radical`, but different assigned names and values to the various variables or attributes describing it. Consequently

`radical-1`	*is-a-member-of*	`radical`
`radical-2`	*is-a-member-of*	`radical`

3. Specification Semantics

Specification mappings, described by the following six semantic relationships, are used to specify the value of attributes of various modeling elements. They express, (a) binary relations, such as whole/part links, (b) communication lines among modeling objects, or (c) the value of simple describing properties.

a. Semantic Relationship 5: "is-composed-of." It defines the link between a modeling object and other modeling objects in which the former is decomposed. For example, to represent a `free-radical-reaction`, one may want to disaggregate it to the level of `initiation-reaction, propagation-reaction,` and `termination-reaction`. In such case

free-radical-reaction	*is composed-of*	{initiation-reaction; propagation-reaction; termination reaction}

b. Semantic Relationship 6: "is-part-of." This relationship defines the link between a modeling object and the modeling object that is containing it. Continuing with the example presented above, we can specify that

    initiation-reaction   *is-part-of*   free-radical-reaction

In particular, this semantic relation is established when the attribute "parent" of the modeling element `initiation-reaction` receives the value `free-radical-pathway`. The *"is-part-of"* relationship deals with structural containment—the *"is-a"* relationship describes conceptual containment. It is obvious that one of the properties of the relation is *transitivity*, and that *"is-composed-of"* and *"is-part-of"* are symmetrical relationships.

c. Semantic Relationship 7: "is-attached-to." This is the first of two semantic relationships used to elucidate the structural connectivity among specific modeling elements, such as *atoms* to form *molecules*. A typical example is

    oxygen-1   *is-attached-to*   (carbon-1; carbon-2)

The *is-attached-to* semantic relationship establishes linkages between different modeling elements, allowing flow of information from one to the other(s). Furthermore, since atoms know their membership into various functional groups, auxiliary links are automatically established; thus LCR creates automatically the following semantic connection:

    oxygen-1   *is-attached-to*   oxo-1

or

    oxygen-1   *is-attached-to*   carboxyl-1

d. Semantic Relationship 8: "is-connected-by." This is the symmetric relationship of *is-attached-to*. Using the previous examples we have

        oxygen-1   *is-connected-by*   carbon-1
        oxygen-1   *is-connected-by*   carbon-2

e. Semantic Relationship 9: "is-described-by." At a basic level, the state of a modeling object is described by the values of a set of variables. Moreover, its function or behavior is described by mathematical relationships (or procedures). This semantic relationship is used to declare the variables and mathematical relationships which describe the state of a modeling object, for example:

`oxygen-1`	*is-described-by*	"hybridization-oxygen-1"
S_n2	*is-described-by*	$K\text{-}get\text{-}sites\text{-}S_n2$.

The first example indicates that the object oxygen-1 is described by the value of the attribute "hybridization", and the second example shows that the behavior of object, S_n2 (an instance of class `reaction`) is described by the procedure $K\text{-}get\text{-}sites\text{-}S_n2$.

f. Semantic Relationship 10: "is-describing." This is the inverse of the previous relationship:

"hybridization-oxygen-1"	*is-describing*	`oxygen-1`
$K\text{-}get\text{-}sites\text{-}S_n2$	*is-describing*	S_n2

LCR has the appropriate mechanisms to keep symmetrical semantic links properly updated, when one of them changes (e.g., is created, deleted, or modified).

g. Semantic Relationship 11: "is-characterized-as." This semantic relationship specializes a class by defining the character (i.e., value) of some of its properties. It represents the semantic relation between a modeling object and an attribute of its description. For example, if the class `pertinent-occupied-molecular-orbital` is described as homolytic, and nucleophilic, semantic links like

POMO	*is characterized-as*	`homolytic`
POMO	*is-characterized-as*	`nucleophilic`

are established between the modeling objects and the attributes of its description. This semantic relationship is particularly important in defining the context of a particular pathway. For example, after a pathway is constructed, the user may wish to change the characterization of an intermediate (e.g., Lewis acid versus nucleophile) so that the implication of the change can be investigated. This is accomplished easily with the semantic relationship *is-characterized-as*. The axioms of commutativity and merging are supported by this relation.

4. Aggregation / Disaggregation Semantics

Aggregation (or, abstraction) is the process by which details of a modeling object are left unspecified in favor of a less cluttered description of its structure. Disaggregation (or refinement) is the opposite process. Thus, abstraction maps a set of modeling objects into an object described by a simpler modeling element.

a. Semantic Relationship 12: "is-disaggregated-in." This semantic relation exists between modeling elements located in different contexts and responds to the need of breaking down systems into smaller, more tractable components, where additional detail can be added. For example, a theoretically feasible but unsubstantiated pathway can be characterized by a set of assumptions that lead to the structural character defining the constituent intermediates. This same pathway may be characterized by another set of assumptions, at a later stage, that leads to a different set of intermediates. Different contexts allow us to encapsulate the distinct representation. However, if we seek to transfer information from the pathway to the species in which it is broken we have to create a communication route. The semantics links

global-pathway-1	*is-disaggregated-in*	initiation-pathway-1
global-pathway-1	*is-disaggregated-in*	propagation-pathway-1
global-pathway-1	*is-disaggregated-in*	termination-pathway-1

provide the communication vehicle.

Notice that the *is-disaggregated-in* link is designed to support communications between objects of the *same type in different contexts*. In this way it is differentiated from *is-composed-of*. The scope of *is-composed-of* is restricted to objects of any type provided they are in the *same context*. The behavior of *is-disaggregated-in* is similar to that of *is-composed-of* with respect to the properties that it supports. This semantic link obeys the axioms of transitivity, commutativity, and merging.

b. Semantic Relationship 13: "is-abstracted-by." It is the inverse of the previous one, and using the same example, we have

{initiation-pathway-1, propagation-pathway-1, termination-pathway-1}

is-abstracted-by   global-pathway-1

5. Properties of LCR's Semantic Relationships

The semantic relations of LCR establish how different objects relate to one another. They were formally defined to obey the following three axioms:

1. *Transitivity*:

   If (O1 "semantic-relation" O2) and (O2 "semantic-relation" O3)
   then
   (O1 "semantic-relation" O3)

   where "semantic-relation" applies to the relations `is-a`, `is-composed-of`, and `is-part-of`, and O1, O2, and O3 are arbitrary modeling objects of this system.

   For example,

   If (`Methyl-1` *is-part-of* `Ethyl-1`) and (`Ethyl-1` *is-part-of* `Ethane-1`)
   then
   (`Methyl-1` *is-part-of* `Ethane-1`)

2. *Commutativity*:

   (THE "*attribute-1*" OF O1 IS A_1)
   ....
   (THE "*attribute-1*" of O1 IS A_{j-1})
   (THE "*attribute-1*" of O1 IS A_j)
   ....
   (THE "*attribute-1*" of O1 IS A_n)

is the same as

   (THE "*attribute-1*" of O1 IS A_1)
   ....
   (THE "*attribute-1*" of O1 IS A_j)
   (THE "*attribute-1*" of O1 IS A_{j-1})
   ....
   (THE "*attribute-1*" of O1 IS A_n)

where "*attribute-1*" is any specific attribute establishing one of the following semantic relationships: "***is-composed-of***," "***is-attached-to***," "***is-connected-by***," "***is-described-by***," "***is-describing***," and "***is-characterized-as***." In other words, the order in which attributes are listed is irrelevant.

3. *Merging*:

   (THE "*attribute-1*" OF O1 IS A_1)
   ....
   (THE "*attribute-1*" of O1 IS A_{j-1})
   (THE "*attribute-1*" of O1 IS A_j)
   ....
   (THE "*attribute-1*" of O1 IS A_n)

is the same as

(THE "*attribute-1*" of O1 IS SET.OF $(A_1 \ldots A_{j-1} A_j \ldots A_n)$)

where "*attribute-1*" is any specific attribute establishing one of the following semantic relationships: "*is-composed-of*," "*is-attached-to*," "*is-connected-by*," "*is-described-by*," "*is-describing*," or "*is-characterized-as*." Hence, attributes of the same concept can be merged. For example,

(THE Neighbor-atoms of Carbon-1 IS Hydrogen-17)
(THE Neighbor-atoms of Carbon-1 IS Nitrogen-21)
(THE Neighbor-atoms of Carbon-1 IS Carbon-2)

is the same as

(THE Neighbor-atoms of Carbon-1
IS THE SET OF
(Hydrogen-1, Nitrogen-21, Carbon-2))

C. SYNTAX OF LCR

Grammars are intended to capture what we may call the *structure* or *syntax* of languages, as opposed to their meaning or semantics. The syntax of a programming language is a strict, precise set of rules that describe the string of symbols that constitute legal statements and specify how a statement breaks down into its constituent parts. The syntax description is done using formal grammar, often referred as *metalanguage* (i.e., special language used to describe other languages). The description language used here is an extended BNF (Backus–Naur form or Backus–normal form), which is more compact than BNF. The BNF (Naur, 1963) is a widely used formal method developed by computer scientists for the precise syntactic description of computer languages. The BNF grammars conventionally have four elements: (1) a *terminal vocabulary* that corresponds directly to the vocabulary of allowed tokens of the language to be defined; (2) a *nonterminal vocabulary*, conventionally enclosed in pointed brackets ($\langle$ and $\rangle$); (3) a *set of production rules* defining way of building up phrases (represented by nonterminal symbols) from terminal and nonterminal vocabulary items; and (4) a *correspondence* indicating which nonterminal symbol is associated with the "master" or "sentence" phrase type of the language. In defining the grammar we will follow the standard practice of writing a production beginning with the "master" phrase type at the top of the grammar.

Two important observations can be made about BNF grammar rules. First, BNF rules can be defined in a recursive manner; that is, the phrase name can appear on both sides of the symbol "::= ." This property can be observed in the rules defining "⟨structural-def⟩," "⟨input-def⟩," "⟨output-def⟩," etc. This recursive property allows an infinite number of statements to be described by a finite number of rules. Second, a BNF description of a grammar is complete when every phrase name has been defined reaching the level of terminal vocabulary.

1. *Explanation of Notation*

   ::= Is defined as.
   [ ] Encloses optional unit.
   { } Indicates mandatory choice.
   ⟨ ⟩ Unit that is described separately.
   ... Indicates repetition of syntactic signs.
   | Separator for alternatives.
   *   What the braces enclose may appear any number of times (including zero).
   +   What the braces enclose may appear any non-zero number of times (must appear at least once).
   [[ ]] Double brackets indicate that any number of the alternatives enclosed may be used, and those may occur in any order, but each alternative may be used at most once unless followed by a star.

2. *Examples of Syntax*

   Word in capital letters: reserved word.
   Lowercase words: metalinguistic variable names.
   ⟨modeling-element⟩::= ⟨modeling-element-name⟩[⟨attribute⟩]*.
   ⟨modeling-element-name⟩::= (⟨modeling-class⟩|⟨atom⟩|
   　　　　　　　⟨bond⟩|⟨reaction⟩|
   　　　　　　　⟨reaction-environment⟩|⟨context⟩)
   　　　　　　　[attribute⟩]*[⟨method⟩]$^+$.
   ⟨attribute⟩::= ⟨attribute-name⟩ IS-ATTRIBUTE-OF
   　　　　　　　⟨modeling-element⟩.
   ⟨method⟩::= ⟨method-name⟩ IS-METHOD-OF ⟨modeling-element⟩.
   ⟨class⟩::= (⟨class-name⟩ IS-A {[⟨modeling-class⟩]$^+$|[⟨class⟩]$^+$}).
   ⟨instance⟩:: +(⟨instance-name⟩ IS-MEMBER-OF {[⟨class⟩]$^+$}.

⟨abc⟩::= ⟨abc-input⟩.
⟨abc-input⟩::= be-matrix|atom-table bond-table.
⟨abc⟩::= ⟨group⟩|⟨ion⟩|⟨radical⟩|⟨molecule⟩.
⟨group⟩::= [atom]$^+$[bond]$^+$.
⟨ion⟩::= [atom]$^+$[bond].
⟨radical⟩::= [atom]$^+$[bond].
⟨molecule⟩::= [atom]$^+$[bond]$^+$.
⟨cb⟩::= {[⟨behavior⟩]$^+$}.
⟨behavior⟩::= (IS-CHARACTERIZED-BY {⟨Internal-Influences⟩}
            [⟨External-Influences⟩]).
⟨Internal-Influences⟩::= (IS-CHARACTERIZED-BY
                    {⟨Ground-state-effects⟩
                     ⟨Excited-state-effects⟩}).
⟨Ground-state-effects⟩::= (IS-CHARACTERIZED-BY
                    {⟨PUMO⟩|⟨POMO⟩|⟨SOMO⟩}).
⟨Excited-state-effects⟩::= (IS-CHARACTERIZED-BY {$\sigma^*|\pi^*$}).
⟨External--Influences⟩::= (IS-CHARACTERIZED-BY
                    {[⟨solvation⟩][⟨hydrogen-bonding⟩]
                     [⟨Van der Waals⟩]}.
⟨composite-operator⟩::= ⟨inputs⟩⟨enabling-conditions⟩
                    ⟨abinitio-operator⟩⟨output⟩.
                    [⟨composite-operator⟩].
⟨input⟩::= ⟨react-cond⟩⟨abc⟩⟨cb⟩.
⟨output⟩::= ⟨abc⟩.
⟨abc⟩::= {[(Instance-of-abc)]$^+$}.
⟨cb⟩::= {[(Instance-of cb)]$^+$}.
⟨reaction-cond⟩::= {(Instance-of reaction-conditions)}.
⟨enabling-conditions⟩::= {[⟨conditionals⟩]$^+$}.
⟨conditionals⟩::= {[[(Instance-of user preferences)]$^+$
            {[(Instance-of physical requirements)]$^+$}]}.
⟨abinitio-operator⟩::= {[⟨Kai⟩]$^+$}.
⟨K_{ai}-operator⟩::= Bond formation|Bond Cleavage|
            Single Electron Transfer|Electron Excitation|
            Electron Decay.
⟨K_{ai}⟩::= ⟨K_{ai}-operator⟩⟨ − ⟨K_{ai}-input⟩⟨K_{ai}-enabling-cond⟩⟨output⟩.
⟨K_{ai}-input⟩::= {[⟨reaction-center⟩]$^+$}.
⟨reaction-center⟩::= {[⟨atom⟩]$^+$}.
⟨K_{ai}-enabling-cond⟩::= {[⟨K_{ai}-condition⟩]$^+$}.
⟨K_{ai}-condition⟩::= {[physico-chemical-requirements]$^+$}.
⟨physical-requirements⟩::= (IS-ABSTRACTING
                    ⟨physio-chemical-requirements⟩).

III. Formal Construction of Representations for Chemicals and Reactions

The nine basic modeling elements of LCR, although they offer a fairly limited vocabulary, are generic enough to allow (1) *infinite extensibility* of LCR's vocabulary through a finite set of rules, and (2) *representation and analysis* of any potential reaction pathway.

In this section we will examine the mechanisms that LCR possesses to achieve these objectives.

A. Extension of LCR's Modeling Objects

In Section II.B we discussed the modeling subclasses emanating from the four basic elements: atom-bond-configuration, chemical-behavior, ab-initio-operator, and composite-operator. Let us examine what mechanisms LCR possesses for such linguistic extensibility, which is imperative for the representation of the widely diverse knowledge of chemistry.

1. Extension of the atom-bond-configuration Hierarchical Tree

Suppose it is advantageous to reason about cyclic aliphatic species directly. For this purpose, we create the modeling class cyclic-aliphatic and link it to the parent class aliphatic (Fig. 7), thereby establishing a set membership:

cyclic-aliphatic   *is-a*   aliphatic

Since aliphatic is a specialization of the class hydrocarbon, it is advantageous to create a link between aliphatic and hydrocarbon (Fig. 7):

aliphatic   *is-a*   hydrocarbon

This allows the attributes describing hydrocarbon to be inherited by aliphatic. Similarly, the link established between aliphatic and cyclic-aliphatic allows this information to be accessed from cyclic-aliphatic directly. Having created these links, any attribute or method residing in the class aliphatic or any of its superclasses is inherited by cyclic-aliphatic. For example, if methods *cyclic-p*,

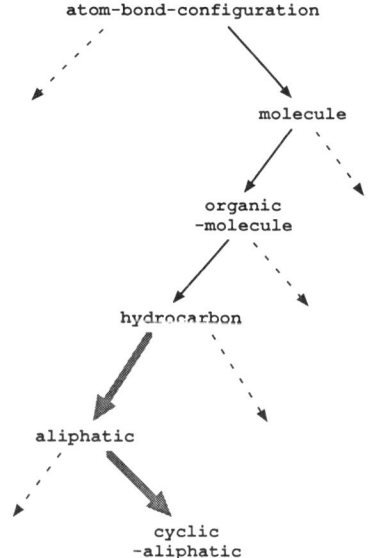

FIG. 7. Extending the atom-bond-configuration tree of subclasses.

compute-nonbonded-strain, and *compute-torsional-strain* reside in atom-bond-configuration and methods *compute-skeleton-chains* and *compute-longest-carbon-chain* reside in organic-molecule, each of these methods as well as any additional descriptive attributes are directly accessible from the modeling class cyclic-aliphatic; no duplication of modeling effort is required. How methods and attributes are combined or mixed into the cyclic-aliphatic class is user-controlled, through LCR's semantic relationships.

It may also be desirable to extend the utility of the cyclic-aliphatic class and its descriptors through the addition of class-specific methods and attributes (Table III). For example, the user may wish to include the following attributes in the class description:

ring-strain	*is-attribute-of*	cyclic-aliphatic
number-of-rings	*is-attribute-of*	cyclic-aliphatic
ring-sizes	*is-attribute-of*	cyclic-aliphatic
ring-number	*is-attribute-of*	cyclic-aliphatic
torsional-strain	*is-attribute-of*	cyclic-aliphatic
nonbonded-strain	*is-attribute-of*	cyclic-aliphatic
...		

TABLE III
SELECT ATTRIBUTES AND METHODS FOR cyclic-aliphatic
AND excited-state-effects CLASSES

cyclic-aliphatic	excited-state-effects
cyclic-aliphatic attributes	**excited-state-effects attributes**
identifier (inherited from super class)	identifier (inherited from super class)
old-identifier (inherited from super class)	old-identifier (inherited from super class)
name (inherited from super class)	bonding-orbital (inherited from super class)
parent-abc (inherited from super class)	anti-bonding-orbital (inherited from super class)
progenitors (inherited from super class)	dominate-behavior (inherited from super class)
...(inherited from super class)	competitive-behavior (inherited from super class)
ring-strain	...(inherited from super class)
number-of-rings	S_0
ring-sizes	S_1
ring-number	S_2
torsional-strain	S_3
nonbonded-strain	T_0
bridged-atoms	T_1
...	T_2
cyclic-alphatic methods	...
compute-identifier (inherited from super class)	**excited-state-effects methods**
compute-old-identifier (inherited from super class)	compute-identifier (inherited from super class)
compute-parent-abc (inherited from super class)	compute-old-identifier (inherited from super class)
compute-progenitors (inherited from super class)	compute-bonding-orbital (inherited from super class)
...(inherited from super class)	compute-anti-bonding-orbital (inherited from super class)
compute-name (super class method modified)	compute-dominate-behavior (inherited from super class)
compute-ring-strain	compute-competitive-behavior (inherited from super class)
compute-number-of rings	...(inherited from super class)
compute-ring-sizes	compute-S_0
compute-ring-number	compute-S_1
compute-torsional-strain	compute-S_3
compute-nonbonded-strain	compute-T_0
compute-bridged-atoms	compute-T_1
monocyclic-ring-p	compute-T_2
bicyclic-ring-p	compute-T_3
fused-ring-p	assess-energy-gap
...	allowed-transition-p
	forbidden-transition-p
	fluorescence-p
	phosphorescence-p
	internal-conversion-p
	internal-system-crossing-p
	...

The values of these attributes are then established by the methods that have been associated with `cyclic-aliphatic`. Methods indicative of this class include the following:

compute-ring-strain	*is-method-of*	`cyclic-aliphatic`
compute-number-of-rings	*is-method-of*	`cyclic-aliphatic`
compute-ring-sizes	*is-method-of*	`cyclic-aliphatic`
compute-nonbonded-strain	*is-method-of*	`cyclic-aliphatic`
compute-bridged-atoms	*is-method-of*	`cyclic-aliphatic`
monocyclic-ring-p	*is-method-of*	`cyclic-aliphatic`
bicyclic-ring-p	*is-method-of*	`cyclic-aliphatic`
fused-ring-p	*is-method-of*	`cyclic-aliphatic`

In these methods, those with the "compute-" prefix determine attribute values; those with the "-p" suffix are predicate methods returning Boolean values. The remaining methods are invoked on instances of the class `cyclic-aliphatic` and are designed to raise the abstraction level (i.e., they allow the user to communicate using higher-level concepts). These methods may be called from outside of the class `cyclic-aliphatic` provided they operate on instances of the `cyclic-aliphatic` class. The method *bridged-atoms* illustrates this concept; it calls *bridged-atom-p*, a predicate method of atom, from the class `cyclic-aliphatic`. Likewise, if reactivity assessment requires knowledge about ring structure, we can access that information by applying the semantic relation *is-describing* to an instance of `cyclic-aliphatic` (ICA), as shown below:

bridged-atoms-ICA   *is-describing*   (`atom-1...atom-n`)

Notice also that although the methods, *compute-nonbonded-strain* and *compute-torsional-strain* were inherited by `cyclic-aliphatic`, more sophisticated versions of these methods were supplied to `cyclic-aliphatic`. This would have been unnecessary if the original versions of these methods supported fully *all* cyclic configurations (i.e., a robust method given at any ring configuration). We have found that the ability to incrementally improve methods accelerates model development.

2. Extension of the `chemical behavior` *Hierarchical Tree*

Suppose that we choose to embellish the modeling class `chemical-behavior` by developing the subclass `excited-state-effects`. We begin by developing a subclass architecture consistent with the existing organization: characterization of the electronic state. In this spirit, we have specialized the class `internal-influences` into the modeling

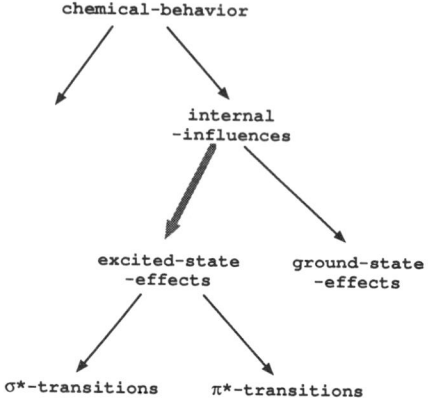

FIG. 8. Extending the chemical-behavior tree of subclasses.

subclasses ground-state-effects and excited-state-effects. We do this using the semantic relation *is-a* (Fig. 8):

excited-state-effects   *is-a*   internal-influence

Further, we specialize excited-state-effects according to the accompanying electronic transitions: transitions to σ^* and transition to π^*; and link these subclasses to excited-state-effects as shown below (see also Fig. 8):

σ^*-transitions   *is-a*   excited-state-effects
π^*-transitions   *is-a*   excited-state-effects

The addition of descriptive attributes and methods to each of these new modeling classes further characterizes each class. To characterize the modeling class excited-state-effects, we add attributes and methods descriptive of singled (S) and triplet (T) states. The states are refined further by energy level. For example ground states (S_0, T_0) are differentiated from their excited states (S_1, T_1) as well as from their higher states $(S_2, T_2,$ and $S_3, T_3)$. As before, we link these attributes using the semantic relationship *is-attribute-of*:

S_0 (or, S_1, S_2, S_3)   *is-attribute-of*   excited-state-effects
T_0 (or, T_1, T_2, T_3)   *is-attribute-of*   excited-state-effects

Additional attributes may also be added. The degree to which it is advantageous to add additional class descriptors (e.g., attributes, methods) is defined by the scope of the modeling efforts.

Methods characterizing excited-state-effects are linked to their associated class using the semantic relation *is-method-of*. These methods can be grouped according to whether the result is used by the attribute's compute method:

compute-S_0 (or, -S_1,-S_2,-S_3) *is-method-of* `excited-state-effects`
compute-T_0 (or, -T_1,-T_2,-T_3) *is-method-of* `excited-state-effects`

or by methods used to derive additional properties of the attribute space. The latter group may include generic methods such as *assess-energy-gap*, which computes the energy difference between two energy states (e.g., S_1 and S_2 or S_1 and T_1); or predicate functions that access feasibility. Several of these are shown below:

assess-energy-gap	*is-method-of*	`excited-state-effects`
allowed-transition-p	*is-method-of*	`excited-state-effects`
forbidden-transition-p	*is-method-of*	`excited-state-effects`
fluorescence-p	*is-method-of*	`excited-state-effects`
internal-conversion-p	*is-method-of*	`excited-state-effects`
internal-system-crossing-p	*is-method-of*	`excited-state-effects`

Together these elements give the modeling class' utility. Notice that the capability of moving electrons between energy levels requires that `atom` or `bond` have attributes that describe atomic and molecular orbitals. Given these attributes, the movement of an electron from one energy level to another can be orchestrated by methods residing in `atom`, `bond`, and `ab-initio-operator` (e.g., *ionization* or *single-electron-transfer*). In the latter case, a well-defined abstraction barrier is established to help maintain program modularity. Clearly, if the modeling effort does not require analysis of electron movement, the functionality need not be included. If it is required, LCR has the capability to expand and accommodate the modeling effort regardless of scope. A state description of `excited-state-effects`, after attribute and method addition, is shown in Table III.

Similarly, the subclasses of `excited-state-effects`, i.e., σ^*-transitions and π^*-transitions, are described by the transitions which characterize them. These include σ to σ^*, n to σ^*, n to π^*, and π to π^* transitions. The corresponding attributes and methods are linked to their respective classes as described earlier. Very briefly, they are shown below:

$\sigma \to \sigma^*$	*is-attribute-of*	σ^*-`transitions`
$n \to \sigma^*$	*is-attribute-of*	σ^*-`transitions`
...		
$\pi \to \pi^*$	*is-attribute-of*	π^*-`transitions`
$n \to \pi^*$	*is-attribute-of*	π^*-`transitions`
...		

and

compute-σ → σ*-transition	**is-method-of**	σ*-transitions
*compute-*n *→ σ*-transition*	**is-method-of**	σ*-transitions
...		
compute-π → π*-transition	**is-method-of**	π*-transitions
*compute-*n *→ π*-transition*	**is-method-of**	π*-transitions
...		

We can also specialize a modeling class by characterizing several of its dominant or critical properties. For selected operations, such as composite or ab initio operations, specialization can further refine the search space and afford greater (search) efficiency.

The semantic relationship *is-characterized-as* is used for this purpose. It provides a window for viewing the more interesting behaviors/structures associated with a class, such as

excited-state-transition	*is-characterized-as*	$\pi \to \pi^*$
excited-state-transition	*is-characterized-as*	$n \to \pi^*$

or

electronic-state	*is-characterized-as*	excimer
electronic-state	*is-characterized-as*	exciplex

Using this semantic relation, the user may investigate alternative pathways by specifying a different characterization. Consistency is maintained by the multiple-context modeling utility of LCR.

3. Extension of the `ab-initio-operator` *Hierarchical Tree*

Our attention now turns to the operators: methods that perform the actual chemical transformations. Bond formation, bond cleavage, electron excitation, and electron decay are a few such examples. These operators lie at the heart of LCR.

We will illustrate the development of these operators by creating the ab initio modeling element, `covalent-bond`. We will establish in this development: (1) the communication protocol between `covalent-bond` and `ab-initio-operator`, (K_{ai}); (2) the criteria on enabling conditions; (3) how criteria can be relaxed, tightened, augmented, or supple-

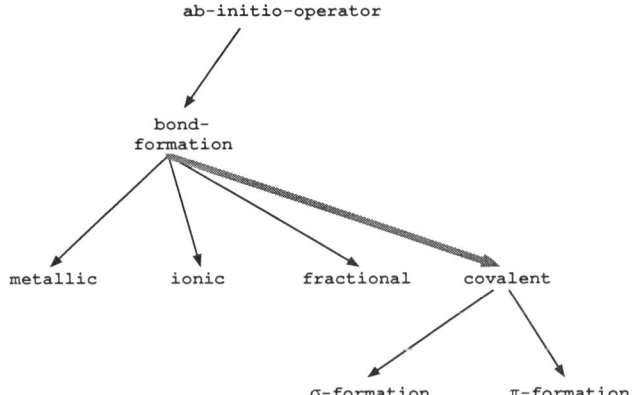

FIG. 9. Extending the ab-initio-operator tree of subclasses.

mented; and (4) how additional functionality can be added to covalent-bond (reflecting advances in our understanding of the bonding mechanism).

We initiate the development of covalent-bond by creating the modeling element and establishing the normal superclass links using the semantic relationship, *is-a* (Fig. 9):

covalent-bond	*is-a*	ab-initio-operator
σ-formation	*is-a*	covalent-bond
π-formation	*is-a*	covalent-bond

However, we will not specify the attributes of covalent-bond, σ-formation, or π-formation as we did previously, where the attributes were used for purposes of bookkeeping. How the attributes are chosen and ultimately used depends on the character and generality of the supporting encoded methods.

To effect a transformation, covalent-bond requires the use of specific methods. In particular, we need to establish generic methods that are responsible for primary transformations. The dominant method of covalent-bond is *make-covalent-bond*. *Make-covalent-bond* establishes a covalent-bond between two reaction centers when enabling conditions are achieved. It also performs all the necessary bookkeeping to ensure the bond is recognizable by the instance of abc. Expressed in pseudocode convention, these methods take the following form:

METHOD: make-covalent-bond
```
        input
        initialize
        if      feasible-p
                then enable-transformation
                return
        end
        return
```

where the method *enable-transformation* performs the transformation and the method, *feasible-p*, represents the enabling conditions, $\langle K_{ai}\text{-enabling-cond}\rangle$, of the transformation.

The syntax for the definition of the subclass was given in Section II.B and is reproduced by the following seven (7) lines in Backus–Naur form (BNF):

1. $\langle K_{ai}\rangle ::= \langle K_{ai}\text{-operator}\rangle\langle K_{ai}\text{-input}\rangle\langle K_{ai}\text{-enabling-cond}\rangle\langle \text{output}\rangle$
2. $\langle K_{ai}\text{-operator}\rangle ::=$ Bond formation | Bond Cleavage | Ionization | Single Electron Transfer | Electron Excitation | Electron Decay
3. $\langle K_{ai}\text{-input}\rangle ::= \{[\langle \text{reaction-center}\rangle]^+\}$
4. $\langle \text{reaction-center}\rangle ::= \{[\langle \text{atom}\rangle]^+\}$
5. $\langle K_{ai}\text{-enabling-cond}\rangle ::= \{[\langle K_{ai}\text{-condition}\rangle]^+\}$
6. $\langle K_{ai}\text{-condition}\rangle ::= \{[\text{physicochemical-requirements}]^+\}$
7. $\langle \text{physical-requirements}\rangle ::=$ (IS-ABSTRACTED-BY $\langle \text{physicochemical-requirements}\rangle$)

Line 1 indicates that the definition of K_{ai} depends on the following four modeling objects: $\langle K_{ai}\text{-operator}\rangle$, $\langle K_{ai}\text{-input}\rangle$, $\langle K_{ai}\text{-enabling-cond}\rangle$, and $\langle \text{output}\rangle$. In turn, $\langle K_{ai}\text{-operator}\rangle$ is defined by a set of attributes (line 2), and $\langle K_{ai}\text{-input}\rangle$ is defined by a list of $\langle \text{reaction-centers}\rangle$ (see line 3) with each element of the list defined as an instance of the basic modeling element, atom.

Continuing with our task, we recognize that covalent-bond requires the specification of $\langle K_{ai}\text{-enabling-cond}\rangle$, $\langle K_{ai}\text{-input}\rangle$, and $\langle \text{output}\rangle$. We will focus on $\langle \text{output}\rangle$ and $\langle K_{ai}\text{-enabling-cond}\rangle$ since the specification of $\langle K_{ai}\text{-input}\rangle$ is transparent (i.e., a list of one or more reaction centers).

The $\langle \text{output}\rangle$ of covalent-bond is an object that represents a covalent bond between two reaction centers. This object (i.e., instance of covalent-bond) is not a single (independent) entity. It knows membership and has pointers that connect it to the new abc, of which it is a part. The new abc recognizes it as a covalent bond. The situation of a nonnil $\langle \text{output}\rangle$ being returned is dependent on the properties of the reaction centers and level of sophistication encoded into $\langle K_{ai}\text{-enabling-cond}\rangle$.

The new `abc` recognizes it as a covalent bond. The situation of a nonnil ⟨output⟩ being returned is dependent on the properties of the reaction centers and level of sophistication encoded into ⟨K_{ai}-enabling-cond⟩.

Specific criteria must be satisfied for a covalent bond to be formed between two atoms. These criteria are based on the physical laws of nature and tempered by our knowledge of the bonding system. They are encoded into the procedure ⟨K_{ai}-enabling-cond⟩.

The general algorithm for ⟨K_{ai}-enabling-cond⟩ is

⟨K_{ai}-enabling-cond⟩:
  input
  initialize
  if feasible-p
    then true
    return
  end
  return

where *feasible-p* represents a list of predicates procedures specifying the criteria for feasibility. The pseudocode procedural form of ⟨K_{ai}-enabling-cond⟩ for the modeling class, `covalent-bond`, is given below:

⟨covalent-bond-enabling-cond⟩
  input: reaction-center-pair
  initialize
    reaction-center-1 ← first element of reaction-center-pair
    reaction-center-2 ← last element of reaction-center-pair
  if favorable-interatomic-distance-p and
    potentially-stable-bond-formation-p and
    available-bonding-electron-p and
    proper-orbital-symmetry-p and
    conservation-laws-upheld-p
    then true
    return
  end
  return

The feasibility criteria, following the "`if`" construct, are implemented as predicate methods. They perform the following evaluations: the method *favorable-interatomic-distance*-p assesses the interatomic distances between the two reaction centers to determine whether bond formation is likely; the method *potentially-stable-bond-formation*-p verifies that the steric repulsion between the reactions centers is less than the potential energy of the system and that the entropic term for bond formation is less than the enthalpy term for bond formation; *available-bonding-electron*-p checks the availability of bonding electrons; *proper-orbital-symmetry*-p determines whether the bonding orbitals are in phase; and *conservation-law-upheld*-p validates whether mass, energy, and momentum have been conserved. These methods are associated with `covalent-bond` using the semantic relationship *is-method-of*, as we have shown previously, specifically:

favorable-interatomic-distance-p	*is-method-of*	`covalent-bond`
potentially-stable-bond-formation-p	*is-method-of*	`covalent-bond`
available-bonding-electron-p	*is-method-of*	`covalent-bond`

Notice that the method *conservation-laws-upheld*-p has not been associated with `covalent-bond`. This method has been associated with K_{ai}, allowing it to be accessed by all subclass modeling elements of K_{ai}.

We can also specialize these methods according to physical requirements and chemical requirements. This provides an abstraction barrier between what is true (i.e., the physical laws) and what is believed to be true (i.e., current theory). For example, suppose that the methods *favorable-interatomic-distance*-p and *potentially-stable-bond-formation*-p are physical requirements whereas the methods *available-bonding-electron*-p and *proper-orbital-symmetry*-p are chemical requirements. We represent this to `covalent-bond` by associating the top-level methods (i.e., methods associated with K_{ai}), chemical requirements, and physical requirements, to `covalent-bond` directly (i.e., we override the method inherited by the mother model class). This is accomplished using the semantic relationship **is-method-of**, as was demonstrated earlier:

physical-requirement	*is-method-of*	`covalent-bond`
chemical-requirement	*is-method-of*	`covalent-bond`

We then tailor these methods by associating the methods of `covalent-bond` to them. This is achieved using the semantic relationship *is-composed-of*:

physical-requirement *is-composed-of* (favorable-interatomic-distance-p,
potentially-stable-bond-formation-p)
chemical-requirement *is-composed-of* (available-bonding-electron-p,
proper-orbital-symmetry-p)

4. Extension of the `composite-operator` Hierarchical Tree

Suppose now that we wish to add an additional operator to the model class `composite-operator`, K. For example, consider the addition of radical-disproportionation, a transformation that captures transformations among radicals:

$$\cdot C_2H_5 + \cdot C_2H_5 \longrightarrow C_2H_4 + C_3H_8$$
$$\longrightarrow C_2H_6 + C_3H_8.$$

As before, the addition of the subclass, $K_{radical-disproportionation}$, to the model class hierarchy requires that links be established between `composite-operator` and `bimolecular-operator`, `bimolecular-operator` and `radical-operator`, and `radical-operator` and `radical-disproportionation`:

bimolecular-operator	*is-a*	composite-operator
radical-operator	*is-a*	bimolecular-operator
radical-disproportionation	*is-a*	radical-operator

Like K_{ai}, the modeling element K performs transformations. This necessitates that procedures be developed and associated with K. Unlike K_{ai}, the procedures of K are based on known transformations. To identify the procedures required of $K_{radical-disproportionation}$, we return to the BNF syntax presented in Section II.D and reproduce the following links:

1. ⟨composite-operator⟩::= ⟨inputs⟩⟨enabling-conditions⟩
   ⟨ab-initio-operator⟩⟨output⟩
   [⟨composite-operator⟩]
2. ⟨input⟩::= ⟨react-cond⟩⟨abc⟩⟨cb⟩.
3. ⟨output⟩::= ⟨reaction⟩,
4. ⟨abc⟩::= {[(instance-of abc)]$^+$}.
5. ⟨cb⟩::= {[(instance-of cb)]$^+$}.
6. ⟨reaction-cond⟩::= {(instance-of reaction-conditions)}.
7. ⟨enabling-conditions⟩::= {[⟨conditionals⟩]$^+$}.
8. ⟨conditionals⟩::= [[[(instance-of user preferences)]$^+$
   {[(instance-of physical requirements)]$^+$}]].

The syntax of K identifies the elements of any subclass modeling element (e.g., $K_{radical-disproportionation}$). These are elements ⟨inputs⟩ ⟨enabling-conditions⟩ ⟨ab-initio-operator⟩ ⟨output⟩. Each of these is defined by simpler entities, as shown above. (The decomposition of ⟨ab-initio-operator⟩ was given in the previous section.)

In addition, we have seen that a composite operator constructs new reactions through the execution of three methods, K_f, $K_{\text{get-sites}}$, and K_t. In the context of radical disproportionation reactions:

$$R_1 + R_2 \rightarrow R'_1 + R'_2,$$

these three procedures do the following:

1. K_f evaluates the feasibility of radical disproportionation (e.g., R_1 and R_2 are both radicals; energy in the reaction environment is less than that required to cleave the newly formed bond; relative stability of products, etc.)
2. $K_{\text{get-sites}}$ identifies potential reaction sites in R_1 and R_2.
3. K_t supplies a sequence of ab initio operators and applies these operators to the reaction sites identified by $K_{\text{get-sites}}$ to obtain the desired transformation.

The feasibility predicate, K_f, for $K_{\text{radical-disproportionation}}$ is similar in spirit to $\langle K_{\text{ai-enabling-cond}} \rangle$ with the exception that the user can encode personal preferences. For example, suppose for safety reasons that the user's interest is focused on a substituted product. This preference could be encoded into K_f by passing in a predicate test, e.g., *substituted-p*, that would be applied to reaction-centers.

METHOD: K_f
    input: substrate-list
    when substrates are radicals
        collect radical centers into potential-reaction-centers
        evaluate potential-bonds between potential-reaction-centers
            for each bond in potential-bonds evaluate bond-energy
                when bond is stable in reaction-environment
                and user-preferences satisfied
                      collect potential-reaction-center-pair
                            into potential-reaction sites
                      return potential-reaction-sites
            return
        return
    end
    return

Notice the usage of inputs (i.e., ⟨abc⟩, ⟨cb⟩, and ⟨reaction-condition⟩) in K_f. Substrates and substrate-list are sets of ⟨abc⟩'s; radicals are identified using ⟨cb⟩; bond stability is assessed in the context of ⟨reaction-condition⟩ (i.e., the values of the instance of `reaction-environment`). Additionally note that multiple products may be found since there may be several favorable sites; potential-reaction-sites are collected on the basis of thermodynamic stability relative to the reaction-environment.

The method $K_{\text{get-sites}}$ is responsible for producing products, given a set of potential reaction sites. It applies K_t to potential reaction center combinations and assesses the relative stabilities of the products. The procedure for $K_{\text{get-sites}}$ is given below:

>METHOD: $K_{\text{get-sites}}$
>   **input:** potential-reaction-sites
>   **for** each reaction-center-pair in potential-reaction-sites
>   apply K_t to reaction-center-pair
>   **collect** potential-structure-pair into potential-structures
>   **return** potential-structures
>   **sort** potential-structure-pair by stability
>   **when** stability-acceptable-p of potential-structure-pair
>   **collect** potential-structure-pair into product-pairs
>   **return** product-pairs
>   **end**
>   **return**

The call to K_t enables the desired transformation. This transformation (encoded below by K_t) begins by cleaving the bond connecting a mobile moiety (e.g., H) to a parent atom adjacent to reaction-center-1. Products are then produced by creating new bonds between these centers using generic methods (e.g., *make-covalent-bond* and *cleave-bond*) to achieve bond formation and bond cleavage. K_t uses these generic methods to access knowledge contained in $\mathbf{K}_{ai}$.

>METHOD: K_t
>   **input**: reaction-center-pair
>   **initialize**   reaction-center ← first element of reaction-center-pair
>   reaction-center ← second element of reaction-center-pair
>   cleave-bond mobile moiety adjacent to reaction-center-1
>   biradical ← reaction-center-1 structure
>   product-1 ← apply make-covalent-bond to mobile moiety
>   and reaction-center-2
>   product-2 ← apply make-covalent-bond to biradical
>   **end**
>   **return**

By changing the transformation code and by adding selector functions we can enhance K_t to enable additional disproportionations. These may include transformations depicting disproportionation of such radicals as R_3CO_2 and R_2HCO_2.

Finally, we associate the procedures K_f, K_t, and $K_{\text{get-sites}}$ to the modeling class **K**$_{\text{radical-disproportionation}}$ using the semantic relation **is-method-of**:

K_f	*is-method-of*	radical-disproportionation
$K_{\text{get-sites}}$	*is-method-of*	radical-disproportionation
K_{ft}	*is-method-of*	radical-disproportionation

We also establish a second (redundant) link using the semantic relationship *is-described-by*. In addition, LCR establishes automatically the following three semantic links, which are the symmetrical relationships of the preceding three:

radical-disproportionation	*is-described-by*	K_t
radical-disproportionation	*is-described-by*	$K_{\text{get-sites}}$
radical-disproportionation	*is-described-by*	K_f

Now, the modeling class, **K**$_{\text{radical-disproportionation}}$, is fully recognized and usable by **K**. An evaluation of **K**$_{\text{radical-disproportionation}}$ on R_1 and R_2 will return an instance of the class `reaction` or set of instances (reflecting multiple favored products). Each `reaction` constructed will contain values for the attributes (e.g., reactants, products, stoichiometry, reaction-environment, enabling conditions) as specified by the modeling element `reaction` (see Section II.A).

B. The "Model Class Decomposition Digraph" (MCDD)

As the examples of the previous section have indicated, the modeling objects in LCR's hierarchies are built from simpler ones by linking them into composite entities. The mechanism for the linkage is the declaration of the semantic relationship between two modeling entities. Thus, starting from one of the nine basic modeling classes (the mother-class), we can create a modeling subclass and endow it with new specialized attributes and methods. Each new attribute could be modeled as an instance of an existing or a new modeling class. This mechanism can be applied recursively until desired representation has been achieved. The assumptions that guide the subclassing of a new modeling element from an existing one, and the specification of its attributes form the scope of the *modeling context*.

Let a modeling object be represented by a node, and let an *edge* represent its semantic link to another modeling object (i.e., another node). Then, the gradual and modular definition of a model can be represented by an evolving directed graph, which is called *model class decomposition digraph* (MCDD). The MCDD represents the net result of the model class decomposition taking into account all the components created at that time. The basic idea is simple and is schematically depicted in Fig. 10. At state 1, the model root A has only one constituent entity called B. At state 2, the entity C has been incorporated into the substructure of B; but C is not a simple entity, it is a composite object itself (e.g., a covalent bond *is-described-by* a set of mathematical components). Consequently, its whole structure is pasted into B's structure. Stage 3 shows that B's substructure has been further specified by incorporating an additional composite object called D. Following with B's definition, stages 4–6 show that a new constitutive part, named E, is being gradually created. This presents a contrast with the previous incorporation of entities C and D; they had predefined substructures. This process will continue until A's structure has been completely specified.

The modularity of the model class definition process allows the modeler to make use of preexisting classes or to gradually define new ones. This feature combined with the communicative properties of the semantic relations makes possible the modification or upgrading of a model class with the minimum of effort.

MCDDs are made up of nodes and edges. The digraph's *nodes* have the particularity of being classes; each one of them corresponds to a constituent entity of the composite object that the model class is representing at the time. Pairs of nodes in a digraph are linked by *edges*. In a MCDD edges describe the semantic relations that victualed the entities represented by the nodes. Any edge of a MCDD can be associated with any of the following specification semantic relations: ***is-composed-of, is-connected-by***, and ***is-described-by***. For example,

A MCDD is a digraph $G = (X, U_R)$, where

(a) $X = \{x_1, x_2, \ldots, x_n\}$ is the set of nodes representing classes of modeling objects.
(b) $U_R = \{(x_i, x_j) \in X \times X / x_i R x_j\}$ is the set of edges connecting nodes x_i and x_j through the three specification oriented semantic relationships, mentioned above.

A MCDD can be characterized as an acyclic digraph whose starting node, S, has a nil predecessor [i.e., $\Gamma^-(S) = 0$] labeled with the name of the system the model is intended to represent. The successor nodes of S, $\Gamma^+(S)$, are labeled with the names of the first-level components of the model. Each of these first-level constituent entities may itself be decomposed into second-level entities, which will be represented by new successor nodes in the digraph, i.e., $\Gamma^+[\Gamma^+(S)]$. Thus, the model of any node, x_i, is the

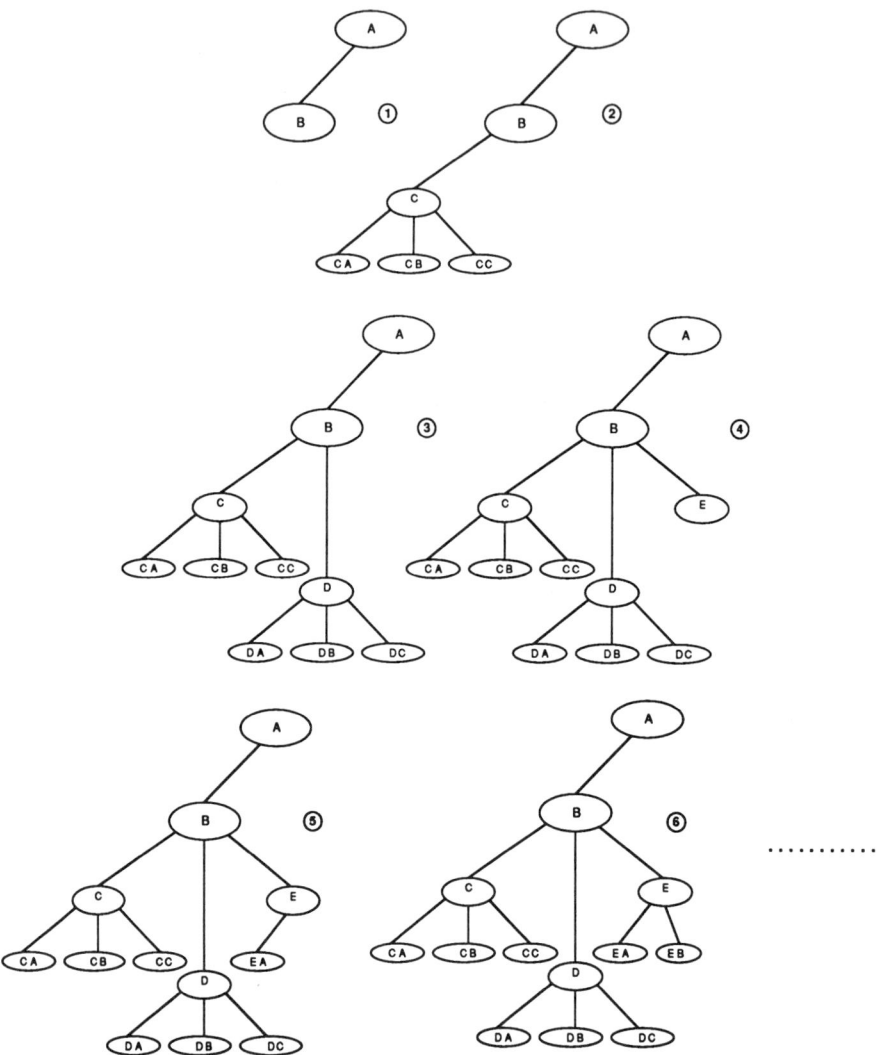

FIG. 10. The evolution of an MCDD's structure. (Reprinted from Stephanopoulos, G., Henning, G., and Leone, H. MODEL.LA A modeling language for process engineering. Part I, *Comp. Chem. Eng.* **14**, Page 813, Copyright 1990, with kind permission from Elsevier Science Ltd, The Boulevard, Langford Lane, Kidlington 0X5 1GB, UK.)

subgraph of the MCDD generated by x_i and its descendants of $(\Gamma^+(x_i) \cup \{x_i\})$;

$$\text{MCDD}(x_i) = \left(\{\hat{\Gamma}^+_{\text{MCDD}}(x_i) \cup \{x_i\}\}, U_R(x_i)\right),$$

where $\hat{\Gamma}^+_{\text{MCDD}}(x_i)$ is the set of descendants of x_i in MCDD and $U_R(x_i)$ is given by

$$U_R(x_i) = \left\{(x_a, x_b) \in \{\hat{\Gamma}^+_{\text{MCDD}}(x_i) \cup \{x_i\}\} X \{\hat{\Gamma}^+_{\text{MCDD}}(x_i) \cup \{x_i\}\} / x_a R x_b\right\}.$$

Formally, a MCDD represents a hierarchical decomposition of a model into constituent entities. Since this decomposition must ultimately terminate, a MCDD should not have directed circuits. Thus, MCDDs are digraphs that have the following properties:

Strict hierarchy. No component node can appear in the path that has as initial end-point one of its constituent entity nodes. Consequently, no component can have a decomposition which eventually contains a constituent entity of the same type.

Uniformity. Any two nodes with the same label have attached to them the same constituent entities. This property implies that any two components of the same type have the same isomorphic decomposition structures.

These properties are particularly important because they guarantee the development of a reasonable model and easy access to its different views, components, etc. An additional property, resulting from the uniformity axiom is that all MCDDs have the capability to specify the digraphs that are associated with the instances of the model class. The instantiation of a MCDD generates a *model decomposition digraph* (MCD). MCDs, like MCDDs, do not have directed circuits. They are finite graphs, obtained as a result of the instantiation of another finite graph.

C. Generation and Representation of Reaction Pathways

In the previous sections we discussed how we can use LCR's mechanisms to add new modeling elements, or to generate representations from existing modeling elements. In this section we will show how to use LCR for the construction of reaction pathways. We will employ examples from the domain of the *free-radical chemistry*, such as oxidation of butane.

LCR provides a method for generating all possible reactions and identifying the resulting pathways, given a set of substrates and a reaction environment. The keyword arguments (i.e., arguments with a colon prefix) accepted by this method are given below:

(find-all-pathways :substrates :operators :override-environment
                    :initiator-p)

The method tests for an override reaction environment allowing the user to investigate the influence of various environments without manually resetting the environment attribute of each substrate. [Recall that an `abc` (i.e., a substrate) "knows" its environment.] The user may also identify whether an initiator is present or believed to be present. This allows the user to investigate the effect of various reaction trajectories without knowing specifics concerning the initiator mechanism. The keyword *:operations* allows the user to specify the types of transformation to be used; the default value is `composite-operator`. The user may specify other operators as well. This focuses the elucidation of pathways. For example, by supplying $K_{free-radical}$ to the keyword argument the user can investigate pathways of free-radical origin.

Alternatively, the user may wish to investigate all theoretically feasible pathways subject to a set of preferences. This is accomplished by supplying *:operators* with K^*, a modified composite operator, which has encoded the user preferences. Recall that K^* as an instance of the `composite-operator`, generates reactions by executing the methods K_t, K_f, and $K_{get-sites}$. K^* contains no encoded transformations within K_t but rather applies the complete set of instances of ab-initio-operator (i.e., K_{ai}) directly. The K_f procedure reflects user preferences. This allows the user to investigate pathways having prespecified features ($\Delta G < 10$ kcal/mol, stereocenters, etc.). Generation of these pathways is accomplished as follows:

(find-all-pathways *:substrates* (`butane oxygen`) *:operators* K^*
                    :initiator-p nil)

Similarly, if there are no preferences, theoretically feasible reactions can be generated by calling find-all-pathways with *:operators* $K_{ab-initio}$ directly:

(find-all-pathways *:substrates* (`butane oxygen`) *:operators* $K_{ab-initio}$
                    :initiator-p nil)

This functionality allows the user to control the scope for the identification of potential pathways. Pathways connecting two states are identified using the method *find-all-pathways-connecting*. Its general form is

(find-all-pathways-connecting :*substrates*   :*desired-products*
                              :*operators*   :*initiator-p*
                              :*strategy*)

where substrates may be a class allowing the user to investigate pathways leading from a desired product to a class or classes of substrates. Alternatively, :*substrates* can be left unspecified. Once a control (e.g., synthesis) strategy has been specified, the method, *find-all-connecting-pathways*, chains on K_f and K_t methods of K, using the semantic relation **is-described-by**, and appropriate selector functions. The current implementation of LCR allows chaining only in the forward direction (i.e., synthetic direction) when $K_{ab-initio}$ or K^* is supplied to :*operators*.

Let us return now to the oxidation of butane. We will illustrate the generation of feasible pathways using K^*. In this example, K^* was restricted to thermodynamically viable free-radical pathways; products were restricted to their homologous series. The generation call was

(find-all-pathways   :*substrates* (butane oxygen)   :*operators* K^*
                     :*initiator* true)

Figure 11a presents several of the pathways identified when *s*-BuOH is formed during the initiation process. The attributes of the initiation reaction, initiation-1, for the oxidation of butane are presented in Fig. 12a, where objects are denoted by #⟨object-name⟩. Note that each of these objects can in turn be expanded or disaggregated using the semantic relationships. Figure 12b illustrates one of the many pathways, pathway-1, generated during the oxidation of butane into *s*-BuOH and AcEt. The expansion of the object, #⟨initiation-1⟩, an element in the value of the attribute, "composing-reactions," results in the description shown in Fig. 12a. The user has access to this information at every step of the synthetic process using the semantic relationships provided by LCR (see Section III.E).

By changing the values of the object, reaction-environment, one can generate new trajectories (denoted in Fig. 11b by dashed lines), corresponding to a higher-energy environment. Figure 11c presents the reaction pathways resulting from yet another reaction-environment. The various pathways stemming from these environments are combined in Fig. 13. The assumption set and conditions specifying each environment are managed by the context generation utility of LCR, discussed in the next section.

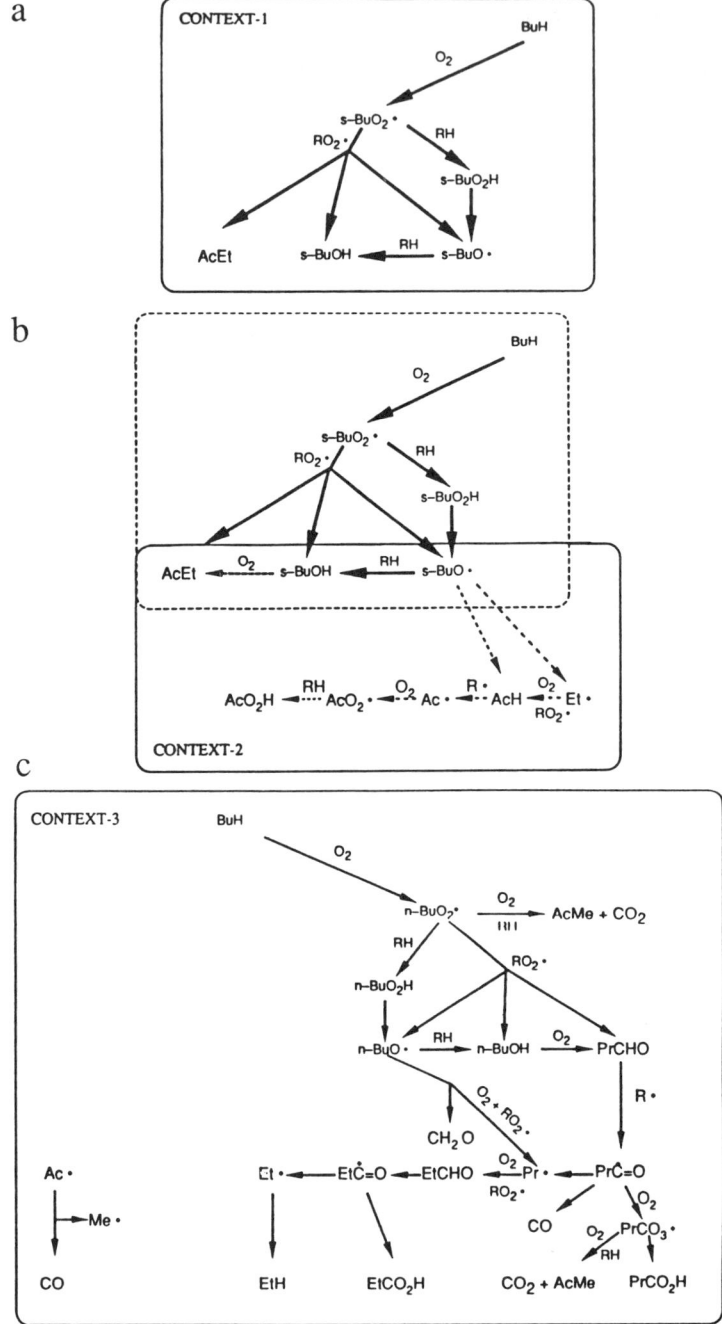

FIG. 11. Three distinct abstractions in describing the oxidation of butane.

a

Reaction Object
identifier:	*unbound*
name:	initiation-1
reactants:	(#<$C_4H_{10}O_2$>)
products:	(#<C_4H_9O> #<HO>)
stoichiometry:	((#<$C_4H_{10}O_2$> . -1)
	(#<C_4H_9O> . +1)
	(#<HO> . +1))
reaction-environment:	#<reaction-environment-1>
enabling-conditions:	K_f
composing-transformations:	K_t
composing-reactions:	*unbound*
rate-expression:	#<rate-expression-1>
equilibrium-constant:	#<equilibrium-constant-1>
context:	#<context-1>

b

Pathway Object
identifier	*unbound*
name	pathway-1
reactants	(# <C_4H_{10}> #<O_2>)
products	(#<$CH_3COCH_2CH_3$>)
stoichiometry	unbound
competing-pathways	(<pathway-2>
	#<pathway-3>)
composing-reactions	(#<initiation-1>
	#<initiation-2>
	#<abstraction-1>
	#<disproportionation-1> ...)
global-rate-expression	#<composite-rate-exp-1>
global-equilibrium-constant	*unbound*

FIG. 12. The description of (a) a reaction object and (b) a pathway object during the oxidation of butane.

By indexing and storing new reactions in their entirety, i.e., by storing the complete objects making up reactions, chemical structures, and chemical conditions, whether they were generated by K, or discovered through the use of K^* or K_{ai}, we have given LCR the ability to learn. Since a context is associated with each reaction, retrieval of specific reactions (from the ever-expanding library of chemical reactions) and incorporation in the evolving generation of pathways can be focused to the desired context, in order to suit specific needs.

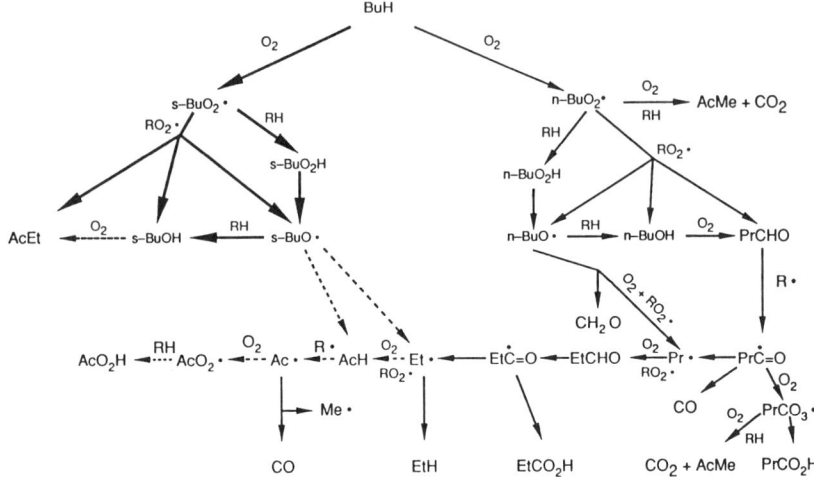

FIG. 13. Linking distinct contexts used for the representation of butane oxidation.

D. Creation of Contextual Reaction Models

The generation and representation of molecular structures, reaction, and pathways is always done within the scope of some "context." In Fig. 11 we see three (3) distinct networks of reactions, occurring during the oxidation of butane, and corresponding to three (3) distinct "contexts" of operating conditions. In general, the representation of any object within LCR depends on a set of assumptions that can be classified into the following three categories:

1. *Assumptions on structural components.* They declare component/subcomponent relations or level combinations of subcomponents: Chemical A has been conjectured as a `cyclic-aliphatic` molecule; bond between atoms A_1 and A_2 is assumed to be `covalent-bond`; pathways P is assumed to be a serial concatenation of pathways, P_1, P_2, and P_3.
2. *Assumptions on behavior or functionality.* These are relations defining: reactivity of various reaction centers in a molecule, heats of reaction, bond strength between two atoms, etc.
3. *Assumptions on the values of variables*, describing attributes of molecular structures or/and reactions.

These assumptions are expressed in terms of generalized constraints, which represent Boolean, qualitative, semiquantitative, or quantitative relationships among the modeling elements. The set of generalized constraints forms the *"context"* within which a model is valid, and are viewed as hard constraints taking on a Boolean character; thus, they are either satisfied or violated (never satisfied to a certain degree).

Let us assume that a model X developed within CONTEXT-1, is to be modified. CONTEXT-2 is created as a "child" of CONTEXT-1, that will be referred to as the "parent-context." In principle, CONTEXT-2 "inherits" all the assumptions that are valid in CONTEXT-1. Then, in order to account for the modifications of model X it is possible to change the inherited assumptions with additions and deletions. The changes are covered by the following rule:

All assumptions in CONTEXT-1 are "inherited" by CONTEXT-2 unless overridden in CONTEXT-2.

Generalizing this rule, we can say that:

The assumptions true in a given context are all the assumptions that are true in the parent-context, minus the assumptions that have been specifically deleted in it, plus any assumption that has been specifically added in it.

Following with this hypothetical case, let us assume that a different representation of X is required. Now, two options are given to us: (1) if the new representation is seen as an alternative to the one already introduced in CONTEXT-2, a new context called CONTEXT-3, should be created as a child of CONTEXT-1; (2) If the modification is intended to change model X built in CONTEXT-2, CONTEXT-3 should be created as a child of CONTEXT-2. The first choice is analogous to the situation found in butane oxidation. The assumptions corresponding to the free-radical oxidation of butane are encapsulated in CONTEXT-1 (Fig. 11a). Given that two distinct global pathways have to be analyzed, CONTEXT-1 gives rise to CONTEXT-2 and CONTEXT-3, which encapsulate the assumptions that generate the representations of Fig. 11b and 11c, respectively. CONTEXT-2 and CONTEXT-3 siblings both have information in common with CONTEXT-1, but there is no direct relation between them. The second situation corresponds to many multistep syntheses. During the synthesis of a complex molecule the chemist makes assumptions regarding how the molecule should be constructed, and these assumptions give rise to new assumptions. For example, assumptions regarding the specification of a carbon skeleton with correct key group orientation may be kept in CONTEXT-A. This context may give rise to CONTEXT-B placing restrictions (i.e. assumptions) on the latent functionality of the skeleton. These restrictions, in turn, may give rise to CONTEXT-C, i.e., assumptions regarding the need for certain protective groups.

Generalizing, we can state that contexts are always arranged in hierarchical manner. The relations among them are established in a direct graph, generated by the context relation (CR) on the set of contexts, C. The relation CR is read as *"gives-rise-to,"* and the digraph is called MODEL ABSTRACTION TREE (MAT). It is a tree because each node representing a context has only one predecessor. A MAT is defined by

$$\text{MAT} = (C, U_R),$$

where

$$C = \{\text{CONTEXTS}\},$$

$$U_R = \{(C_i, C_j) \in C \times C / C_i CR C_j\}.$$

This tree of contexts is similar to the "class precedence list" used in hierarchical inheritance mechanisms. An example of such tree is presented in Fig. 14, where CONTEXT-1 gives rise to CONTEXT-2 and CONTEXT-3, and CONTEXT-2 gives rise to CONTEXT-4. All contexts represent different reaction pathways for the same overall reaction.

1. Communication between Contextual Models

Contexts can be related as (a) siblings (e.g., CONTEXT-2 and CONTEXT-3 in Fig. 14) or (b) "parent"–"child" (e.g., CONTEXT-1 and CONTEXT-2 in Fig. 14).

In the first case, the contexts do not need to communicate with each other, but in the second case, information must pass from the parent-context to the child-context. The operations associated with the modification of the parent-context to produce the child-context, presently available in LCR are

1. *MODIFY*: an operation applied to an individual assumption in order to alter its value.
2. *DELETE*: an operation applied to a single assumption in order to alter its value.
3. *ABSTRACT*: an operation that is applicable to the set of assumptions representing an object model. The representation of this object is abstracted to obtain a less detailed description of its structure.
4. *DISAGGREGATE*: is the opposite to ABSTRACT. It is applied to the assumption set, representing a model, but its purpose is to introduce more refinement.

Whereas MODIFY and DELETE are simple operations, ABSTRACT and DISAGGREGATE are quite complex. Whichever of the last two operations is performed during the modification of a new context, it should allow the user to bind together components that are conceptually related in the two contexts. Thus, the ABSTRACT and

MODELING LANGUAGES

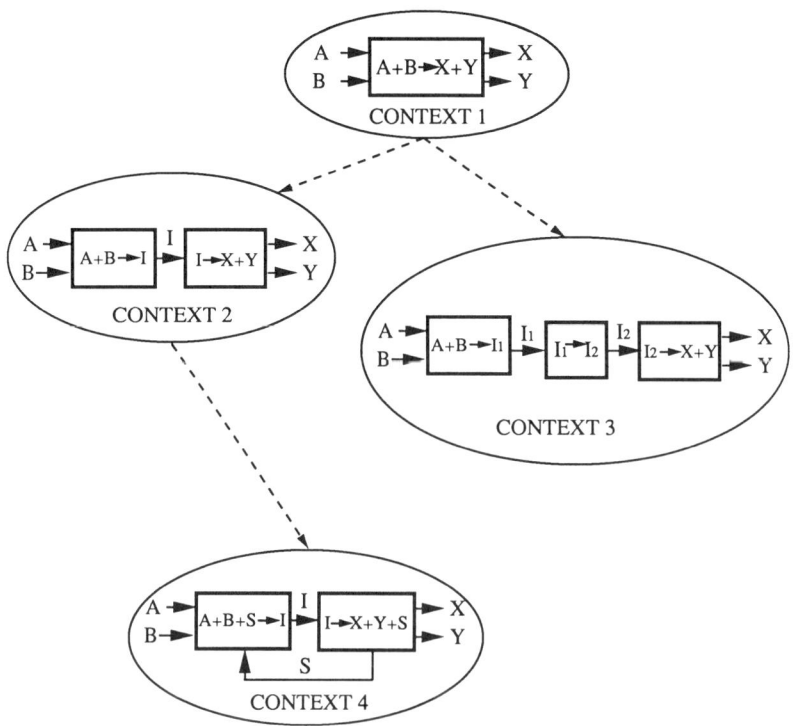

FIG. 14. A tree of contextual inheritances.

DISAGGREGATE operations should provide the framework for the communication across contexts. The symmetric operations of abstraction and disaggregation can be described by an algorithm that considers them in a very general sense, and can be summarized as follows:

```
ABSTRACTION (NEW-CONTEXT NEW-UNIT OLD-UNITS)
    ESTABLISHED-STRUCTURAL-COMPATIBILITY
    (NEW-CONTEXT NEW UNIT OLD-UNITS)
    ESTABLISH-BEHAVIORAL-COMPATIBILITY
    (NEW-CONTEXT NEW-UNIT OLD-UNITS)
END
DISAGGREGATION (NEW-CONTEXT NEW-UNITS OLD-UNIT)
    ESTABLISH-STRUCTURAL-COMPATIBILITY
    (NEW-CONTEXT NEW-UNITS OLD-UNIT)
    ESTABLISH-BEHAVIORAL-COMPATIBILITY
    (NEW-CONTEXT NEW-UNITS OLD-UNIT)
END
```

Since contexts have only one parent there is no need to specify the name of the context where the old-units are located. With the name of a new context, its parent context will be uniquely specified. The details of the above algorithms will be discussed in the following paragraphs, where we will make extensive use of the semantic links, *is-disaggregated-in* and *is-abstracted-by*.

2. Structural Compatibility among Contextual Models

Structural compatibility analysis operates on pathways that are located in two different contexts, bound together by a CR relation. During the disaggregation operation, a pathway that exists in the parent context is substituted by a set of pathways in the next context. The structural compatibility operation will establish the following semantic link:

P_{old} *is-disaggregated-in* (SET.OF($P_1,\ldots,P_n,$)) in NEW-CONTEXT

Since a parent-context may have several children, it is always necessary to specify the name of the child-context when the disaggregation occurs. For example, during the oxidation of butane, the context butane-oxidation-reaction-environment gives rise to the contexts butane-oxidation-termination-1 and butane-oxidation-termination-2. So, the relation

butane-pathways *is-disaggregated-in*
 (SET.OF (initiation, propagation, termination))
  in butane-oxidation-termination-1

provides an unambiguous link and avoids any confusion with models located in the other contexts (initiation, propagation,..., etc.)

The abstraction procedure is just the inverse operation:

(SET.OF ($P_1,\ldots,P_n,$)) *is-abstracted-by* P_{new} in NEW-CONTEXT

Now, a set of P values are abstracted in a single pathway. Figure 15a shows an example of the abstraction and disaggregation processes between CONTEXT-i and CONTEXT-j. If CONTEXT-j is created as a child of CONTEXT-i, the structural-compatibility operation will generate the link

P *is-disaggregated-in*   (SET.OF (P_1 P_2))   in CONTEXT-j

On the other hand, if CONTEXT-i is a child of CONTEXT-j, the structural compatibility operation would generate the following link:

(SET.OF (P_1, P_2))   *is-abstracted-by*   P in CONTEXT-i

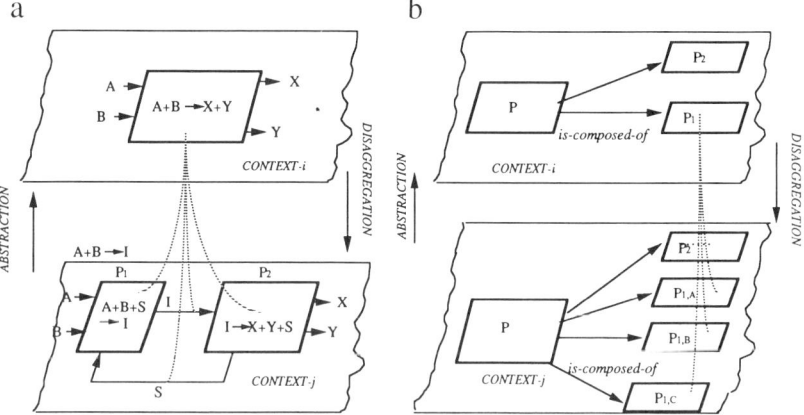

FIG. 15. Aggregation and disaggregation of representations of reacting systems.

However, there are more complex situations that need further analysis. Specifically, the case of pathways that are being disaggregated or abstracted and that themselves are not simple pathways from a structural point of view, but are also "part-of" other pathways in the parent context. What happens with these semantic links in the new context? A typical situation is depicted in Fig. 15b, where pathway P_1, which is *part-of* pathway P in CONTEXT-i, is disaggregated in pathways $P_{1,A}$, $P_{1,B}$ and $P_{1,C}$ in CONTEXT-j. The structural-compatibility procedure, apart from establishing *is-disaggregated-in* links across the contexts, should also establish *is-part-of* links between pathways $P_{1,A}$, $P_{1,B}$ and $P_{1,C}$, and pathway P.

More formally, the structural-compatibility procedure should perform the following checking operation every time a pathway P_{old} is disaggregated in a new context.

IF P_{old} *is-part-of* ?PATHWAY-X in OLD-CONTEXT
THEN ?PATHWAY-Y *is-part-of* ?PATHWAY-X in NEW-CONTEXT

where ?PATHWAY-Y is one of the pathways in which P_{old} is disaggregated, i.e., the following link exists between them:

P_{old} *is-disaggregated-in* ?PATHWAY-Y in NEW-CONTEXT

In order to avoid the generation of redundant links that the checking operation can produce due to the transitivity axiom, ?PATHWAY-X, will be restricted to be the predecessor of P_{old} that is related to P_{old} by means of a *is-composed of-* link. This condition is met by the only member of the following set

$$\{x \mid x \in A \wedge \{\{y \mid x \text{ is-composed-of } y\} \cap A\} = \phi\}$$

where

$$A = \{x \mid x \text{ is-composed-of } P_{old}\}$$

Let us now consider the inverse situation, depicted in Fig. 15b, where pathways $P_{1,A}$, $P_{1,B}$, and $P_{1,C}$, that are part of the pathway P in CONTEXT-j are abstracted in pathway P_1 in CONTEXT-i. In this case, the structural-compatibility procedure will establish *is-abstracted-by* links across the contexts and a *is-composed-of* link between pathways P and P_1 in CONTEXT-i. In order to consider situations like the one presented above, the following checking operation is also carried out every time a set of pathways $\{P_i, i = 1, \ldots, n\}$ *is-abstracted-by* the pathway P_{new} in a new context.

IF ?PATHWAY-X *is-composed-of* $\{P_i, i = 1, \ldots, n\}$ in OLD-CONTEXT
THEN ?PATHWAY-X *is-composed-of* P_{new} in NEW-CONTEXT

where $\{P_i, i = 1, \ldots, n\}$ *is-abstracted-by* P_{new} in NEW-CONTEXT.

For the same reasons expressed above, that is, to avoid the creation of redundant links, ?PATHWAY-X will be restricted to be the pathway that is a predecessor of the members of the set $\{P_i, i = 1, \ldots, n\}$, and which is also related to them by a *is-composed-of* link. Like before, ?PATHWAY-X is the only member of the set

$$\{x \mid x \in P \wedge \{\{y \mid x \text{ is-composed-of } y\} \cap P\} = \phi\}$$

where

$$P = \{x \mid x \text{ is-composed-of } \{P_i, i = 1, \ldots, n\} \text{old}\}$$

E. Case Study: Ethane Pyrolysis

To demonstrate the utility of LCR and the interaction between various modeling elements, semantic relationships, and supporting methods, let us first consider the pyrolysis of ethane forming principally ethylene and hydrogen. Although this example is relatively simple, it highlights the functionality of LCR and underscores various issues that require resolution for computer implementation to be successful.

1. Initialization

We begin by initializing the instance of atom-bond-configuration, which represents ethane (Table IV). Each attribute in Table IV whose value is an object can in turn be expanded, using the semantic relationships of LCR. For example, the attribute "empirical-formula" contains a list of ab-atom instances, whose expansion makes accessible those properties of an atom, which are independent of its environment (Table V), and include electronegativity, valence electrons, and atom-weight. Similarly, expansion of the instances atom in the attribute "atoms" describes substrate-dependent properties of an atom (Table VI), which include "bonds," "hybridization," "neighbor-atoms," "neighbor-groups," etc. Similarly, Fig. 16 shows the attributes of a particular instance of bond, describing the characteristics of a specific bond in ethane, while Table VII shows the attributes describing the group "methyl-1".

2. Generation of Pathways

The identification of various free-radical pathways and underlying elementary reactions involved in the pyrolysis of ethane are evaluated by applying the procedure, FIND-ALL-PATHWAYS, to the substrates, and associating with that call the designated instance of reaction-environment and the group of composite operators to be used (e.g., K, K^*, $K_{\text{ab-initio}}$). The call to FIND-ALL-PATHWAYS is shown below:

(FIND-ALL-PATHWAYS
:substrates (ethane) :*operators* $K_{\text{free-radical}}$
:*override-environment* reaction-environment)

The method FIND-ALL-PATHWAYS begins by calling the composite operators that constitute the domain of free-radical operations on the substrates specified by the keyword argument :*substrates*. These operators loop over the substrate list and evaluate properties specified by $K_{\text{free-radical}}$. Homolytic dissociation of ethane is initiated when FEASIBLE-P, a compound predicate method of $K_{\text{initiation}}$ ($K_{\text{initiation}}$ is an operation of $K_{\text{free-radical}}$), is evaluated on the sites identified by $K_{\text{get-sites}}$. These sites are passed to FEASIBLE-P, which searches the reaction-environment to establish potential radical forming processes: thermal cleavage, photochemical cleavage, and oxidation–reduction processes. Attributes describing reaction-environment establish thermal cleavage as a likely initiation mechanism; photochemical cleavage

TABLE IV
Attribute Values of an Instance of Ethane

identifier:	"C2H6-T0779"
name:	"Ethane"
atoms:	(#⟨ATOM C-T0764⟩ #⟨ATOM H-T0765⟩
	#⟨ATOM H-T0766⟩ #⟨ATOM H-T0767⟩
	#⟨ATOM C-T0768⟩ #⟨ATOM H-T0769⟩
	#⟨ATOM H-T0770⟩ #⟨ATOM H-T0771⟩)
bonds:	(#⟨BOND b-T0772⟩ #⟨BOND b-T0773⟩
	#⟨BOND b-T0774⟩ #⟨BOND b-T0775⟩
	#⟨BOND b-T0776⟩ #⟨BOND b-T0777⟩
	#⟨BOND b-T0778⟩)
empirical-formula:	((#⟨DB-ATOM 414000575⟩.6) (#⟨DB-ATOM 414001073⟩.2))
empirical-formula-string:	"C2H6"
molecular-weight:	30.07
charge:	0.0
terminal-skeleton-atoms:	(#⟨ATOM C-T0764⟩ #⟨ATOM C-T0768⟩)
equivalent-atoms:	((#⟨ATOM C-T0768⟩ #⟨ATOM C-T0764⟩)
	(#⟨ATOM H-T0771⟩ #⟨ATOM H-T0765⟩
	#⟨ATOM H-T0766⟩ #⟨ATOM H-T0767⟩
	#⟨ATOM H-T0769⟩ #⟨ATOM H-T0770⟩))
equivalent-bonds:	((#⟨BOND b-T0775⟩)
	(#⟨BOND b-T0778⟩ #⟨BOND b-T0772⟩
	#⟨BOND b-T0773⟩ #⟨BOND b-T0774⟩
	#⟨BOND b-T0776⟩ #⟨BOND b-T0777⟩))
weakest-bond:	(#⟨BOND b-T0775⟩)
weakest-bond-strength:	82.6
weakest-bond-strength-ratio:	1.0
ordered-eq-bonds:	((#⟨BOND b-T0775⟩)
	(#⟨BOND b-T0778⟩ #⟨BOND b-T0772⟩
	#⟨BOND b-T0773⟩ #⟨BOND b-T0774⟩
	#⟨BOND b-T0776⟩ #⟨BOND b-T0777⟩))
groups:	(#⟨GROUP methyl-1⟩#⟨GROUP methyl-2⟩
	#⟨GROUP terminal-sp3-methylene-1⟩
	#⟨GROUP terminal-sp3-methylene-2⟩)
group-bonds:	#⟨BOND b-T0775⟩ #⟨BOND b-T0772⟩
	#⟨BOND b-T0776⟩)
meta-groups:	#⟨GROUP ethyl-1⟩ #⟨GROUP ethyl-2⟩)
methyl-carbon:	NIL
primary-carbons:	(#⟨ATOM C-T0764⟩ #⟨ATOM C-T0768⟩)
secondary-carbons:	NIL
tertiary-carbons:	NIL
terminal-carbons:	(#⟨ATOM C-T0764⟩ #⟨ATOM C-T0768⟩)
backbones:	(#⟨ATOM C-T0764⟩ #⟨ATOM C-T0768⟩)
backbone-length:	2
progenitor:	NIL
environment:	#⟨reaction-environment-1 11235723⟩
⋮	⋮

TABLE V
DATA FROM AN EXTERNAL DATABASE FOR atom "CARBON"

Name:	"Carbon"
atomic-symbol:	"C"
atomic-number:	6
atom-weight:	12.001
valence:	4
row:	"1"
column	"4a"
orbitals:	*unbound*
valence-electrons:	4
electronegativity:	*unbound*

and initiation by oxidation–reduction processes are eliminated because their attribute values are unbound or nil (i.e., nontrue).

Using this knowledge, $K_{initiation}$ constructs a list of potentially-cleavable-bonds by applying the method IDENTIFY-WEAKEST-BONDS to the bond representation of each substrate contained in the substrate-list (i.e., ethane). IDENTIFY-WEAKEST-BOND returns a list of bonds,

TABLE VI
SELECT ATTRIBUTES OF AN ATOM

identifier:	"C-T0764"
old-identifier:	NIL
free-valence:	0
diradical-p:	NIL
formal-charge:	0.0
electron-withdrawing-substituent-p:	NIL
connectivity-number:	4
hybridization:	"sp3"
conjugated-p:	NIL
p-orbitals:	NIL
open-approach-p:	T
parent-molecule:	#⟨ORGANIC-MOLECULE C2H6-T0779⟩
progenitors:	NIL
parent-groups:	(#⟨GROUP methyl-1⟩
	#⟨GROUP terminal-sp3-methylene-1⟩
	#⟨GROUP ethyl-1⟩ #⟨GROUP ethyl-2⟩)
neighbor-atoms:	(#⟨ATOM C-T0768⟩ #⟨ATOM H-T0767⟩
	#⟨ATOM H-T0766⟩ #⟨ATOM H-T0765⟩)
neighbor-groups:	(#⟨GROUP methyl-2⟩
	#⟨GROUP terminal-sp3-methylene-2⟩)
bonds:	(#⟨BOND b-T0775⟩ #⟨BOND b-T0774⟩
	#⟨BOND b-T0773⟩ #⟨BOND b-T0772⟩)
DB-atom:	#DB-ATOM 414001073⟩

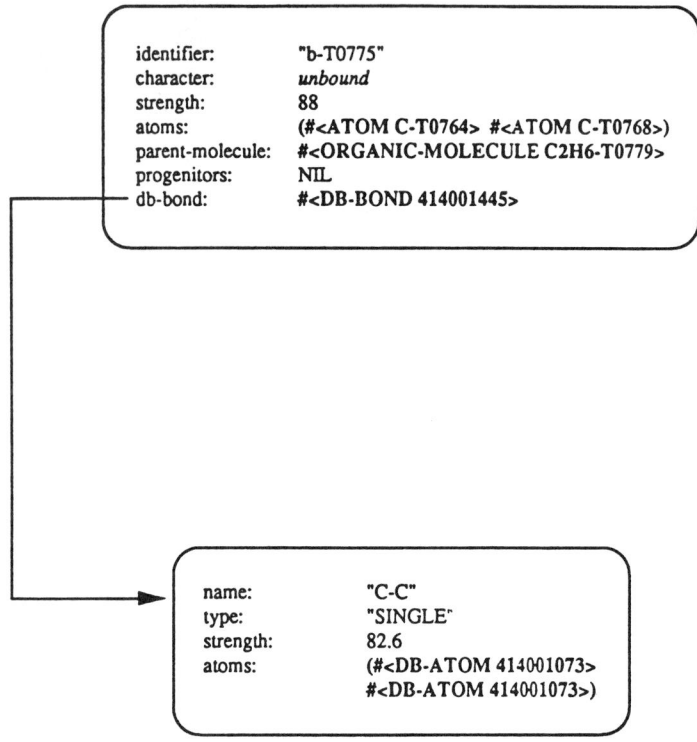

FIG. 16. Description of a bond with reference to data from an external database.

TABLE VII
SELECT ATTRIBUTES OF METHYL GROUP

identifier:	"methyl-1"
string-identifier:	"-CH3"
type:	*unbound*
neighbor-groups:	(#⟨GROUP methyl-2⟩
	#⟨GROUP terminal-sp3-methylene-2⟩
	#⟨GROUP ethyl-1⟩ #⟨GROUP ethyl-2⟩)
comprising-atoms:	(#⟨ATOM C-T0764⟩ #⟨ATOM H-T0767⟩
	#ATOM H-T0766⟩ #⟨ATOM H-T0765⟩)
connecting-bonds:	#⟨BOND b-T0775⟩
connecting-atoms:	#⟨ATOM C-T0768⟩
group-weight:	15.035001
parent-molecule:	#⟨ORGANIC-MOLECULE C2H6-T0779⟩
DB-group:	#⟨DB-GROUP methyl⟩

arranged in increasing strength. This is shown below:

IDENTIFY-WEAKEST-BOND applied to `bond-abstraction` ⇒
 ((`bond-1`) (`bond-2 bond-3 bond-4 bond-5 bond-6 bond-7`))

where

`bond-1` = carbon-carbon single bond connecting `methyl-1` to `methyl-2`
`bond-2` to `bond-4` = carbon-hydrogen single bond on `methyl-1`
`bond-3` to `bond-7` = carbon-hydrogen single bond on `methyl-2`

Bond selection is based on relative strength. A default energy difference of 10 kcal/mol is used by IDENTIFY-WEAKEST BOND, but this value can be changed by the user. Bonds of equivalent strength are listed together.

For each system of reactants, IDENTIFY-WEAKEST-BOND identifies the weakest substrate `bond`, and compares it with others in the system to ensure minimal global bond strength. Selected instances of `bond` are then appended to potentially-cleavable-bonds.

The weakest system `bond` is identified by applying the method IDENTIFY-ABSOLUTE-WEAKEST to the bond representation of `ethane`. The value of *weakest-bond* is the carbon–carbon single bond connecting `methyl-1` to `methyl-2`. Specifically

   IDENTIFY-ABSOLUTE-WEAKEST-BOND
   applied to potentially-cleavable-bonds `bond-1`

Once selected, it is associated with the global queue *weakest-bond*. Specifically

   weakest-bond `bond-1`

This queue prevents $K_{initiation}$ from processing the weakest bond of individual molecules when the energy difference between those bonds is greater than the default value. It also ensures that the initiation process is focused on substrates that are most likely to cleave in `reaction-environment`. For example, in a complex molecule or set of species where one particular bond is substantially weaker than others in the system (e.g., a peroxy bond), knowledge of *weakest-bond* focuses the initiation process on that bond. Once potentially-cleavable-bonds and *weakest-bond* become known to $K_{initiation}$, the method, FEASIBLE-P evaluates the energy in `reaction-environment` and determines whether there is sufficient energy for bond cleavage. The method SUFFICIENT-THERMAL-ENERGY-P performs this evaluation. When FEASIBLE-P evaluates true, $K_{initiation}$ calls K_t on the bond selected for cleavage. Application of bond cleavage operations, operators of $K_{ab-initio}$, to the weakest bond of ethane produces a microhomolysis-reaction. Each micro-reaction has associated with it reactants, products, and reaction stoichiometry. In addition, the bond selected for cleavage is explicit, as are various other properties of the reaction. Methods of kinetic operator are used to identify these properties; they include COLLECT-PRODUCTS, COMPUTE-STOICHIOMETRY, COMPUTE-BOND-FAVORABILITY, ASSESS-THERMODYNAMIC-FAVORABILITY, and COMPUTE-EQUILIBRIUM-CONVERSION.

Information pertinent to the transformation is associated with micro-reaction to make it easily accessible by methods and operations external to micro-reaction. Information essential to the external environment includes:

micro-homolysis-reaction
    reactants: (ethane)
    products: (methyl-radical-1 methyl-radical-2)
    stoichiometry: ((ethane.-1)
                   (methyl-radical-1.+1)
                   (methyl-radical-2.+1))
    bond-cleaved: bond-1
    environment: reaction-environment

Since multiple bonds may reside in potentially-cleavable-bonds, each possibly leading to a micro-reaction, $K_{initiation}$ creates a global-reaction that acts as a place holder for managing high-level information: bond-queue and micro-reactions. Global-reactions maintain a record of this information and the association of a micro-reaction corresponding to a particular bond cleavage:

    global-homolysis-reaction
       bonds-to-be-cleaved: (bond-1...bond-n)
       weakest-bond: (bond-1)
       micro-reaction: (micro-homolysis-1...micro-homolysis-n)

K_t, the function responsible for performing a particular transformation, calls methods contained in $K_{ab-initio}$. Principal methods invoked by K_t to effect homolytic bond cleavage of ethane are GENERAL-SCISSION and CLEAVE-BOND. The methods often utilize helping functions, contained in various modeling elements (e.g., IDENTIFY-BOND-TYPE, HOMOLYTIC-P, IDENTIFY-CHARGE-DISTRIBUTION), to assist the transformation process. After a molecular bond has been selected for cleavage, it is passed to GENERAL-SCISSION by $K_{ab-initio}$. Operations composing GENERAL-SCISSION identify the bonds parent-structure, perform the cleaving function, and manage fragments resulting from the cleavage process.

Fragment management is facilitated by ABSTRACT-GROUPING and ABSTRACT-ATOM-GROUPING. These methods partition in abc, given a set of starting points (e.g., atoms). Normally, these points are specified by the endpoints (e.g., connecting atoms) of the reaction center identified by $K_{\text{get-sites}}$. In ethane disassociation, the reaction center endpoints are the two carbon atoms associated with the carbon–carbon single bond. As a consequence, in the application of CLEAVE-BOND to weakest-bond (i.e., the element specified by *weakest bond*), two abc's are identified. These are grouped by ABSTRACT-ATOM-GROUPING and become radicals (i.e., methyl-radical-1 and methyl-radical-2 on instantiation by abc).

The (abc) instantiation process evaluates, updates, and classifies the abc as an instance of radical. During instantiation, abc invokes MAP-OLD-ATOM-TO-NEW-ATOM, a method that maintains pointers within the chemical system so that atoms constituting a substrate know where they came from. This is accomplished using attribute's identifier and old-identifier. During the dissociation of ethane the value of carbon-1 identifier is "C-T0764," whereas its old-identifier has value nil, reflecting external creation for ethane (i.e., database for user specification of the abc). However, the attribute value of old-identifier for carbon-3, the carbon making up new methyl radical, methyl-radical-1, reflects its progenitor carbon-1:

carbon-3		carbon-1	
identifier:	"C-T0790"	identifier:	"C-T0764"
old-identifier:	"C-T0764"	old-identifier:	nil
...	...	...	...

Since the value of old-identifier describing carbon-3 has the same value as carbon-1's identifier value, these pointers enable construction of a substrate's complete history together with the operators that enabled each transformation.

A history trace is constructed by chaining through the values of abc attribute progenitors. Similarly, the values of enabling-conditions, an attribute describing reaction, is traced to identify the various reactions constituting a particular pathway. These utilities, in combination, allow every attribute describing a modeling element (e.g., abc) to be logged and accessed, in chronological order. This schema is one way in which LCR affords multifaceted and multilevel description of a modeling element throughout its history, and allows, for example, a multilevel description of a particular chemical species and the reaction pathways in which it participates throughout its life cycle.

Once a composite operation has successfully been executed, FEASIBILITY-P returns true and ab initio operations making up the sequence of transformations contained in K_t return viable transformations; unique products identified by reaction are appended to the global queue denoted *potential-reactant-queue*. This queue contains a complete listing of potential reactants throughout the reaction cycle. Management of potential reactants in this manner allows the behavior of a system as it moves toward an equilibrium state, to be simulated.

Continued application of composite operations, invoked by FIND-ALL-PATHWAYS on the substrates contained in *potential-reaction-queue* identifies additional instances of transfer, propagation, and termination reaction. By tracing the history of these instances of reaction, free-radical-reaction (i.e., free-radical pathways), one can construct a free-radical reaction. Methods of free-radical-reaction provide the utility for searching through a set of chemical species, using the value of the progenitors attribute and MOLECULE-EQUIVALENT-P, to identify substrate loops within the propagation reaction network. Propagation reactions are then identified and the attribute value established.

However, with the myriad products capable of being formed in termination sequences, application of $K_{free-radical}$ to substrates in the *potential-reactant-queue* may not terminate. This potential exists because new reactants, resulting from coupling-reactions and combination-reactions, are constantly appended to *potential-reactant-queue*. For example, two ethyl-radicals can combine to form butane, and hydrogen abstraction of butane can form butyl, which may combine to form octane. This cycle can, in principle, continue indefinitely. To prevent such cycles, substrates are removed from *potential-reactant-queue* when they have been consumed, or when a homologous series of a reactant has been previously examined. The latter is accomplished using HOMOLOGOUS-SERIES-P, with the register function responsible for appending potential reactants to *potential-reactant-queue*.

Pathways of the overall reaction sequence are constructed using instances of free-radical-reaction. Information associated with the individual instances initiation-reaction, transfer-reaction, propagation-reaction, branching-reaction, and termination-reaction facilitates this task. With this information, free-radical-reaction is able to construct individual free radical pathways.

IV. MODEL.LA.: A Modeling Language for Process Engineering

MODEL.LA. is a high-level, special-purpose language, which was developed to support the modeling activities of a broad range of process engineering tasks; process design, simulation, operations planning, diagnosis, configuration of control systems, and others. In this regard, MODEL.LA. is quite distinct in both scope and capabilities from other modeling languages such as ASCEND, MODASS, and OMOLA. The technical details of MODEL.LA.'s structure and implementation can be found in Stephanopoulos *et al.* (1990a, b) and Henning and Leone (1990).

MODEL.LA. shares many common features with LCR. Both languages share a common set of semantic relationships and a common syntax. They differ in their modeling elements, which reflect the different vocabularies of chemistry and process engineering. Nevertheless, the modeling elements of each language are organized into subsets, which depict the *structural* and *behavioral* knowledge of the corresponding domains. In this section we will provide an overview of MODEL.LA.'s structure, and in Section V we will discuss the utilization of MODEL.LA. within the scope of engineering problems.

A. Basic Modeling Elements

To account for all representational needs in process engineering, MODEL.LA uses six (6) basic classes of modeling elements, and fairly rich modeling hierarchies of subclasses, emanating from the basic modeling elements.

1. Elements Defining Structures

To represent topological structures of processing systems, MODEL.LA employs the following modeling elements:

a. Modeling-Element 1: `generic-unit`. This is used to capture a system at any level of detail. Thus, an overall plant, a plant-section, a processing unit, a part of a processing unit (e.g., tubes of a heat exchanger), a phase in a processing vessel (e.g., liquid 1 in a two-liquid phase reactor), are all represented as instances of `generic-unit` (or, its specialized subclasses, as we will see in subsequent section). Also, sensors, actuators, control loops, and information processing systems are all represented and instances of `generic-unit`. Each instance of a generic unit encapsulates its own boundary, internal structural components, modeling relationships, and associated assumptions.

b. Modeling-Element 2: `port`. These are special objects that the generic units use to pass information to each other. Thus, the boundary of a generic unit is in essence defined by the set of ports through which it communicates with the rest of the world. Flow of materials, energy, or information into or from a generic unit takes place *only* through a port.

c. Modeling-Element 3: `stream`. These objects relate the ports of connected generic units, and are the conduits through which material, energy, information, or other quantities pass from one generic unit to the next. The type of a specific stream is always the same as the type of the ports that is associated with it.

2. Elements Defining Behavior

The following three modeling elements are used to capture the behavioral characteristics of materials and processing systems.

a. Modeling-Element 4: `constraint`. This class is used to capture any piece of knowledge associated with a declarative relationship among variables and parameters, such as a logical, qualitative, or quantitative relationship. An instance of `constraint` (or, its subsidiary subclasses) contains information about the form of the relationship it represents, the terms and variables that compose it, the "meaning" and significance of the relationship, as well as the range of its applicability.

b. Modeling-Element 5: `generic-variable`. Instances of this class (or its subsidiary subclasses) constitute the building blocks for the construction of modeling relationships. They represent parameters and variables describing physical quantities, engineering terms, design or operating specs, etc. An instance of `generic-variable` encapsulates information about the physical significance, value, range of possible values, trends over time, units, and other parameters of the quantity it represents.

c. *Modeling-Element 6:* `modeling-scope`. This object is a list of consistent instances of `constraint`, which represent the declarative relationships that apply to all components of a model, i.e., assumptions, simplifications, conjectures expected to be true, and modeling relationships. It is clear that `modeling-scope` is a *redundant* modeling element (e.g., all information in an instance of `modeling-scope` is contained in a set of instances of `constraint`). The importance, though, of maintaining a high-level container of the context in which the model was developed is sufficient to warrant the element's elevation into a basic element of the modeling language, allowing easier and direct manipulation of the modeling context.

B. Semantic Relationships

MODEL.LA. possesses the same set of 13 semantic relationships as LCR, with the same semantic implications:

1. *Semantic-Relationship 1: is-attribute-of.* Indicates that an entity is an attribute of a particular class of objects.

    "entity"   *is-attribute-of*   `object-class`

2. *Semantic-Relationship 2: is-method-of.* Indicates that a particular algorithmic procedure is a method of a specific class of objects.

    PROCEDURE   *is-method-of*   `object-class`

3. *Semantic-Relationship 3: is-a.* Relates a subclass to the mother class:

    object-subclass   *is-a*   object-class

4. *Semantic-Relationship 4: is-a-member-of.* Relates an instance to the class that generated it:

    instance-of-class-*x*   *is-a-member-of*   class-*x*

5. *Semantic-Relationship 5: is-composed-of.* Relates a modeling object to its component parts:

    jacketed-CSTR   *is-composed-of*   (jacket; stirred-tank)

6. *Semantic-Relationship 6: :is-part-of.* It is the semantic relationship symmetrical to the previous one.

7. *Semantic-Relationship 7: is-attached-to.* This relationship is used primarily to define the boundary of a generic unit, by defining the ports

attached to a specific unit:

> part-x   *is-attached-to*   unit-y

8. *Semantic-Relationship 8: is-connected-by.* It is the inverse of the previous one.
9. *Semantic-Relationship 9: is-described-by.* It is used to indicate the variables or relationships used to describe the behavior of a generic unit:

> unit-*x*   *is-described-by*   (variable-*y*; equation-*z*; rule-*w*)

10. *Semantic-Relationship 10: is-describing.* This is the inverse of the previous one.
11. *Semantic-Relationship 11: is-characterized-as.* It is used to specialize a modeling class by specifying a fixed, default value for a given attribute of the class; for example, the following statement

> Reactor-x   *is-characterized-as*   "isobaric"

puts a constant value in the attribute, "pressure", of the class Reactor-x. All instances of reactor-*x* will have constant pressure.
12. *Semantic-Relationship 12: is-disaggregated-in.* As in LCR, this semantic relationship associates a modeling element, located in a given context, with its constituent modeling components that are located at a different (more detailed) context. For example,

> plant-*x* *is-disaggregated-in* (reaction-section; separation-section)

13. *Semantic-Relationship 13: is-abstracted-by.* This is the inverse of the previous one.

The semanic relationships of MODEL.LA. obey the same axioms as those of LCR, namely *transitivity*, *monotonicity*, *commutativity*, and *merging*. For more details, see Section III.D.

C. HIERARCHIES OF MODELING SUBCLASSES

Six basic modeling classes presented in Section IV.A are the root-classes for expanded trees of modeling subclasses. Whereas the six basic elements are designed to possess generic attributes and methods, independent of the particular domain they represent, the hierarchical trees emanating from

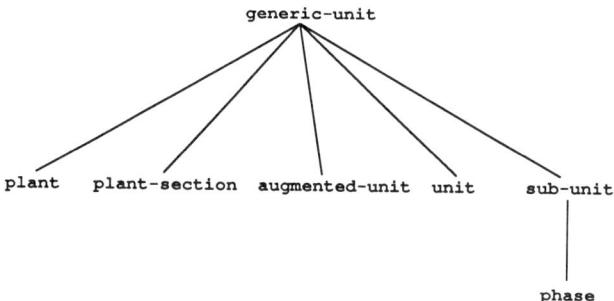

FIG. 17. The tree of generic-unit modeling subclasses.

them include modeling subclasses that are specialized to *reflect the vocabulary of process systems engineering*:

1. *The tree of* generic-unit *classes* (Fig. 17). The subclass generic-processing-unit, encompasses the various abstractions of processing systems: plant, plant-section, augmented-unit, sub-unit. The sub-unit has a specific subclass, phase, to capture the specialized attributes of a material with uniform thermodynamic state.

2. *The tree of* port *classes.* Four subclasses of ports emanate from the basic modeling element; convective-port, material-port, energy-port, and information-port. Each is specialized to express the characteristics of the quantity that flows through it.

3. *The tree of* stream *classes.* The number of stream subclasses is equal throughout and reflect similar structure of attributes as the port subclasses.

4. *The tree of* modeling-scope *classes.* It has two subclasses: model and context. The first captures all those relationships that reflect assumptions, hypotheses, and conjectures, all of which define the contextural scope of a model. The second captures the list of mathematical relationships describing the behavior of a generic-unit.

5. *The tree of* constraint *classes.* (Fig. 18). The assignment subclass is used to represent constraints, which are *solved* with respect to a particular variable. The relationship subclass is used to represent *unsolved* constraints.

6. *The tree of* generic-variable *classes.* This is composed of two subclasses, the variable and the term. The most important subclass is the term, which represents a compound quantity, made up from the

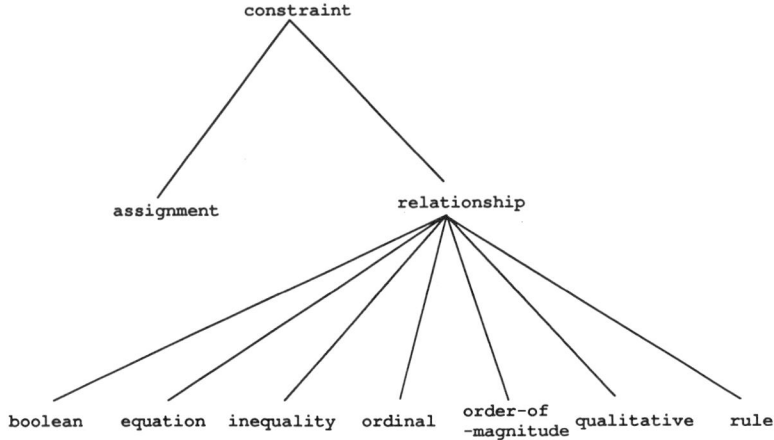

FIG. 18. The tree of constraint modeling subclasses.

combination of other elementary variables and parameters. It is used to model; terms appearing in modeling equations (e.g., enthalpy of stream-i, accumulation), dimensionless numbers [e.g., Re, Pr, Nu (Reynolds, Prandtl, Nusselt numbers)], design quantities [e.g., L/mG (liters per milligauss) of absorbers].

D. SYNTAX

MODEL.LA., like LCR, uses the BNF grammar rules for the construction of syntactically correct modeling sentences. For more details on the syntactic characterization of new models, the reader is referred to Stephanopoulos et al. (1990a, b), and MODEL.LA.'s manual (Henning and Leone, 1990).

V. Phenomena-Based Modeling of Processing Systems

MODEL.LA. is a language with infinite extensibility of its vocabulary, enabled by a fixed set of six modeling hierarchies and a fixed set of 13 semantic relationships. In Section IV.C, we discussed the hierarchies of modeling subclasses emanating from the six basic modeling elements. What is far more important for the modeling power of MODEL.LA. is its ability to capture all aspects of chemical engineering science, using the physical and chemical phenomena as the basis for the automatic generation of models for processing systems, using as input the declared physical and chemical characteristics of a processing system. For example, the user (or, another program) declares the following about a reactor-x:

Reactor-x is a two-liquid-phase jacketed CSTR.

MODEL.LA. is capable of parsing the sentence and of automatically creating the following "interpretation":

1. Reactor-x is composed of two subsystems: jacket and continuous stirred-tank.
2. The reacting mixture in the stirred-tank is made up of two liquid phases.
3. There is mass transfer between the two liquid phases.
4. There is heat transfer between reacting mixture and jacket.
5. Materials enter and leave the reactor through convective flows.
6. Enthalpy enters and leaves the reactor through convective flows.

Once this "interpretation" has been established, MODEL.LA. (a) generates all the requisite modeling elements and (b) constructs the modeling relationships, such as material balances, energy balance, heat transfer between jacket and reactive mixture, mass transport between the two liquid phases, equilibrium relationships between the two phases, estimation of chemical reaction rate, estimation of chemical equilibrium conditions, estimation of heat generated (or consumed) by the reaction, and estimation of enthalpies of material convective flows. In order to automate the above tasks, MODEL.LA. must possess the following capabilities:

1. Rich hierarchies of modeling elements, which can be used to represent any conceivable quantity of interest in chemical engineering science.
2. A series of procedures, which can automatically generate the complete set of modeling equations representing the behavior of a processing system.

A. THE "CHEMICAL ENGINEERING SCIENCE" HIERARCHIES OF MODELING ELEMENTS

The modeling hierarchies of `constraint` and `generic-variable` (see Section IV.C) have been expanded to a series of subclasses, which have been specialized to capture the knowledge of chemical engineering science, and use it in an explicit manner during the construction of models for processing systems.

1. Classes of Variables

Class `variable` is a subclass of the basic class `generic-variable`. From the class `variable` emanates the tree of subclasses, a partial view of which is shown in Fig. 19. Unlike other modeling approaches, MODEL.LA. does not represent variables through their values alone, but it provides an extensive structure that includes many additional attributes in the description of a variable. The additional attributes allow MODEL.LA. to reason about these variables and not just acquire their values. Thus, a set of methods in the class `variable` allow any of the subclasses to monitor their values, react with predefined procedures when the value of the variable changes, invoke values from external databases, and so on.

2. Classes of Terms

The class `term` is a subclass of the `generic-variable`, and one of the most important modeling elements in MODEL.LA. It is used to represent a very broad spectrum of compound physical quantities, which appear in modeling relationships. Figure 20 shows a partial view of that

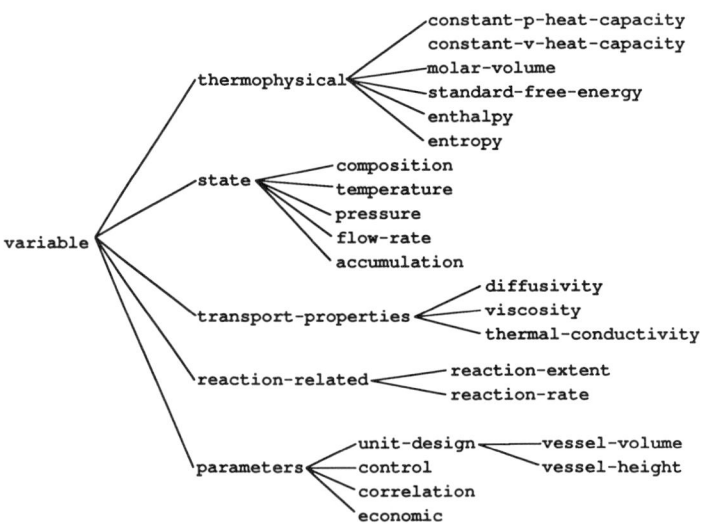

FIG. 19. Partial view of the `variable` hierarchy of modeling subclasses.

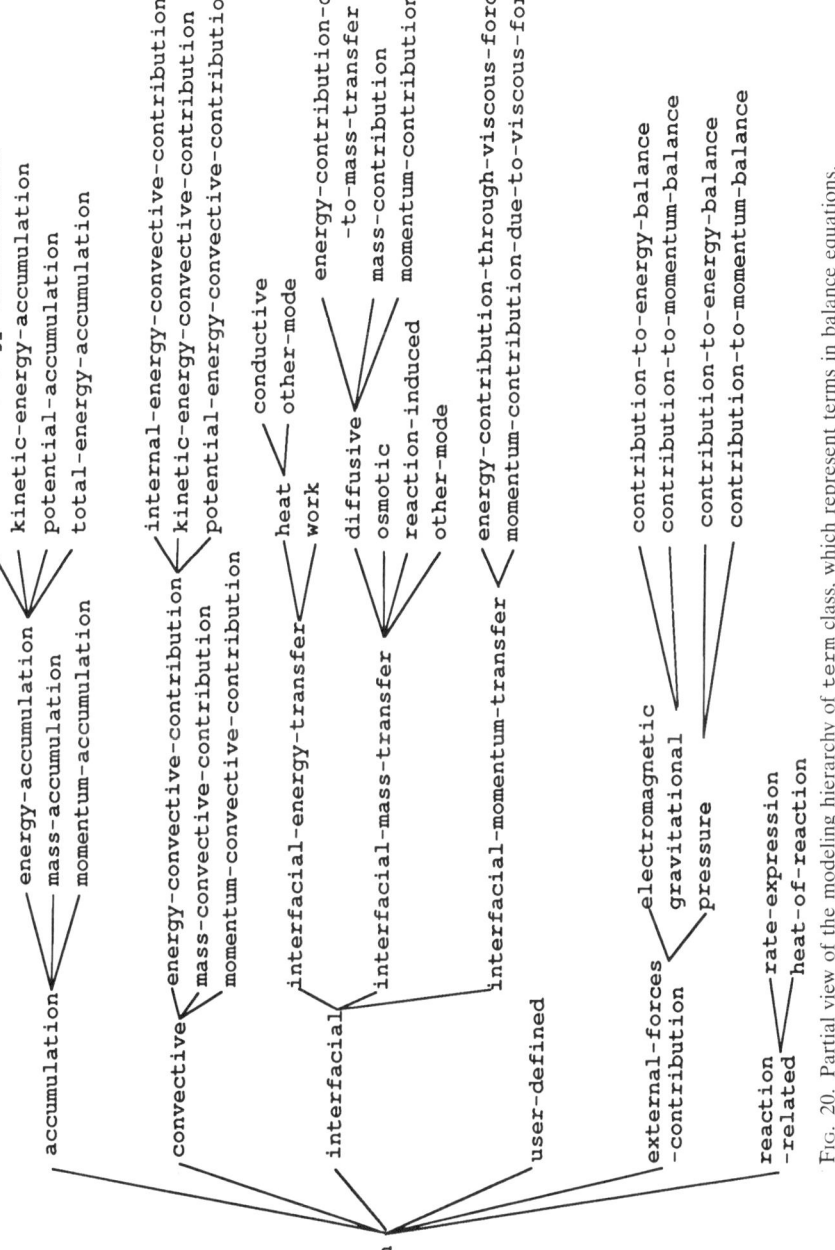

FIG. 20. Partial view of the modeling hierarchy of term class, which represent terms in balance equations.

part of the hierarchy of classes emanating from `term`, which are involved only in the definition of balance equations during the modeling of processing units. The complete structure of the hierarchy of terms is fairly extensive and covers the bulk of terms appearing in balance equations, transport phenomena, and equilibrium relationships.

3. Classes of Equations

All these modeling elements emanate from the class `equation` (Fig. 18), and a partial view is shown in Fig. 21. Each subclass captures all the information about a given equation, including its significance, preconditions for its correct use, and implications on the physics of the modeled system.

B. Formal Construction of Models

Every model constructed through MODEL.LA. is represented through a MCDD (see Section III.B), with the modeling elements playing the role of nodes and the semantic relations among the modeling elements representing the edges of the digraph. Figure 22 shows the MCDD associated with the model of a continuous-stirred-tank reactor (CSTR) without a jacket. Note the nodes representing topological elements (e.g., the ports, `input1-vessel` and `input2-vessel`), variables (e.g., `pressure`, `temperature`, `composition`), modeling equations (e.g., `mass-lumped-balance`, `energy-lumped-balance`, `phase-equilibrium-equation`). The element `vessel-surroundings-heat-transfer` is not specified. If it is specified to be a jacket, then the MCDD of the jacket is appended to that of the CSTR and produces the composite MCDD of Fig. 23, which represents the model of a *jacketed CSTR*. This type of modular construction of processing models provides MODEL.LA. with infinite flexibility in representing any processing systems. Furthermore, predefined models can be used as components in future representations, making the next modeling effort no more difficult than previous ones.

C. Multifaceted Modeling of Processing Systems

Consider the three abstractions of the plant shown in Fig. 1. How could one use the capabilities of MODEL.LA. to generate consistent represen-

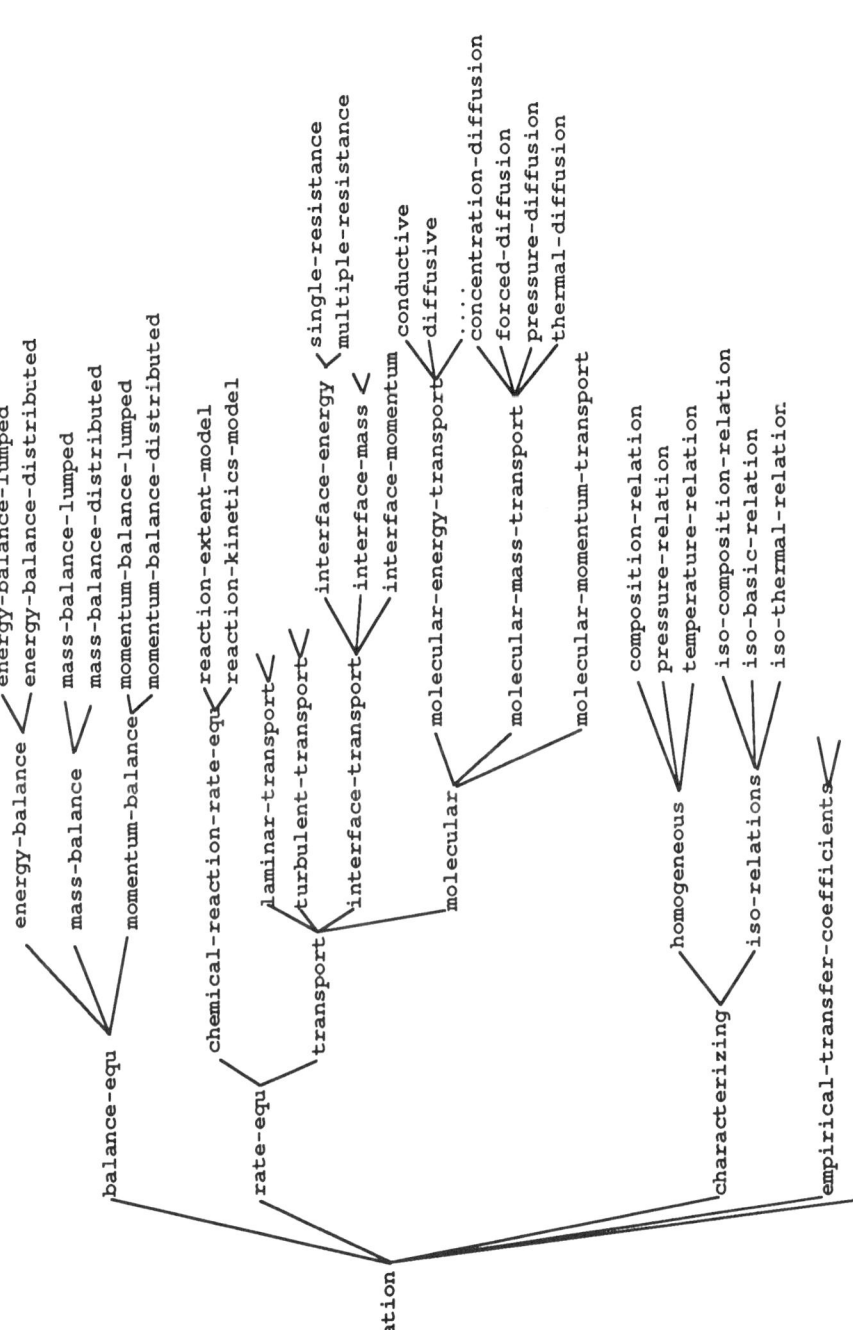

FIG. 21. Partial view of the hierarchy of modeling classes representing various types of equations in chemical engineering science.

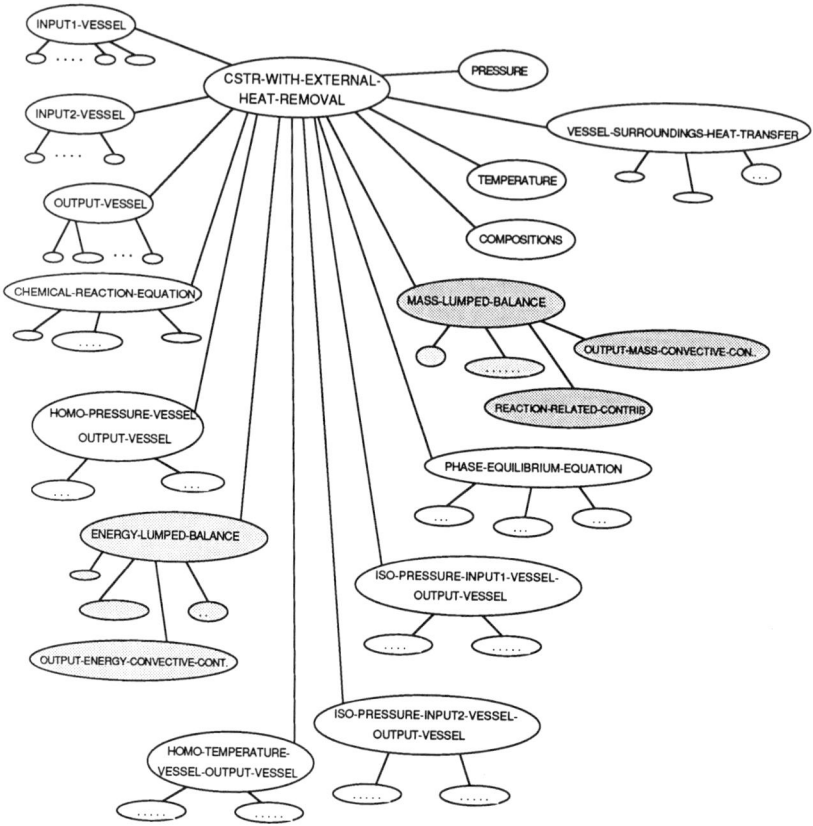

FIG. 22. The MCDD associated with the model of a CSTR without jacket. (Reprinted from *Comp. Chem. Eng.* **14**, Stephanopoulos, G., Henning, G., and Leone, H. MODEL. LA A modeling language for process engineering. Part I, Page 813, Copyright 1990, with kind permission from Elsevier Science Ltd, The Boulevard, Langford Lane, Kidlington 0X5 1GB, UK.)

tations of the three views of the same plant? On the other hand, the two versions of the plant shown in Fig. 2 have many modeling elements in common. How could one use MODEL.LA. to maintain two distinct facets of the same plant at the same level of detail? Such multiviewing of process models, along with the corresponding multilevel and multicontext requirements, define the scope of the so-called *multifaceted modeling of processing systems.*

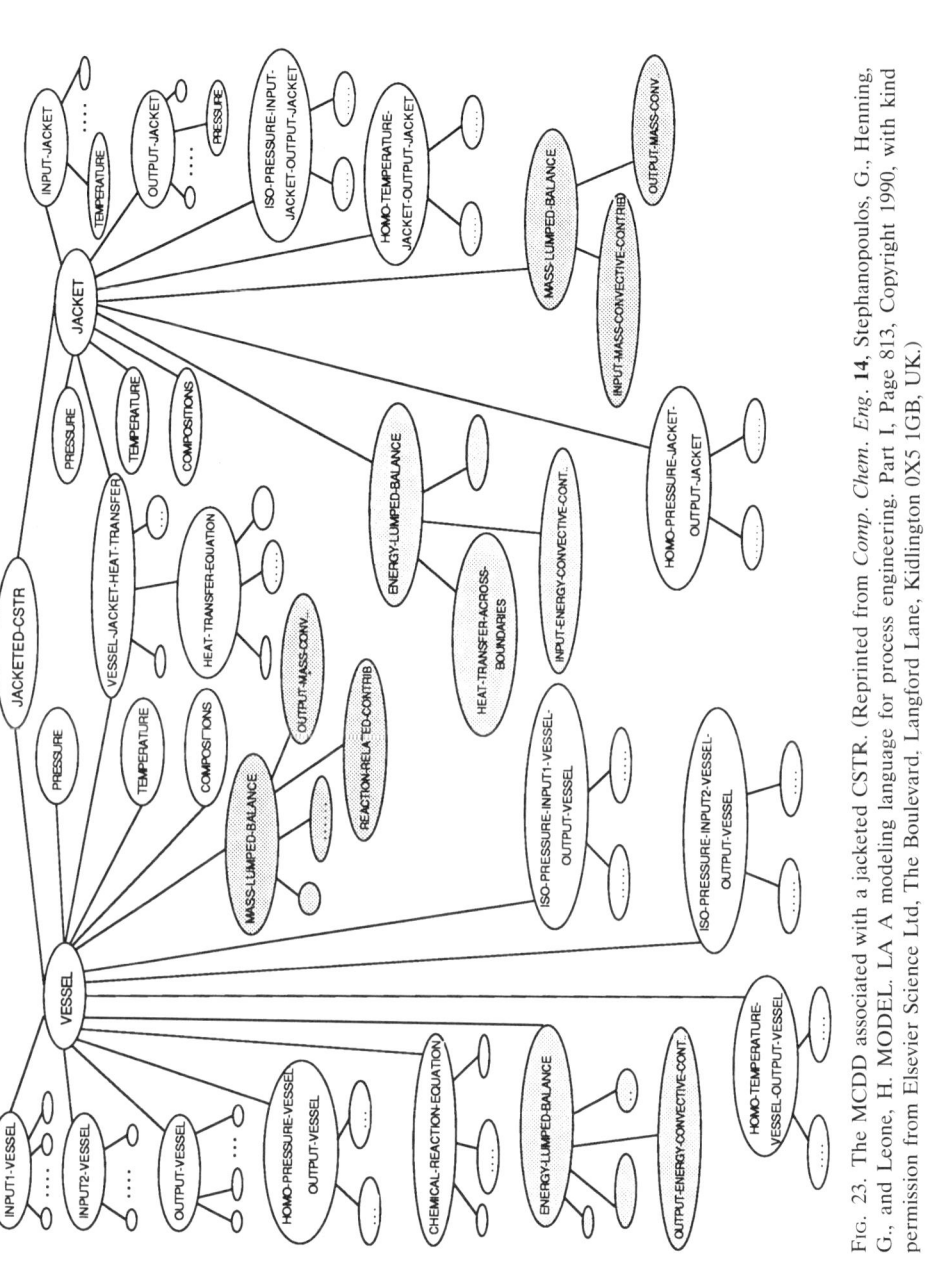

FIG. 23. The MCDD associated with a jacketed CSTR. (Reprinted from *Comp. Chem. Eng.* **14**, Stephanopoulos, G., Henning, G., and Leone, H. MODEL.LA A modeling language for process engineering. Part I, Page 813, Copyright 1990, with kind permission from Elsevier Science Ltd, The Boulevard, Langford Lane, Kidlington 0X5 1GB, UK.)

1. Concurrent Models at Multiple Abstractions

Consider again the "overall process" of Fig. 1a. It can be represented by an instance of the modeling element, `generic-unit` (or, `plant`, if we want to take advantage of its special attributes). MODEL.LA. can derive automatically the necessary modeling instances for the refined plant of Fig. 1b, through the invocation of the following semantic relationship:

overall-process *is-disaggregated-in* (generalized-reaction-section; generalized-separation-section)

If the `generalized-reaction-section` and `generalized-separation-section` are defined as instances of the class `plant-section`, then the correspondence between the various ports in Fig. 1a and 1b is automatically established. Thus, all the "external" ports `port-A`, `port-B`, `port-P`, and `port-BP` in Fig. 1b acquire the same information that their "external" counterparts possess in Fig. 1a. Furthermore, MODEL.LA. automatically establishes instances of the streams `Gas-Recycle` and `Liquid-Recycle` and associates them with the corresponding ports of the two plant sections in Fig. 1b. The automatic generation of models for the refined representation of the same plant as in Fig. 1b is carried out by a series of procedures that MODEL.LA. activates when an instance of the `generic-unit` (or its subclasses) is (are) refined through the *is-disaggregated in* semantic. Specifically, these procedures carry out automatically the following tasks:

1. *Establish structural compatibility* between the abstract and the set of refined generic units.
2. *Establish topological compatibility* between the ports and streams of the abstract entity and those of the refined entities generated after the disaggregation process.
3. *Establish compatibility between the physicochemical phenomena* occurring at the abstract and refined contexts of the plant.
4. *Propagate values of variables* from the abstract to the more detailed context (or, vice versa), using the compatibility links established above.

For the technical details of the procedures, carrying out the above tasks, the reader is referred to Stephanopoulos *et al.* (1990a, b).

2. Concurrent Models at the Same Abstraction

MODEL.LA. maintains concurrent models of the same process at the same level of abstraction using the notion of *context* (see also Section

III.D). For example, let CONTEXT-1 signify the set of (1) topological relationships characterizing the structure in Fig. 2a and (2) modeling relationships describing the functionality and behavior of the generalized reaction and separation sections. Then, we can create CONTEXT-2 as the child of CONTEXT-1, by changing a few of its relationships so that the new context can capture the "splitter" and the "purge-stream" of Fig. 2b. Since CONTEXT-2 (1) inherits all relationships from CONTEXT-1, (2) overrides or eliminates those that changed, and (3) adds new modeling elements and relationships, the two contexts share the common modeling elements (without duplication) and maintain their distinctive character through their private modeling elements or/ and relationships. For more technical details on the MAT, see Section III.D; and for the knowledge sharing of concurrent models at the same level of abstraction, see Stephanopoulos *et al.* (1990a, b).

D. COMPUTER-AIDED IMPLEMENTATION OF MODEL.LA.

MODEL.LA. was implemented within IntelliCorp's KEE, using Common LISP (list processing language) as the low-level programming language. KEE's frame system provided a fairly flexible, object-oriented programming environment. To take advantage of a number of facilities and to be integrated with an advanced, computer-aided engineering framework, MODEL.LA. was incorporated into the DESIGN-KIT (Stephanopoulos *et al.*, 1987).

1. Organization

The different objects comprising the modeling system are grouped into several knowledge bases:

- The basic modeling elements of the language, and the subclasses that emanate from them, are stored in a knowledge base (KB) called NETWORK-MODEL.
- When the composite objects representing new model classes are generated, their parts constituent are created as more specialized subclasses of the modeling elements of the NETWORK-MODEL KB. However, all the specialized entities constituting model classes are stored in the LIBRARY KB.
- When a model class is instantiated to represent a given model situation, all the entities used to generate the composite object that represents the model instance are created in the WORKING KB.

The rationale behind this division is that managing large knowledge bases is difficult. As knowledge bases grow in size and complexity, they strain the capacities of software tools for knowledge editing and maintenance. This division is based on the different functionality of the several KBs. The NETWORK-MODEL KB contains the basic knowledge, and consequently, is almost fixed. It will only be changed if the language is upgraded or modified. The LIBRARY KB will change if the model classes are modified or new ones are defined. Finally, the WORKING KB contains the most volatile type of knowledge, that is, entities that are created to represent very specific situations.

The WORKING KB stores all the instance objects that are used to represent the models employed during a particular process engineering activity. As was shown in previous sections, however, a model may have multiple descriptions of its structure, topology, behavior, etc. These different modeling situations are encapsulated in CONTEXTS. The CONTEXT facility is built on top of KEE worlds. KEE worlds is a set of tools provided by the KEE system to allow the representation of alternative states of the knowledge base, and to facilitate the exploration and comparison of alternative scenarios. The information in a standard KEE knowledge base provides a starting point for modeling hypothetical situations represented as worlds. In our case this information, called the "background," is stored in the WORKING KB. This KB contains the assumptions that are true in every hypothetical situation, and perhaps the problem-specific facts describing the initial modeling situation.

2. Facilities for the Generation of Models

Model classes can be generated in one of two ways:

(a) By using the MODEL.LA. language, a natural language built on top of KEE's Tell And Ask. The Tell And Ask language is a special facility for interacting with a knowledge base, by telling it new facts to incorporate, and asking about the facts it already contains. It is basically a set of KEE functions and a set of linguistic forms that format information for KEE.

(b) By using the MODEL-EDITOR, that is a specially designed mouse-and-menu-style interface. The description of the MODEL-EDITOR interface can be separated clearly in two parts: one governing the appearance and properties of the display itself, and the other defining the logical relations between display and modeling objects.

Our application domain makes it natural to adopt menus as the underlying display metaphor. *Menus* are represented as structured objects organized in *panels*. Menus themselves are constructed of rectangular primitive regions called *boxes* and nonprimitives called *polyboxes*. Boxes (which need not have visible outlines) are used to accept input data (typed or mouse-selected data), to allocate regions for displaying text and annotations of various kinds, etc. The adoption of a menu-driven interface has several advantages, among which we note the following: (1) *error prevention*, which is achieved naturally, because menus give the user the complete list of assumptions that can be specified at any point during the definition of a model; (2) *memorization*, to avoid or minimize the need for reference to an external manual or calls to help facilities; (3) *immediate visual feedback*; and (4) *active menus*, which help ensure that all the assumptions that need to be set for the model to be defined properly are in fact set. The model editor interface has been designed according to a "conditional display" paradigm. The complete definition of a model class can in principle require the user to specify a very large number of individual assumptions. In practice, fortunately, most model classes can be defined supplying only a small set of the possibly relevant data.

3. Facilities for User-MODEL.LA. Interaction

Specially designed tools provide an integrated set of knowledge access facilities serving various purposes such as inspecting the current status of the several KBs, linking the available model classes and their associated assumptions, displaying the structure of models or just partial views of them, listing the assumptions associated with a model instance in several contexts where it is defined, listing the relations and variables associated with a model instance, and listing the variable values.

To the extent that the structure of a model reflects the logical relations among its several constituent entities, models can be verified in a straightforward way, by confirming that the necessary links exist. Thus, specific browsers are used to display the different specification semantic relations existing among the pieces of a model. It is obvious that this browser facility allows a selective access to the different views of a model.

Other types of browsers are employed to show different semantic relations between the components of the language, for example, to display "*is-a*" or "*is-a-member-of*" links, or to show the directed graph that is formed by the contexts and the parent/child relations between them.

Summarizing, we can say that the user-interaction facilities incorporated in this system are governed by the need to provide a smooth, natural, and efficient environment for accessing, analyzing, and debugging knowledge.

References

Andersson, M., An object-oriented modeling environment. *In* "Simulation Methodologies, Languages and Architectures and AI and Graphics for Simulation" (Iazeolla *et al.*, eds.), 1989 European Simulation Multiconference, Rome, pp. 77–82. The Society for Computer Simulation International, 1989.

Corey, E. J., *Pure Appl. Chem.* **14**, 19 (1967).

Dugundji, J., and Ugi, I., An algebraic model of constitutional chemistry as a basis for chemical computer programs. *Top. Curr. Chem.* **39**(19) (1973).

Evans, L. B., Boston, J. F., Britt, H. I., Gallier, P. W., Gupta, P. K., Joseph, B., Mahalec, V., Ng, E., Seider, W. D., and Yazi, H., ASPEN: An advanced system for process engineering. *Comput. Chem. Eng.* **3**, 319 (1979).

Henning, G. P., and Leone, H. P., "Design-Kit Users' Guide." Laboratory for Intelligent Systems in Process Engineering, Massachusetts Institute of Technology, Cambridge, MA, 1990.

Meyer, B., Reusability: The case for object-oriented design. *IEEE Software*, March, pp. 50–64 (1987).

Nagel, C., Identification of hazards in chemical process systems. Ph.D. Thesis, Department of Chemical Engineering, Massachusetts Institute of Technology, Cambridge, MA (1991).

Naur, P., Revised report on the algorithmic language ALGOL 60. *Commun. ACM* **6**, 1 (1963).

Nilsson, B., Object-oriented modeling of chemical processes. Doctoral Thesis, Department of Automatic Control, Lund Institute of Technology (1993).

Pantelides, C. C., SPEEDUP—recent advances in process simulation. *Comput. Chem. Eng.* **12**, 745–755 (1988).

Perkins, J. D., and Sargent, R. W., SPEEDUP: A computer program for steady-state and dynamic simulation and design of chemical processes. Selected topics on computer-aided process design and analysis. *AIChE Symp. Ser.* **214**, 78 (1982).

Piela, P. C., ASCEND: An object-oriented computer environment for modeling and analysis. Ph.D. Thesis, Carnegie-Mellon University, Pittsburgh, PA (1989).

Piela, P. C., Epperly, T. G., Westerberg, K. M., and Westerberg, A. W., ASCEND: An object oriented computer environment for modeling and analysis: The modeling language. *Comput. Chem. Eng.* **15**(1), 53 (1991).

Sørlie, C. F., A computer environment for process modeling. Dr. Ing. Thesis, Department of Chemical Engineering, Norwegian Institute of Technology (1990).

Stephanopoulos, G., Artificial intelligence: What will its contributions be to process control? *In* "The Second Shell Process Control Workshop" (D. M. Prett, C. E. Garcîa, and B. L. Ramaker, eds.), p. 591. Butterworth, London, 1990.

Stephanopoulos, G., Johnston, J., Kriticos, T., Lakshmanan, R., Mavrovouniotis, M., and Siletti, C., DESIGN-KIT: An object-oriented environment for process engineering. *Comput. Chem. Eng.* **11**(6), 655 (1987).

Stephanopoulos, G., Henning, G., and Leone, H., MODEL.LA. A modeling language for process engineering. I. The formal framework. *Comput. Chem. Eng.* **14**(8), 813 (1990a).

Stephanopoulos, G., Henning, G., Leone, H., MODEL.LA. A modeling language for process engineering. II. Multifaceted modeling of processing systems. *Comput. Chem. Eng.* **14**(8), 847 (1990b).

Vernin, G., and Chanon, M., eds., "Computer Aids to Chemistry." Ellis Horwood, Chichester, 1986.

Wipke, W. T., Heller, S. R., Feldmann, R. J., and Hydes, E., eds., "Computer Representation and Manipulation of Chemical Information." Wiley, New York, 1974.

Woods, E. A., The Hybrid phenomena theory: A framework integrating structural descriptions with state space modeling and simulation. Dr. Ing. Thesis, Department of Engineering Cybernetics, Norwegian Institute of Technology (1993).

Woodward, R. B., *in* "Perspectives in Organic Chemistry" (A. Todd, ed.), p. 155. Wiley (Interscience), New York, 1956.

Woodward, R. B., *in* "Pointers and Pathways in Research" (G. Hofteizer, ed.), p. 23. Ciba of India, Ltd., Bombay, 1963.

AUTOMATION IN DESIGN: THE CONCEPTUAL SYNTHESIS OF CHEMICAL PROCESSING SCHEMES

Chonghun Han and George Stephanopoulos

Laboratory for Intelligent Systems in Process Engineering
Massachusetts Institute of Technology
Cambridge, MA 02139

James M. Douglas

Department of Chemical Engineering
University of Massachusetts, Amherst, MA 01003

I. Introduction	94
A. Conceptual Design of Chemical Processing Schemes	96
B. Issues in the Automation of Conceptual Process Design	98
II. Hierarchical Approach to the Synthesis of Chemical Processing Schemes: A Computational Model of the Engineering Methodology	103
A. Hierarchical Planning of the Process Design Evolution	104
B. Goal Structures: Bridging the Gap between Design Milestones	107
C. Design Principles of the Computational Model	117
III. HDL: The Hierarchical Design Language	122
A. Multifaceted Modeling of the Process Design State	123
B. Modeling the Design Tasks	129
C. Elements for Human–Machine Interaction	134
D. Object-Oriented Failure Handling	138
E. Management of Design Alternatives	138
IV. ConceptDesigner: The Software Implementation	139
A. Overall Architecture	139
B. Implementation Details	143
V. Summary	144
References	145

If you really know how to carry out an engineering task, then you can instruct the computer to do it automatically. This self-evident truism can be used as litmus test of whether a human "really" knows how to, say, design an engineering artifact. Experience has shown that in very few instances, engineers have been able to automate the process of design, thus demonstrating the presence of serious flaws in (1) their understanding of how to do design or/and (2) their ability to clearly articulate the design methodology, both of which can be traced to the inherent difficulty of making the "best" design decisions. The pivotal element in automating the design process is *modeling the design process itself*, which includes the following modeling tasks: (1) modeling the *structure of design tasks* that can take you from the initial design specifications to the final engineering artifact; (2) representing the *design decisions* involved in each task, along with the assumptions, simplifications, and methodologies, needed to frame and make the design decisions; and (3) modeling the *state of the evolving design*, along with the underlying rationale.

In this chapter, we will show how one can use ideas and techniques from artificial intelligence, such as symbolic modeling, knowledge-based systems and logic, to construct a computer-implemented model of the design process. By using the Douglas hierarchical approach as the conceptual model of the design process itself, this chapter will show how to generate models of the structure of design tasks, design decisions, and the state of design, thus leading to automation of large segments of the synthesis of chemical processing schemes. The result is a *human-aided, machine-based* design paradigm, with the computer "knowing" how the design is done, what the scope of design is, and how to provide explanations and the rationale for the design decisions and the resulting final design. Such paradigm is in sharp contrast with the traditional *computer-aided, human-based* prototype, where the computer carries out numerical calculations and data fetching from files and databases, but it has no notion on how the design is done, i.e., knowledge resting exclusively in the province of the individual human designer. In addition, we will argue that the human-aided, machine-based design is the paradigm that will characterize future design systems, where rapid conceptualization and prototyping of engineering artifacts will be the source of competitive edge.

I. Introduction

The engineering design of products and/or processing systems is a dialectic process (Stefik *et al.*, 1982) between goals (i.e., what is desired) and possibilities (i.e., what is actually realizable), aimed at the satisfaction of functional and performance specifications. No general theory exists for the systematic and rigorous development of a procedure that leads to the design of the desired engineering artifacts. This is due to two inescapable facts: (1) any design is a knowledge-intensive task; the more and better-quality knowledge, the more efficient the design and better the quality of the designed artifact and (2) the knowledge required for a particular class of design problems is specific to the class of problems. Thus, various attempts to formalize the overall design procedure as a large combinatorial optimization problem, have not provided a *generic theoretical framework for design*, but have simply underlined the fact that every design problem is a complex decision process with real-valued and integer decisions. On the other hand, combinatorial optimization techniques based on branch-and bound strategies (Edgar and Himmelblau, 1988), when applied to narrowly defined specific domains, have been shown to be fairly effective in selecting good design solutions through the implicit enumeration of many alternatives (Grossmann, 1985, 1989; Kocis and Grossmann, 1989). Nevertheless, even in such cases, the successes are an indication of the efficiency of the implicit enumeration techniques rather than a manifestation of a theory for design.

In the absence of a general theory on how design is done, research in the filed of artificial intelligence has been addressing the following distinct but complementary areas of inquiry:

1. *Axiomatic theory of design*, with the objective to establish a theoretically firm ground for the definition of design and thus bring it into the realm of "science," rather than "art," where it presently stands. Efforts in this direction have led to a number of approaches for the representation of cognitive models of the design process, but no significant breakthroughs have been achieved.

2. *Engineering science of knowledge-based design* (Tong, 1987; Tong and Sriram, 1992), aiming at the development of a rational framework for organizing, evaluating and formulating knowledge-based models on how the design is done. Efforts in this direction have been more successful. They have led to a number of new representation schemes (e.g., frames, scripts, streams, actors; see Rich and Knight, 1991; Quantrille and Liu, 1991), which have broken previous limitations in articulating, representing, and utilizing all forms of available knowledge; a particularly important requirement for a successful design approach. In addition, theoretical work has led to the development of new mathematical frameworks, which can manipulate these forms of knowledge representation (e.g., algebra of approximations, logic inferencing through semantic networks, analysis of logic clauses). Consequently, today we can effectively use all forms of

available knowledge within a computer-based environment; this is a significant addition to the numerical algorithms of the traditional computer-aided design tools.

Computer-aided design environments should support the designer in reaching better solutions than in the past, but in shorter periods of time. *We cannot afford to have the designs of tomorrow take as long as the designs 20 years ago.* Here is where the new artificial intelligence (AI)-based programming styles (e.g. object-oriented programming), the development of design-oriented languages, and object-oriented databases have already made significant contributions to the design process, by allowing the development of highly complex, design-oriented software systems in remarkably short periods of time. The existing applications, especially in other areas of engineering, are very impressive and do not allow room for disputing this fact. In particular, the complexity of the so-called *cooperative design environments* (or equivalently, *concurrent engineering*) is of such scale that traditional procedural programming approaches would have rendered the development of such software systems impossible. Thus, *AI is enabling the development of new design environments.* In chemical process design, the corresponding effort in academic institutions is at early stages, although the articulation of the design needs by the industry has been expressed in more mature terms. Lack of education in the developments of modern computer science has been the primary reason for this delay.

A. Conceptual Design of Chemical Processing Schemes

Conceptual process design involves the series of tasks leading to the development of a process flowsheet from the given reaction information and product specifications. This process usually comes at the early stage of a design project. Since all other design activities depend on the results of the conceptual process design, it has been considered the most important stage during the development and deployment of a new process. According to Douglas (1988), while the cost of the conceptual design usually takes 10–20% of the total cost to develop a new commercial process, the decisions committed at this stage fix about 80% of the total project cost.

Conceptual process design is an underdefined problem. Only a very small fraction of the information needed to define a design problem is available from the problem statement. The design decisions of the process designer provide this missing information. For example, the designer makes design decisions about what kind of process units to use, how to interconnect those units, and so on. From this perspective, the conceptual design is a combinatorial problem. It has been reported (Douglas, 1988) that there are 10^4–10^9 ways that we might consider to accomplish the conceptual design. On the other hand, experience indicates that at the conceptual design level, less than 1% of ideas ever prove to be successful, so the emphasis should be placed on the quick screening of numerous process alternatives.

1. Previous Approaches and Their Limitations

The pivotal role of the conceptual process design has motivated significant levels of research on the automation of the conceptual process design. AIDES (Adaptive Initial DEsign Synthesis) (Siirola et al., 1971; Powers, 1972; Rudd et al., 1973) is the pioneering prototype in the automatic synthesis of conceptual process flowsheets. AIDES makes use of a heuristically modified version of the means–ends analysis technique (Quantrille and Liu, 1991, pp. 268–270); given information on raw materials and the desired products, it attempts to bridge the difference between raw material and products through a sequence of operations that are eventually transformed into unit operations. BALTAZAR (Mahalec and Motard, 1977) is another early example on the use of AI techniques in process synthesis. Using techniques from the area of automatic theorem proving, BALTAZAR proposes logic-based approaches to the synthesis of flowsheets.

Douglas (1988) has proposed a hierarchical decision procedure. This procedure decomposes the design problem into a series of design or decision levels, sequenced and solved in a hierarchical order. It first sketches a design that is complete but vague, and then refines the vague parts into more detailed designs until finally the design has been refined to a complete sequence of detailed unit operations. Heuristics are used at all levels to fix the structure of the flowsheet. Economic potential is computed at each level to reduce the number of alternatives that have to be considered. It should be noted that hierarchial decision procedure is a systematic design procedure, not a design automation system.

Synthesis of process flowsheets through a simultaneous structural and parametric optimization, has received significant levels of attention (Grossmann, 1985, 1989; Kocis and Grossmann, 1989). This approach requires a superstructure as an initial structure from which all the designs of interest can be derived. Branch-and-bound algorithms have been used to solve the mixed-integer nonlinear programming (MINLP) formulations, and find the optimal structure and parameters of the unknown flowsheet. The possible design alternatives need to be articulated and modeled, if one is to develop an MINLP formulation that, using an implicit enumeration algorithm, will solve the MINLP problem and find the optimal solution. Clearly, this approach does not give much attention on the design process that we want to automate. However, it is a very promising technique for selecting the best alternative, once the superstructure has been identified.

Knowledge-based expert-system development has been very active during the last decade. After a design methodology has been identified for a given class of design problems, the expert system can be constructed in such a way that it maps uniquely the engineering design methodology into a computer program. The human usually interacts with the program by monitoring the design process, providing decisions and guidance at critical junctures, or the values which the program requests. Example of such programs are *PIP* (Kirkwood, 1987; Kirkwood *et al.*, 1988), *BioSepDesigner* (Siletti, 1988; Siletti and Stephanopoulos, 1992), *ProDesigner* (Kritikos, 1991). The construction of PIP (a program for the synthesis of process flowsheets) led to a rather rigid program with a strict procedural model to emulate the design methodology. Currently, a new object-oriented version is under development with significantly improved flexibility, extensibility and maintenance. *BioSepDesigner*, on the other hand, was developed to be a flexible design environment for the synthesis of separation sequences for the recovery and purification of proteins. Object-oriented in character, it is easily extendible with new knowledge and provides integrated treatment of data, models, and design decisions. Nevertheless, all three of the above systems and several others do not contain an explicit model of the design process and thus they cannot easily accommodate the incorporation of new knowldege as it becomes available.

B. Issues in the Automation of Conceptual Process Design

Extensive efforts over the last 30 years have led to rich computer-aided design environments that support the various tasks during the synthesis, development, and engineering of processing schemes. Figure 1 shows the standard outlay of such a computer-aided design (CAD) environment. A series of tools (e.g., equation solvers, optimization routines, specific design methodologies for localized problems, physical property estimation techniques) are being invoked by the human designer to execute various design tasks, such as design of heat exchanger networks, synthesis of separation systems, design of individual processing units, sizing, and costing. Convenient graphic interfaces make the interaction between designer and computer very smooth and transparent. Sophisticated database management

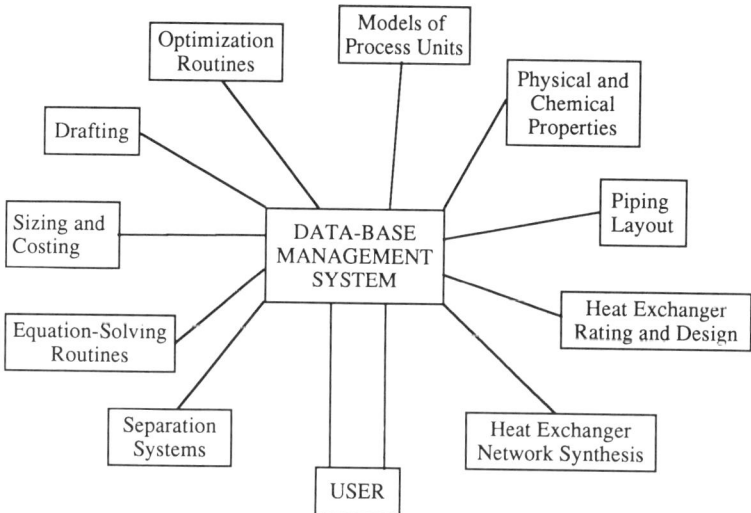

FIG. 1. Typical computer-aided design environment.

systems provide easy means for depositing and retrieving of data, intermediate results, previous designs, etc.

Despite the continuous enrichment of CAD environments, the character of the overall process design procedure remains the same (Mostow, 1985): "the human does the design and the computer provides the support tools, without understanding the design process, its rationale, or the design decisions." The drawbacks of the CAD paradigm are several and, at times, critical. They all stem from the fact that the design procedure is implicit in the designer's mind, for example

(a) Where does a design start from?
(b) What is to be done next?
(c) How are the assumptions, conjectures, and simplifications needed for the design to proceed, to be formulated?
(d) How are major design decisions made?

In other words, the CAD paradigm "knows" nothing about (1) the design agenda and (2) the "context" of the design activities.

Furthermore, the computer structure of tasks during the synthesis of a process flowsheet can be very large, detailed and complex for any human designer to document and mentally carry with him/her. To the extent that we can untangle and make explicit the design procedure, thus emulating the designer's own methodology, the process can be mechanized. But in this case, we are moving towards a *human-aided, machine-*

based design paradigm, where the computer, through human guidance, can carry out significant portions of a design by "knowing" the design procedure itself, its rationale and the reasoning behind a number of design decisions. This is the paradigm whose development and computer implementation has been significantly advanced by research in artificial intelligence, and which we will discuss in subsequent sections. The benefits from the availability of such mechanized models for design are many and diverse:

1. Rapid prototyping and evaluation of processing schemes.
2. Improvements in cost and reliability.
3. Explicit documentation of the design process itself: why certain goals were set during the design and how they were achieved, how design decisions were made, what assumptions and simplifications were involved, what models were used at the various stages of design, what alternative designs were examined, and why certain ones were selected over others.
4. Explicit documentation of the designed process itself, i.e., what are its components and their characteristics, how are they interconnected, what are their functional and performance characteristics, what are the critical design variables and the intrinsic tradeoffs.
5. Easy verification and modification of the resulting design. Having an explicit documentation of the intermediate design tasks, generated alternatives, rationale behind various design decisions, assumptions, conjectures, and simplifications, one can replay the design scenario and easily verify the validity of the derived processing scheme, or modify its design premises for further improvements.
6. The mechanized model of a design methodology offers an excellent depository for the organization of new empirical knowledge and/or the systematic incorporation of new theoretical results and analytic tools. Such inclusion of new knowledge will progressively increase the automation of the process synthesis procedure itself.

1. Modeling the Process of Design

How do you construct a software system, which emulates a specific methodology for the synthesis of processing schemes? Figure 2 shows a generic, conceptual model of the design methodology. It is composed of three distinct components; a *Planner*, a *Scheduler*, and a *Designer*.

a. Planner. Defines a top-level milestones through which the state of the processing scheme is expected to pass. For the synthesis of processing schemes, using the Douglas hierarchical approach, these milestones are (1)

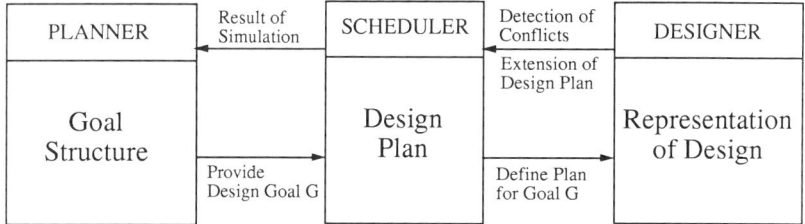

FIG. 2. A generic model for the design process.

design of the multiplant complex configuration, (2) design of the input/output structure for each plant, (3) design of the recycle structure, (4) design of the generalized separation structure, (5) synthesis of the separation subsystems (vapor, liquid, solid), (6) synthesis of heat exchanger networks, and (7) integration of processing subsystems.

b. Scheduler. Determines the sequence of design steps to be taken as one attempts to advance the current state of a processing scheme to the next milestone, defined by the Planner. It embodies a theory of how design goals and associated tasks are created, prioritized, and decomposed; how they interact, and how they are satisfied. As a result, it offers a fairly detailed design plan, by identifying all the requisite engineering design tasks. In subsequent sections, we will see that the *Scheduler* is represented by a tree of design tasks, which emulate the Douglas hierarchical design methodology (e.g., see Figs. 7–10, 12, 13).

c. Designer. It maintains the representation of the processing scheme being synthesized and other domain-specific knowledge. Thus, given a design step (from *Scheduler*), the *Designer* knows what must be done to execute the design step, e.g., reason with a specific set of rules, execute a design algorithm, carry out an optimization procedure, etc. It also updates the state of the evolving flowsheet, detects conflicts, and using domain-specific data, prescribes modifications to the design plan, which are communicated back to the *Scheduler*.

2. High-Level, Design-Oriented Languages

The conceptual model of the design process, discussed above (see also Fig. 2), should be turned into a logically well-defined computational process for the computer. There are high-level programming languages, such as FORTRAN, C, LISP, which can bridge the gap between the engineering model of conceptual design of Fig. 2, and its computer-based counterpart. However, as it has been pointed out by Newell (1982) and Chandrasekaran (1986), this gap is too wide for these languages to bridge in an expressive and efficient manner. Process designers cannot describe their design activities in their own terms, using FORTRAN, C, or LISP. Since the structure of language defines the boundaries of thought and expression, it is clear that the computer-based representation of a design methodology must rely on higher-level linguistic constructs, which allow us to do the following: (1) to represent the evolving structures of processing schemes in significant detail, including all the semantic relationships between the various representational components (processing units, streams, materials, modeling relationships, variables, decision-making procedures, numerical algorithms, etc.) and (2) to represent the design methodology itself, i.e., the intermediate design milestones, the goal structures taking the process from one milestone to the next, the procedures supporting various design tasks, etc.

3. Object-Oriented Representation of Knowledge

As a conceptual process design is a knowledge-intensive activity, its computer-based automation leads to a very large and complex computer program, which represents a very broad range of data models and procedural methodologies. Writing a large computer program is neither a trivial task, nor a straightforward exercise when all the design activities are already known.

The recent advances in object-oriented programming technology make this complex task much simpler by introducing the concepts of the *encapsulation* (or information hiding), *inheritance*, and *polymorphism*. In an object-oriented environment, the structure of the program can be designed in such a way that it represent a one-to-one mapping of the conceptual design methodology itself. For each object in engineering design methodology, a corresponding computational object can be constructed. For each engineering task, a symbolic operation possibly containing numerical computations in the computational model can be defined. With the use of this strategy, extending the model to accommodate new objects or new actions requires no strategic changes to the structure of the program itself. It only needs the addition of the new symbolic analogs of those objects or actions.

Many currently available CAD programs do not fully implement this concept. This is in part due to the unavailability of this technology when those systems were developed, and in part due to the lack of compatibility between the old codes and these new technologies. As a result, it is not a trivial task to modify the existing models of the CAD systems, or add a new model to the design system.

4. Human–Computer Interface

Given that the design process is such a complex process, the human–computer interface is very important for the user, in order to monitor the progress of the evolving design, provide decisions, and guide the direction of design in a user-friendly manner. It is not surprising that many computer-aided simulation environments, such as ASPEN Plus, PRO/II, DESIGN II, and HYSIM, have incorporated advanced graphic user interfaces into their environment. However, to help the process designer control the complexity of the design process, these user interfaces should be designed in a manner consistent to the model of the design process itself. The goal is to make the design process explicit to the user and allow the user to monitor the design process and communicate his/her ideas through user interfaces. When the user needs to make a design decision, the system should provide him or her with a context for the design decision. This context should have minimal information, yet still provide the key information necessary for the decisionmaking. After the user makes a design decision, the design decision should be recorded and made available to the user. The user should be able to see the effect of the design decisions on the process flowsheet, request and receive explanations, and solicit advice for the evolution of design. When the user interfaces satisfy these requirements, the user can take full control of the design process, while large segments of the design process are automated by the computer.

II. Hierarchical Approach to the Synthesis of Chemical Processing Schemes: A Computational Model of the Engineering Methodology

Douglas' hierarchical design approach (Douglas, 1985, 1988) was chosen to be the model of the engineering methodology, whose computer-based automation we seek to achieve. This approach has been used for the design of single-product continuous processes with single or multistep reactions (Douglas, 1985, 1988, 1989), solid processes (Rossiter and Douglas, 1986a, b; Rajagopal *et al.*, 1992), polymer processes (McKenna and Malone, 1990), pharmaceuticals, and specialty chemicals

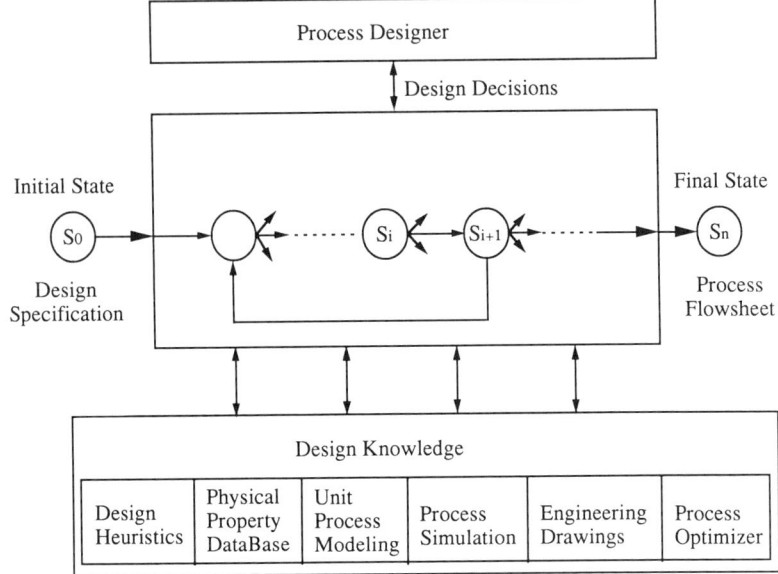

FIG. 3. Schematic diagram of the generic design process.

(Stephanopoulos *et al.*, 1994). It provides a well-defined structure of how the synthesis of conceptual processing schemes proceeds from very abstract, vague, and incomplete to fairly detailed, exact, and quite complete flowsheets. However, when it was developed, it was not intended to be a computational model for the design process itself. As a result, it lacks the requisite formalism and representational exactness needed for the construction of a computational process. In this section, we will provide a formalized restatement of the hierarchical procedure, in an effort to cast it more closely to a computational process for its computer-based automation.

A. Hierarchical Planning of the Process Design Evolution

It is common to characterize a design history as a series of transformations from specification to final design (Fig. 3). However, every design methodology does not handle the gap between "specifications" and "final design" in one sweeping structure of design steps. Instead, it identifies the abstract milestones through which the design is expected to go. In the case of the synthesis of processing schemes, the Douglas methodology has identified the following sequence of intermediate design milestones:

Beginning. Collect project specifications, e.g., chemistry to be used, feedstocks available, products to be produced, desired production specifications (amounts, purity), and economic constraints not to be violated (e.g., required capital $\leq a$, a product unit cost $\leq b$, return on investment $\geq c$, where a, b, and c are available bounding values).

Milestone 1. Synthesize and evaluate *plant complex structure*, i.e., identify the number of abstract "plants" each of which is formalized around a set of reactions, which take place under the same conditions (Fig. 4a).

Milestone 2. Design the *input/output structure* for each "plant," i.e., identify the set of *distinct* input and output streams associated with each "plant" of the overall structure (Fig. 4b).

Milestone 3. Select the *recycle structure* connecting the generalized abstract reaction and separation sections (Fig. 4c). During this phase of the design, the essential design alternatives of the generalized reaction section are also identified, i.e., the types of feasible reactors and their interconnections.

Milestone 4. Select the *generalized separation structure*, i.e., the elements of all-encompassing structure, shown in Fig. 4d, which are needed for the specific output from the reaction section of a given "plant."

Milestone 5. Synthesize the structure of the specific separation subsystems, e.g., vapor recovery, gas separation, liquid separation, or/and solid separation, which are relevant to the generalized separation structure of a given "plant."

Milestone 6. Integrate the process flowsheets, identified for each "plant" of the plant complex structure (Fig. 4a), and carry out consolidation of common processing tasks. The resulting design represents a flowsheet for the whole plant.

Milestone 7. Synthesize energy management systems for the consolidated flowsheet, derived at the last step.

End. Carry out sensitivity analyses of the process flowsheet's economic measures and revisit the critical design decisions, thus giving rise to improved designs.

It is clear from the above discussion that the Douglas approach to the synthesis of conceptual process flowsheets is based on the idea of *abstract refinement* (Mitchell *et al.*, 1981; Stephanopoulos, 1989), which is shown in Fig. 5. Since top-down abstract refinement does not handle goal coupling

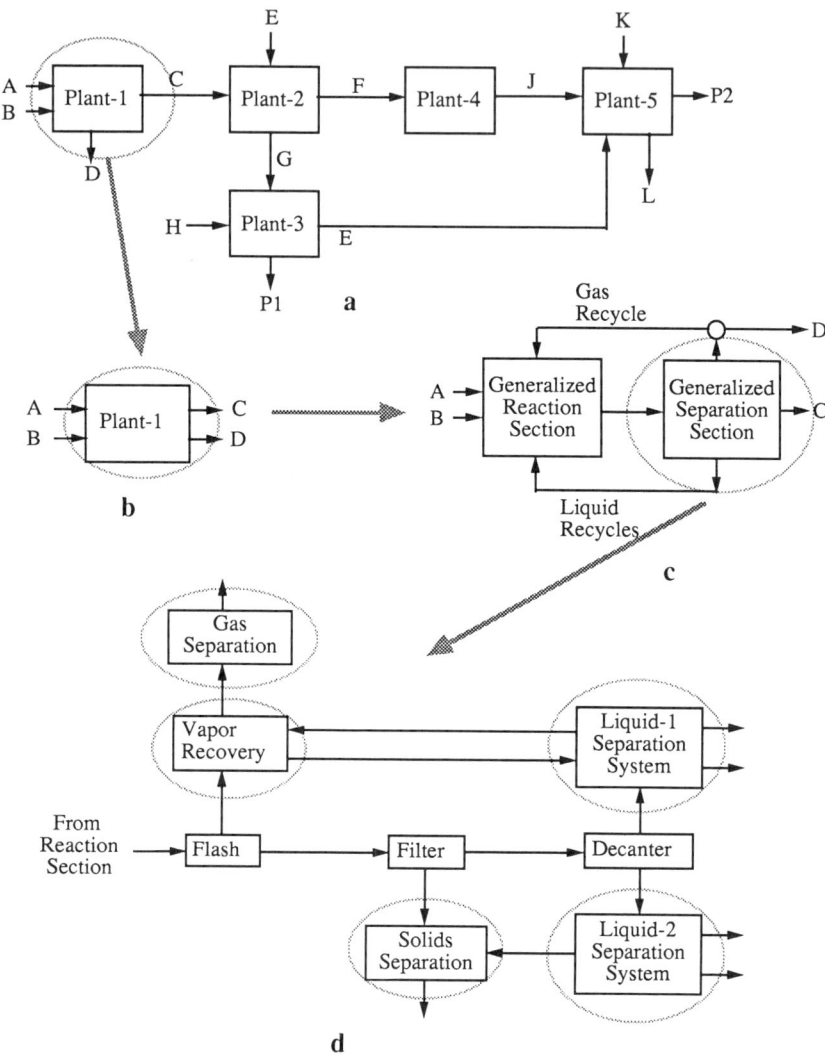

FIG. 4. Hierarchical evolution of conceptual processing schemes.

well, the approach must use, and in fact it does, *constraint propagation* to achieve consistency between the different parts of the overall processing scheme, as it will be discussed in a subsequent paragraph. However, for the time being, it is clear that Fig. 5 outlines the hierarchical planning stages for the conversion of specifications into one or more final process flowsheets.

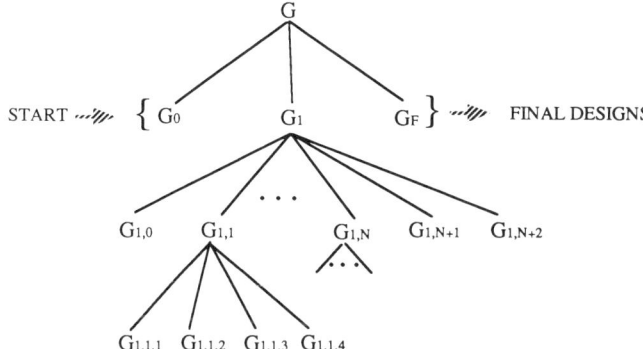

FIG. 5. Tree of goal structures in abstract refinement—(G—design the conceptual processing scheme for a multistep reaction plant; G_0—initialize the project's specifications; G_1—synthesize the process flowsheet for the whole-plant designs; G_F—evaluate process flowsheet and create improved designs; $G_{1,0}$—synthesize the plant complex structure; $G_{1,i}$—synthesize the flowsheet for plant-i; $G_{1,i,1}$—design the input/output structure for plant-i; $G_{1,i,2}$—design the recycle structure for plant-i; $G_{1,i,3}$—design the generalized separation structure for plant-i; $G_{1,i,4}$—design the liquid separation subsystem for plant-i; $G_{1,N+1}$—consolidate the flowsheets of all plant-i, $i = 1, 2, \ldots,$ into one; $G_{1,N+2}$—synthesize energy management systems for consolidated process).

B. Goal Structures: Bridging the Gap between Design Milestones

Moving the synthesis of process flowsheets from one milestone to the next entails a series of design steps with progressively increasing amount of details. These design steps need, nevertheless, to be articulated very explicitly and modeled by specific entities, before they can be automated by a computational procedure. In this section, we will focus our attention on the articulation of design tasks or steps, limiting our view to the synthesis of processes with a single fluid product with no solids present anywhere in the process.

1. Goal Structures and the Transformational Model

While the model of abstract refinement is quite satisfactory for the hierarchical planning of the process synthesis methodology, it is quite cumbersome and possibly incorrect as a model to describe the intricate web of design actions from one milestone to the next. Instead, the *transformational model* (Balzer *et al.*, 1976; Scherlis and Scott, 1983; Stephanopoulos, 1989) provides a more appropriate vehicle. This model converts specifications into designs through a series of correctness-

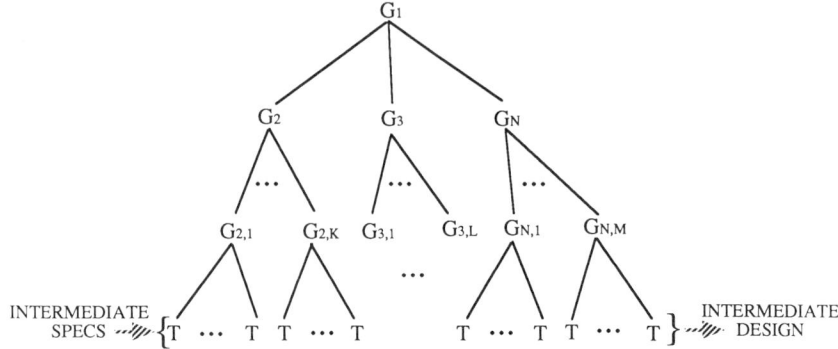

G: Goal   T: Transformation to achieve corresponding goal
Fig. 6. Structure of design tasks in the transformational model.

preserving transformations from one complete description to another (see Fig. 6). Thus, a single transformation may operate on several components of an evolving flowsheet at once. It is more general than the abstract refinement, which is limited to steps that replace a single component with a more detailed description of it. However, the generality comes at the potential cost of increased complexity, since a transformation on a complete description must deal with more information and must tackle all design decisions simultaneously. Nevertheless, since a transformation sequence can be viewed as an executable program for implementing the desired specifications, it is in agreement with the nature of the computational tools available in process engineering, i.e., branch-and-bound algorithms to solve mixed-integer constraint- or optimization-based design problems. Let us see now how we can model the transformation of one design milestone to the next, using Douglas' methodology for the synthesis of conceptual processing schemes. In doing so, in the subsequent paragraphs, we will generate the goal structures and requisite tasks following the general transformational model shown in Fig. 6, which can be easily converted into a computational process.

a. Goal-Structure 1: Initialize Project Definition. Looking back to Fig. 5, we notice that goal G_0, i.e., initialize the project's specifications, is the first to be accomplished. Figure 7 shows the refinement of G_0 into a series of data-acquisition subgoals, all of which are satisfied by user actions. Table I provides an example of project specifications at the outset of the process synthesis for a plant for the hydrodealkylation of toluence (Douglas, 1988).

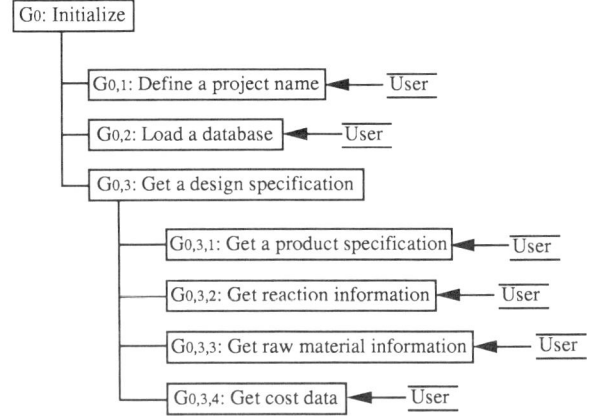

FIG. 7. Goal structure for the goal G_0, "Initialize the project specification."

b. Goal-Structure 2: Specify the Plant Complex Structure. For multistep reaction schemes, the synthesis of the process flowsheet starts by identifying the plant complex structure, i.e., achieving goal $G_{1,0}$ in Fig. 5. Figure 8 presents the transformational model of the goal structure, which identifies the individual "plants" defined around a set of reactions. From the design specification, the methodology can identify the number of simple plants. The reactions that take place at the same temperature, pressure, and catalysts are allocated to the same plant. The input and output streams of

TABLE I

DESIGN SPECIFICATIONS FOR HYDRODEALKYLATION (HDA) OF TOLUENE TO PRODUCE BENZENE[a]

Product specification
   Product: benzene
   Production rate: 265 lb-mol/h
   Product purity: 0.9997 mole fraction
Reaction information
   Toluene + $H_2 \rightarrow$ benzene + CH_4 at gas phase, 500 psia, 1150°F
     Extent of reaction = $\text{prod}/[1 - 0.0036/(1 - X)^{1.5}]$
   2 Benzene $\rightleftharpoons$ diphenyl + H_2 at gas phase, 500 psia, 1150°F
     Extent of reaction = $\text{prod}[0.0036/(1 - X)^{1.5}]/2[1 - 0.0036/(1 - X)^{1.5}]$
Raw material information
   Pure toluene at 15 psia, gas phase
   H_2: 95%, CH_4: 5% at 550 psia, 100°F, gas phase

[a]Key: prod represents the molar flowrate of the product; X represents the conversion of the primary reaction; psia = pounds per square inch absolute.

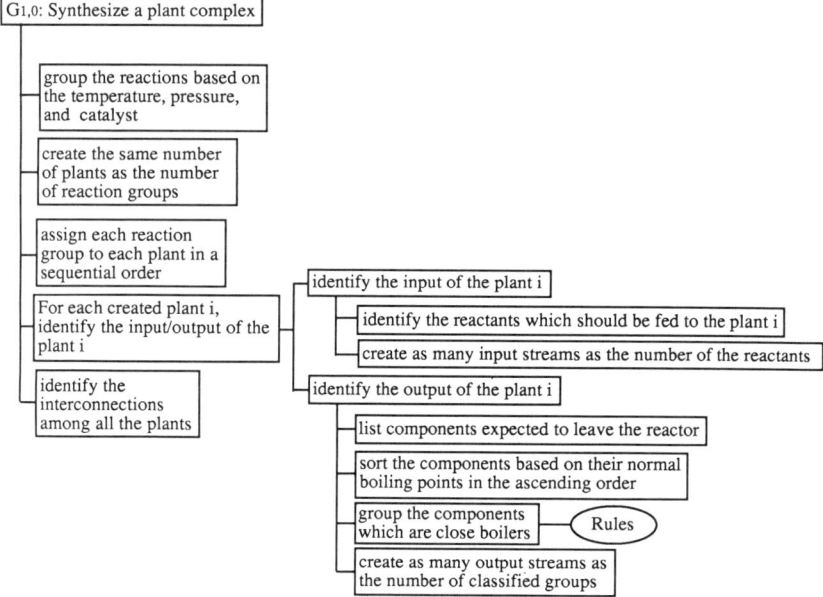

FIG. 8. Goal structure for the goal $G_{1,0}$, "Synthesize the plant complex structure."

each plant are identified. The rules (Douglas, 1988) for deciding the destination code are used to decide the number of output streams and the destinations of output streams. Once input and output streams of each plant have been identified, we can establish the interconnections among the plants using stream matching.

c. Goal-Structure 3: Design the Input-Output Structure. For each "plant-i" identified above, the hierarchical design of a process starts by selecting the structure of input and output streams. The design tasks at this stage are depicted in the transformational model of the goal structure shown in Fig. 9 and described below.

(i) *Synthesis.* When there are impurities in the feed streams, the methodology needs to make a decision for each feed stream about whether it needs feed pretreatment before the plant system.

(ii) *Analysis.* The design variable is a conversion for the primary reaction. If there are multiple reactions, then extents of reactions should be provided as functions of a conversion for the primary reaction, product

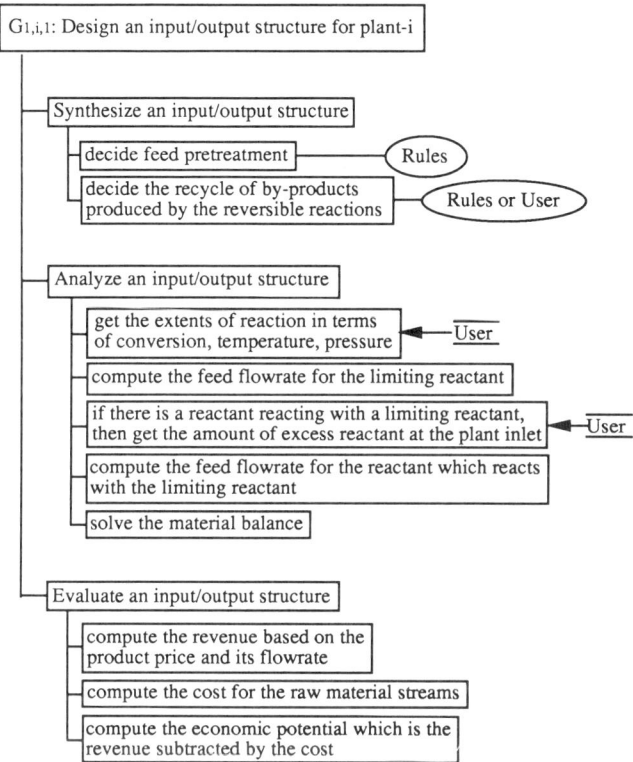

FIG. 9. Goal structure for the goal $G_{1,i,1}$, "Design an input/output structure for plant-i."

flowrate, temperature, and pressure. The extents of reactions represent the selectivity. The excess reactant ratio at the plant inlet is another design variable. It represents the ratio between the amount of the key reactant and that of the limiting reactant. A composition in the gaseous outlet stream can be substituted for the excess reactant ratio at the plant inlet.

(iii) *Evaluation*. Economic potential is computed by subtracting the costs of raw materials and feed pretreatment systems from the revenue from the product and the byproducts.

(iv) *Refinement*. The plant system identified at the input/output structure level is refined at the recycle structure level. The plant system is used as a boundary system at the recycle structure level.

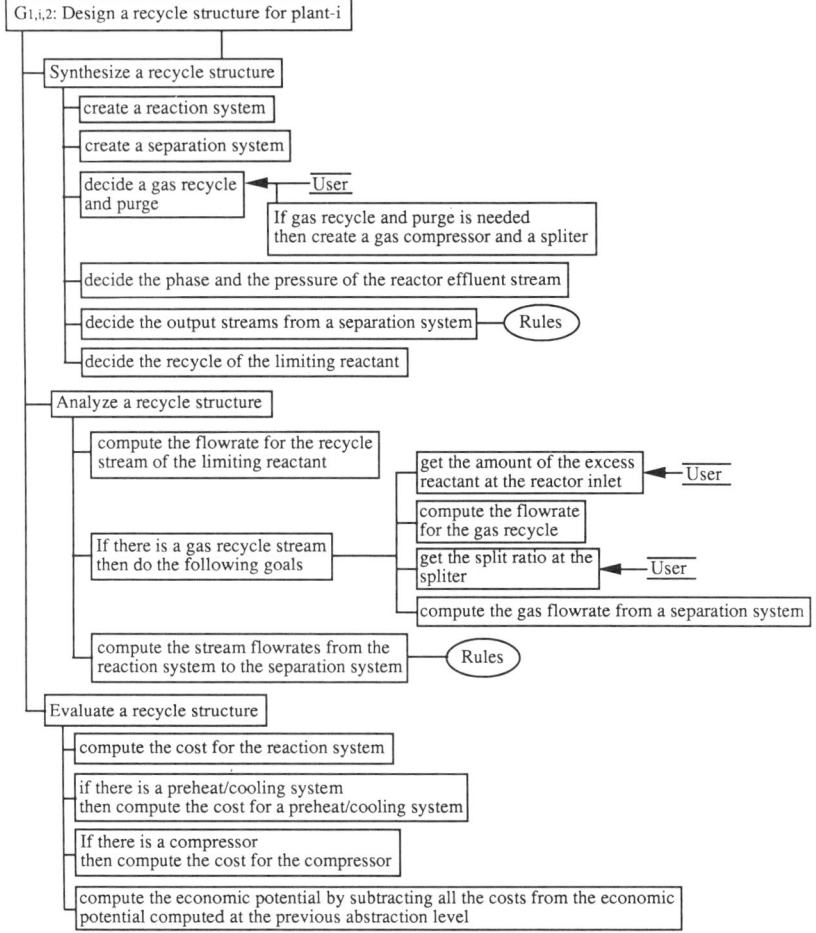

FIG. 10. Goal structure for the goal $G_{1,i,2}$, "Design a recycle structure for plant-i."

d. Goal-Structure 4: Design the Recycle Structure. The input/output structure of "plant-i" is refined into a more structured network of two sections; the generalized reaction and separation subsystems (e.g., see Fig. 4c). Figure 10 shows the goal structure, depicting the tasks needed for satisfying the top-goal, $G_{1,i,2}$, following again the transformational model in Fig. 6. The design tasks are discussed below.

(i) *Synthesis.* It is assumed that the recycle structure consists of a reaction section and a separation section. It is also assumed that 99% recovery is the same as 100% recycle. Thus, the unreacted limiting reactant will be recycled completely. The user needs to make a design decision about either just purging the gas stream that has reactants, or recycling a part of the gas stream and purging the remaining gas stream. As a result of this decision, a gas compressor and a spliter may be added to the flowsheet. When the plant system has a secondary reversible reaction, the user has to make a design decision about whether it is desirable to shift the equilibrium by recycling the byproduct.

(ii) *Analysis.* As the number of design alternatives is small, the material balance can be solved based on the structural decisions done at the synthesis phase. The ratio between the amount of the limiting reactant and that of the key reactant at the reactor inlet is needed as a parametric decision variable. This design variable indicates the excess amount of key reactant at the reactor inlet.

(iii) *Evaluation.* Economic potential is computed by subtracting the costs for the reaction system and gas compressor from the economic potential computed at the input/output structure level. To compute the costs for those units, several assumptions, such as the reactor type and the kinetic model, should be made. More detailed algorithms and example applications are available in Douglas' design text (Douglas, 1988).

(iv) *Refinement.* The separation system identified at the recycle structure level is refined at the general separation structure level. The separation system is used as a boundary system at the general separation structure level.

e. Goal-Structure 5: The General Separation Structure. The refinement of the generalized separation section leads to the superstructure of Fig. 4d, which accounts for all possible configurations of abstract separation subsystems. We will limit our attention to processes with fluids and single liquid phase only, thus eliminating the need for a filter, decanter and the associated subsystems of Fig. 4d. The resulting feasible options are shown in Fig. 11. Figure 12 shows the goal structure of the transformational model, with the corresponding tasks organized as follows.

(i) *Synthesis.* It is assumed that the general separation structure will be one of the three structures shown in Fig. 11. A structure is chosen on the basis of the phase of the reactor outlet stream. When the phase of the reactor effluent stream is not liquid, the user needs to make a design decision about whether it needs a vapor recovery system. If it does, the user should decide the location of the vapor recovery system. The location will be one of "on the recycle stream," "on the spliter outlet stream," or

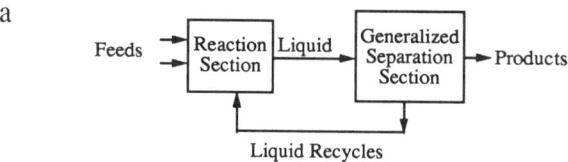

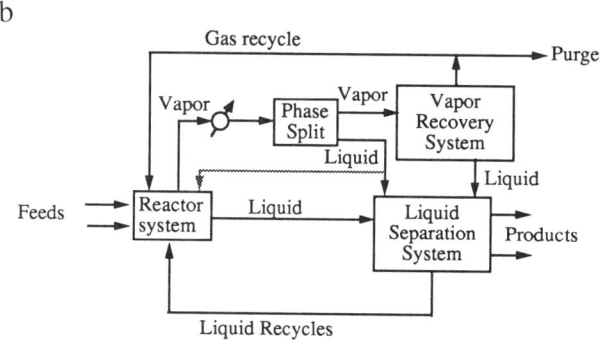

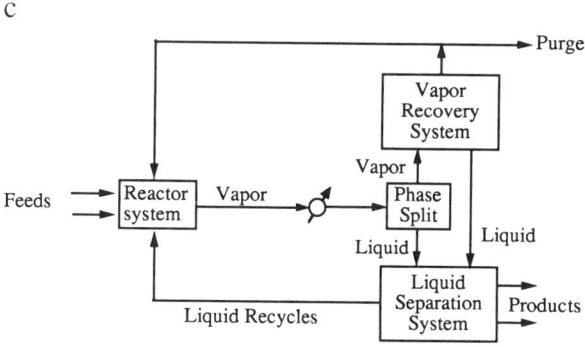

Fig. 11. Three possible general separation structures when reactor exit is (a) liquid (b) vapor, and liquid (c) vapor only [Reproduced from Hierarchical Decision Procedure for Process Synthesis, J. M. Douglas, *AIChE J.*, **31**, 353 (1985), by permission].

"on the purge stream." To help the user make this design decision, the system makes the flash calculation for the spliter and shows the flowrate and compositions of the vapor-phase outlet stream from the spliter. If the user decides to locate the vapor recovery system on the recycle stream to the reaction system, the control automatically goes back to the recycle structure level and locates the vapor recovery system on the recycle stream, and then the design procedure restarts at the general separation structure level.

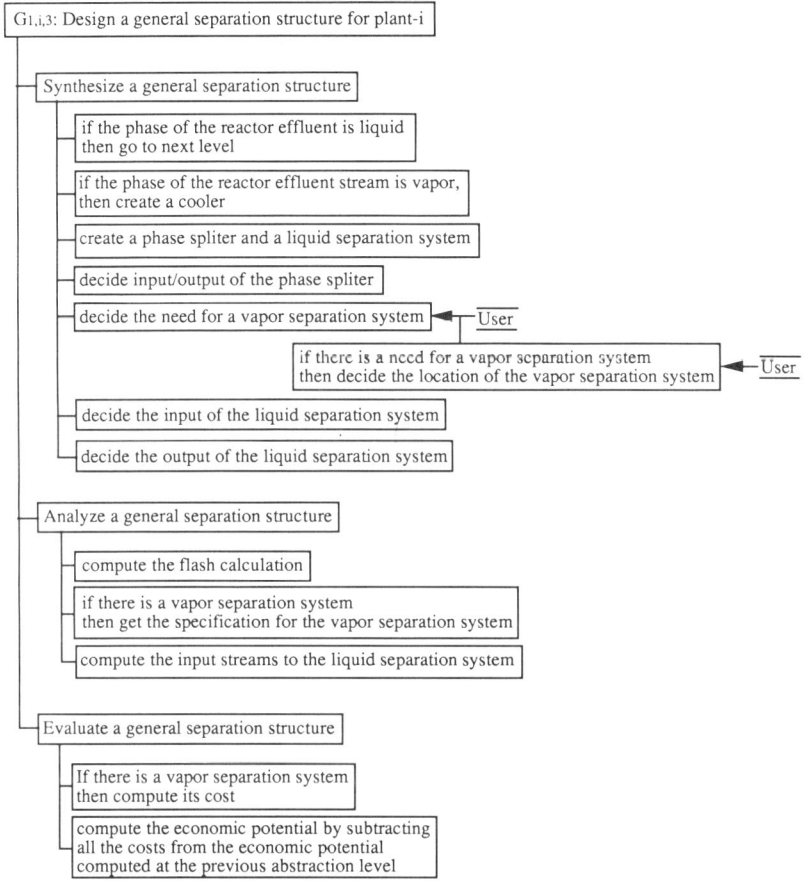

FIG. 12. Goal structure for the goal $G_{1,i,3}$, "Design a general separation structure for plant-i."

(ii) *Analysis*. If a vapor separation system is in a flowsheet, the user should give a specification for the vapor separation system, e.g., the existing components and their compositions at the liquid-phase outlet stream from the vapor separation system. As there are no more degrees of freedom, all the material and energy balance calculations can be done without any interaction with the user.

(iii) *Evaluation*. Economic potential is computed by subtracting the cost of vapor recovery system from the economic potential computed at the previous level. To compute the cost for a vapor separation system, the

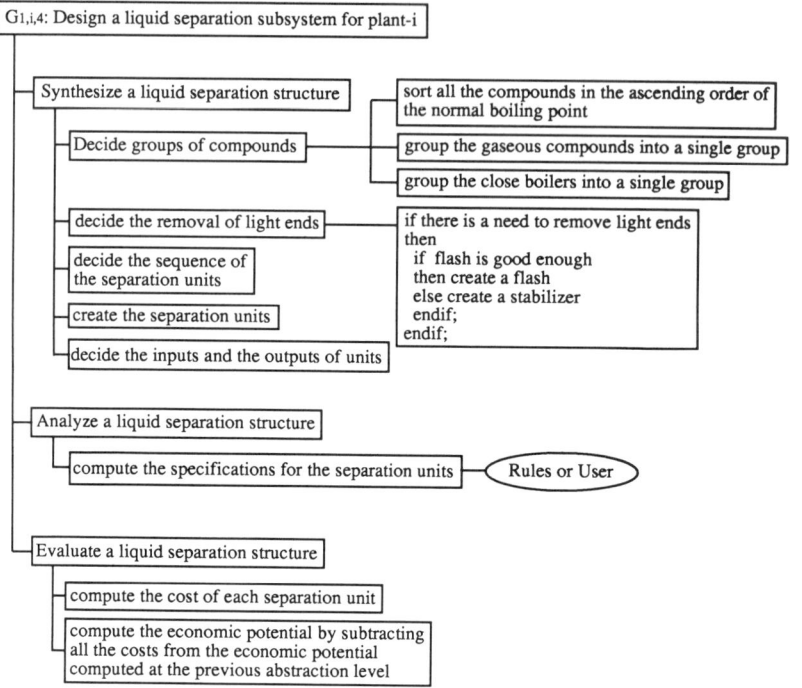

Fig. 13. Goal structure for the goal $G_{1,i,4}$, "Design a liquid separation subsystem for plant-i."

type of the system, e.g., absorption, condensation, should be specified. The algorithm for the absorber costing is available in Douglas' design text (Douglas, 1988).

(iv) *Refinement.* The liquid separation system identified at the general separation structure level is refined at the liquid separation structure level. The liquid separation system is used as a boundary system at the liquid separation structure level.

f. Goal-Structure 6: The Liquid Separation Subsystem. Figure 13 shows the goal structure and associated tasks during the design of the liquid separation subsystem.

(i) *Synthesis.* As the liquid separation system has only liquid feed streams, the only synthesis subproblem is that of the synthesis of the liquid separation sequence. If the input streams to the liquid separation system have light ends, then the methodology checks whether the light ends should be removed to satisfy the product specification. If light ends need to be removed, the methodology considers the flash as a means to remove the light ends. If the flash is not enough, then the design methodology decides to put a stabilizer column as a first column in the separation sequence. For the remaining components, the system decides the separation sequence. When the process has only three components to be separated, then, Eq. (1) is used to select either the direct or indirect sequence (Malone et al., 1985). Whenever the left-hand side of Eq. (1) takes on a negative value, the indirect sequence is preferred.

$$\frac{\delta V}{F} = 1.2 \left\{ \left(\frac{x_B + x_C}{a_{BC}} \right) \left(\frac{x_A x_C}{1 + x_A x_C} \right) + \left(\frac{1}{a_{AC} - 1} \right) \left(\frac{x_C - f x_A + x_A x_C^2}{f(1 + x_A x_C)} \right) - \frac{a_{BC}(x_A + x_B)(f - 1)}{(a_{AC} - a_{BC})f} \right\} - x_A, \quad (1)$$

where a_{xy} is a relative volatility between component x and component y, x_A is a mole fraction of component A in the feed stream, and f is a correction factor, $f = 1 + 1/100 x_B$.

When the process has more than three components, the selection is based on either heuristics (Douglas, 1988), or an implicit enumeration of the alternative sequences.

(ii) *Analysis*. It is assumed that in every separation unit, 99.5% of the light key is recovered in the overhead, and 99.5% of the heavy key in the bottom. It is also assumed that all the components lighter than the light key are taken overhead and that all components heavier than the heavy key leave with the bottom.

(iii) *Evaluation*. Economic potential is computed by subtracting all the costs of the distillation units from the economic potential computed at the previous abstraction level. To compute the cost of each separation unit, Fenske–Underwood–Gilland methods are used. To approximate the behavior of the distillation unit, a user can supply the average relative volatility for the column.

C. Design Principles of the Computational Model

Let us try to summarize the main principles (Han, 1994) on which a formalized computational model (outlined in the previous two sections) of the Douglas conceptual process design methodology was based (Douglas, 1988).

1. Principle 1: Hierarchical Planning of the Design Methodology

The overall design methodology was modeled as a hierarchical planning process. It starts with project specifications, goes through a predefined set of intermediate design milestones, and ends with a number of design alternatives.

2. Principle 2: Successive Refinement of Specifications into Implementations

The principle of hierarchical planning by itself cannot identify the design characteristics of the intermediate milestones. Successive refinement, on the other hand, does. Thus, at each stage of the hierarchical planning, we have a refinement of the specifications with more detailed description of the design characteristics of the next milestone. Figure 4 shows how the implementation specifications are successively refined from those of the input/output structure (first implementation), to recycle structure (second implementation), to generalized separation structure (third implementation), to separation subsystem structures (fourth implementation), etc. The generic *goal structure* of Fig. 5, based on the notion of *abstract refinement*, is the essential model that has been used to capture both the hierarchical planning and the successive refinement of the overall design methodology.

3. Principle 3: Propagation of Constraints

The top–down refinement of the abstract refinement model does not handle goal coupling well. Thus, the refinement of an abstract unit into a network of subunits, leads to design problems which could be solved independently; clearly, a gross violation of the overall system's intended functionality. Consequently, *constraint propagation*, to achieve consistency between the different parts of the design, is essential. Constraint propagation manifests itself through two distinct mechanisms; *induction of new constraints* and *propagation of constraint values*. Let us discuss two mechanisms in more detail since both appear during the synthesis of conceptual process design.

a. Propagation of Constraint Values. Consider the two abstractions of a process, shown in Fig. 14. At the input/output level (Fig. 14a), the input and output streams have flows as shown. At the recycle level (Fig. 14b), the input and output streams through the dashed boundary must have the same values, if the two abstractions are to be consistent. Thus, constraints on the flows of A, B, P, and (BP, A) have been propagated from one abstraction to the next.

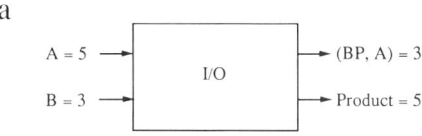

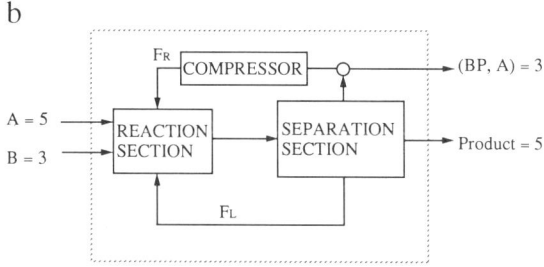

Economic Potential = 10 - Annualized_Compressor_Cost(F_R)
 - Annualized_Reactor_Cost($\overline{F}_R$) > 0

FIG. 14. Schematic used to indicate propagation of constraints from the input/output structure (a) to the recycle structure (b).

b. Induction of New Constraints. The design abstraction of Fig. 14b has revealed the presence of a compressor on the gas recycle stream. Since the economic potential must remain positive, the following constraint on the value of F_R is induced:

$$\text{Economic potential} = 10 - \text{annualized\_compressor\_cost}(F_R)$$
$$- \text{annualized\_reactor\_cost}(F_R) > 0$$

Thus, as design constraints are propagated from abstraction i, to abstraction $i + 1$, they also induce new constraints that must be satisfied by the implementation at level $i + 1$.

4. Principle 4: Hierarchical Decomposition with Coordination

Each refinement consists of two steps: a decomposition and a coordination. Whenever a system is decomposed into a set of subsystems, a specific knowledge about how the current system should be decomposed into a set of subsystems is required. This knowledge should come from the problem context. For example, at the recycle structure level, the knowledge that a plant system will be decomposed into a reaction system and a separation system, is specific only for the decomposition at the recycle structure level. At the coordination phase, the inputs and outputs of all process units and their interconnections are identified. A coordination step is necessary because the decomposition of a system into a set of subsystems introduces additional details to the design. It should be emphasized that the coordination is accomplished through the two constraint propagation mechanisms discussed above.

5. Principle 5: Unified Transformational Design

The model of abstract refinement is satisfactory only for the depiction of the intermediate design milestones. It is unacceptable as the model to describe the design steps at each abstraction level. Instead, we have relied on the transformational model represented by the generic goal structure of Fig. 6. Therein we notice that the generation of an intermediate design is accomplished through a series of tasks, which preserve the correctness (i.e., constraint satisfaction and optimality) of the derived designs. It is important to note that the generic goal structure of Fig. 6 is an abstract notation of the intended transformations. Thus, it does take different forms depending on the specifics of the design methodology employed. For example, the goal structures of Figs. 9, 10, 12, and 13, all of which are manifestations of the transformational model, *are not unique* and simply signify the methodological design steps employed by the Douglas approach. One can envision an implicit enumeration algorithm, where the various tasks correspond to the computational steps of the particular algorithm.

In the present work, we have opted for an explicit description of the intended design goals along with the associated design tasks, thus giving rise to explicit transformational models at each level of abstraction, as shown by the detailed tasks in Figs. 9, 10, 12, and 13.

Furthermore, it is important to underline that the model of the unified transformational design has an internal structure that is fairly generic, and in all likelihood would be present in any specific implementation, namely, the cycle of *synthesis, analysis, and evaluation*. As can be seen from the goal structures in Figs. 9, 10, 12, and 13, synthesis, analysis and evaluation are the three common pivotal goals around which the specific design implementations are expressed.

Synthesis is the first and represents the activity of generating structural designs. When the design specifications are given as an input, the structure is synthesized by heuristic reasoning, algorithmic procedures, or based on the interaction with the user. During the synthesis phase, all the process units and the process streams are identified. Thus, the design is given to the analysis phase.

In the second phase, analysis implies the setting up and solving material and energy balances for the synthesized structure. When the degrees of freedom for the structure exceed zero, or the structure is underdefined, optimization search algorithms or the user supply as many values as the degrees of freedom. These variables are selected by the design heuristics.

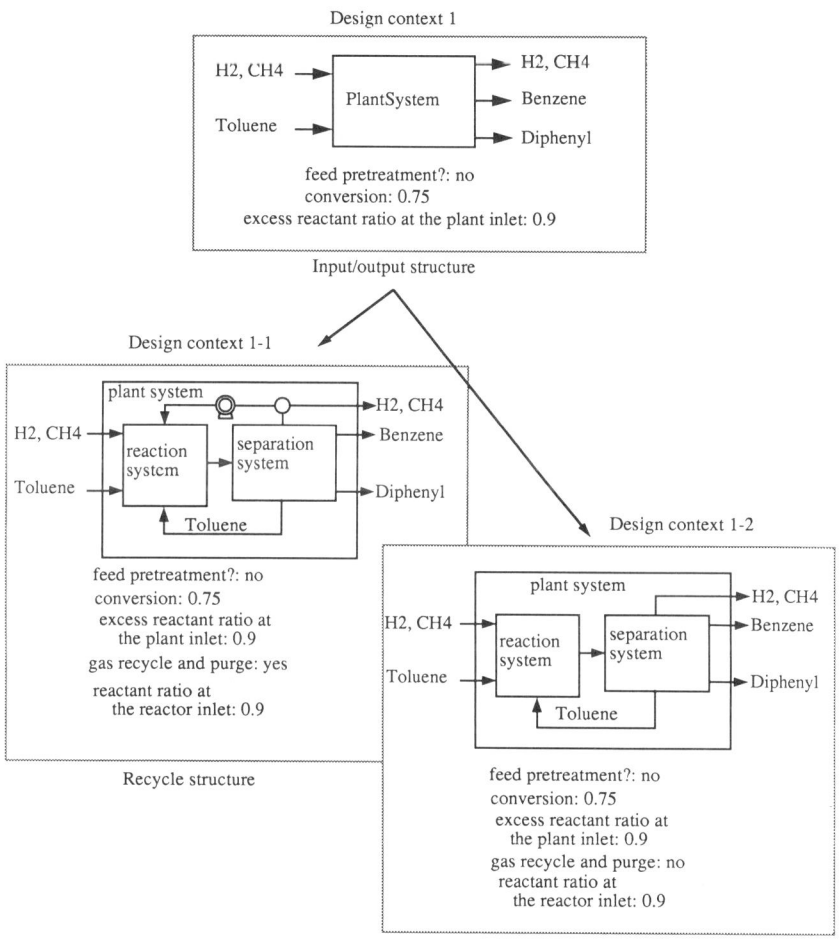

FIG. 15. Context-based design. This figure illustrates the scoping relationships among design contexts. The design decisions at the recycle structure are within the scope of the design decisions at the input/output structure level.

After the analysis phase, all process units and streams have quantitative values (flowrate, compositions, etc.).

Evaluation is the third phase and represents the activity of evaluating the structure in terms of economic measures, economic potential (EP), defined as follows:

$$EP_i = \begin{cases} \text{revenue} - \text{cost}, & \text{at } i = 0, \\ EP_{i-1} - \text{cost} & \text{at } i > 0, \end{cases} \quad (2)$$

where i represents the abstraction level.

Clearly, when implicit enumeration techniques are used, the parametric and structural optimization carries out synthesis, analysis, and evaluation in an integrated, unbroken cycle of simultaneous computations.

6. Principle 6: Context-Based Design

The design at the current abstraction level is used as a specification that will be implemented at the next abstraction level and defines the scoping relationship between the abstraction levels. This relationship is important to understand the mechanism for managing alternative designs. On the basis of this scoping relationship, we can identify the hierarchical dependency relationships among designs at all the abstraction levels. Figure 15 illustrates a design context generated from such relationships. The design decision "feed pretreatment?" has been made at the input/output structure. This design decision affects not only the input/output structure, but also all the designs refined from this input/output. For example, a local decision, "gas recycle and purge," shown in Fig. 15, is within the scope of the global decision, "feed pretreatment".

III. HDL: The Hierarchical Design Language

In the previous section, we discussed the structure of a conceptual model that can be used to represent a design methodology. But the value of this model rests with the effectiveness of the representation schemes that one employs to describe the declarative and procedural components of the model in a way that the computer can "understand." Thus, we are led to the need of defining a design-oriented language for the description of the computational process.

From the discussion in Section II, it is clear that a design-oriented language must address the following two needs:

1. Multifaceted modeling of the various states of the evolving process design.
2. Modeling of the procedural design tasks, as these are described by the goal structures.

HDL (Han, 1994) has been designed to meet the above two classes of modeling needs. It has been influenced by other previous work on process-design-related languages, such as MODEL.LA. (Stephanopoulos *et al.*, 1990a, b; see also first chapter in this volume), ConStruct (Johnston, 1991), SYDERELA (Kritikos, 1991), and ASCEND (Advanced System for Computations in Engineering Design) (Piela, 1989). The structure of HDL conforms with the design principles discussed in the previous section, and the object-oriented modeling, and human–computer interface requirements discussed in Section I. HDL provides a fairly simple and powerful framework of modeling classes, which the user can extend to develop a customized design environment. It should be emphasized that HDL is a formal framework for the development of the computational process that emulates the Douglas methodology for conceptual process design. In the following paragraphs, we will discuss its specific characteristics.

A. MULTIFACETED MODELING OF THE PROCESS DESIGN STATE

As the process flowsheet of a chemical plant is designed, it goes through a series of design states with variable details, with each design state being a snapshot description of the process flowsheet. Such a description encompasses the physical entities that are parts of the design. In addition, it includes the relationships among these variables. Thus, the declarative representation defines what a design or partial design is. Clearly the rationale, expressed in the first chapter in this volume, for a language such as MODEL.LA., finds a perfect example in the needs of HDL. Therefore, it is not surprising that the multi-faceted modeling capabilities of HDL, have drawn heavily from the structure of MODEL.LA.

1. Basic Modeling Elements

When we describe the chemical processes, a model should fully highlight the problem at hand, although not at the expense of excluding other possibly coexisting models. For example, models of chemical processes have modeling primitives corresponding to process units (e.g., reactors, pumps, pipes). Models of chemical processes also have primitives for aggregated units (e.g., reaction system, separation system, plant systems). So, although both design tasks are in the same domain, and both designs ultimately are composed of the vessels, pumps, pipes, etc., different modeling elements are used to highlight the important features of the design. Designs using the appropriate modeling elements convey the necessary design information with greater comprehension, because modeling primitives match intuitive description elements. These primitives can then be used to create an effective declarative representation.

To account for all these requirements, HDL provides a set of basic modeling elements drawn and extended from MODEL.LA (see first chapter in this volume, Stephanopoulos *et al.*, 1990a), which are used to represent the states of evolving process flowsheets at varying levels of abstraction. They also provide an efficient vehicle for the representation of alternative process designs with consistent referencing to common design elements of the alternative processing schemes. The ability to provide hierarchical representations and consistent versioning, allows HDL a truly multifaceted modeling of processing schemes.

Let us present HDL's basic modeling elements. For a pictorial depiction of their character, see Fig. 16. The first three modeling elements are sufficient to represent the "structure" of any processing system. The following two elements are used to represent complex systems which consist of networks of instances of GenericUnit, Port, and Stream.

a. Modeling-Element 1. GenericUnit. This is identical in character to the MODEL.LA.'s modeling element with the same name. It is used to represent an isolated system by defining the boundary that separates it from the surroundings. Thus, any system can be modeled as a GenericUnit, if the system has a clear boundary between itself and the environment. For example, a plant can be modeled as a GenericUnit. A reactor can also be modeled as a GenericUnit. To distinguish a plant from a reactor, the GenericUnit has two subclasses that have more specific properties. They are the GenericSystem and the GenericProcessUnit. The GenericSystem is the unit that can be decomposed into subsystems. For example, within the context of the hierarchical decision procedure, the plant system at the input/output structure level is decomposed into a reaction system and a separation system at the recycle structure level. On the other hand, GenericProcessUnit is the unit that cannot be decomposed into subsystems without losing its physical identity. For example, when a reactor is decomposed into a vessel and a heating coil, a vessel does not have the function "reactor" any more.

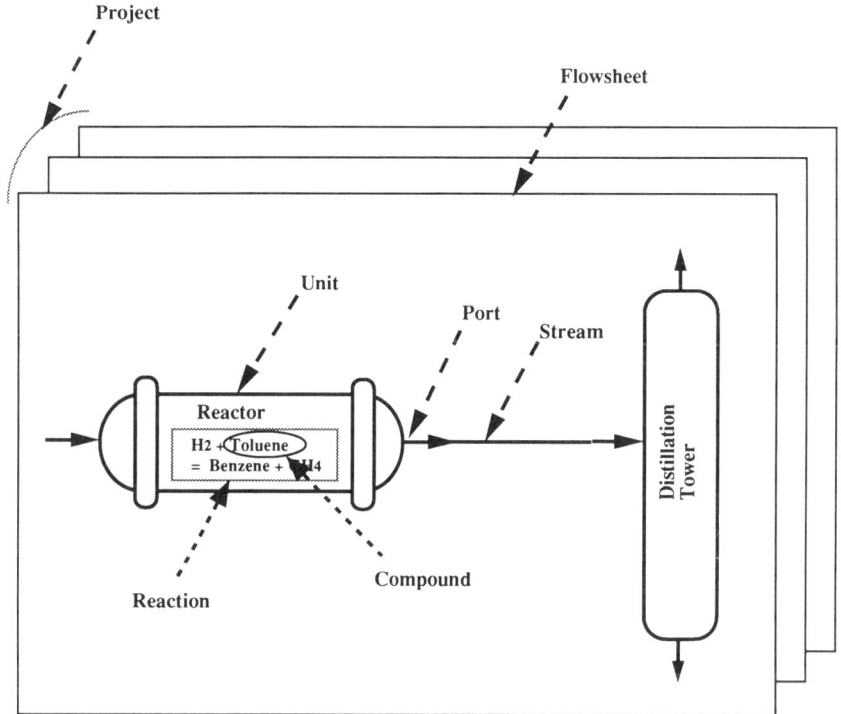

Fig. 16. Basic modeling elements in the HDL (hierarchical design language).

b. Modeling-Element 2: `Port`. As in MODEL.LA., instances of `Port` are used to define the boundaries of instances of `GenericUnit` and to provide the vehicles through which these instances transfer data to each other.

c. Modeling-Element 3: `Stream`. `Stream` is an abstraction that connects a port of an instance of `GenericUnit` to another port of another instance of `GenericUnit`. `Streams` know the `Ports` they are attached to, and symmetrically, each `Port` knows the stream to which it is attached. In logically consistent connections of `GenericUnits`, the flow through a stream from one port is connected to the another port of the same type.

d. Modeling-Element 4: `Flowsheet`. `Flowsheet` is an abstraction for a process flowsheet. It consists of process units and process streams. Except when the `Flowsheet` represents the design state at the highest abstraction level (least detail level), the `Flowsheet` has a boundary system which defines the design scope. The boundary system is transferred from the previous abstraction level. The user can monitor the design process through the Flowsheet and get the information which he/she wants to know from the Flowsheet. Each design alternative at each abstraction level is represented as an instance of Flowsheet.

e. Modeling-Element 5: Project. Project is an abstraction for a design project. An instance of Project keeps all the process flowsheets generated during the design project and handles the requests from a user, e.g., shows the Flowsheet at the recycle structure level. When a process alternative is generated, the alternative is also managed by the Project. Every instance of Project contains a tree-like data structure, in which it stores the state of the evolving Flowsheet. Specifically, as shown in Fig. 20, the root of the tree is the set of design specifications (empty Flowsheet), the first-level nodes represent alternative input-output structures, the second-level nodes the emanating recycle structures, etc. Clearly, the link between two connected nodes represents the design decision(s) generating one Flowsheet from the other.

f. Modeling-Element 6: GenericVariable. GenericVariable is an abstraction for a variable and is used to construct mathematical models for GenericUnits. It encapsulates the following information about a variable: variable name, variable value, possible range of values, and units. This additional information can be used to post constraints. These constraints are used to check whether the variable has a physically feasible value, e.g., a molar composition should be within the range from 0 to 1. The variable name and variable value type can be used to transfer information about the variable to the different environment, e.g. transfer the variable from HDL to Nexpert, which is a shell for the construction of expert systems, or copying a value from Nexpert to HDL.

g. Modeling-Element 7: Compound. Compound is an abstraction for a chemical compound. It has various elementary physical properties and the operations to access them from a database or to estimate derived properties through models and correlations. The elementary properties include normal boiling point, molecular weight, and critical temperature. The derived properties include the heat capacity at the given temperature or the vapor pressure at a given temperature.

h. Modeling-Element 8: Reaction. Reaction is an abstraction for a single reaction. It has information about the reactants, products and their stoichiometric coefficients, reaction temperature, reaction pressure, etc. It

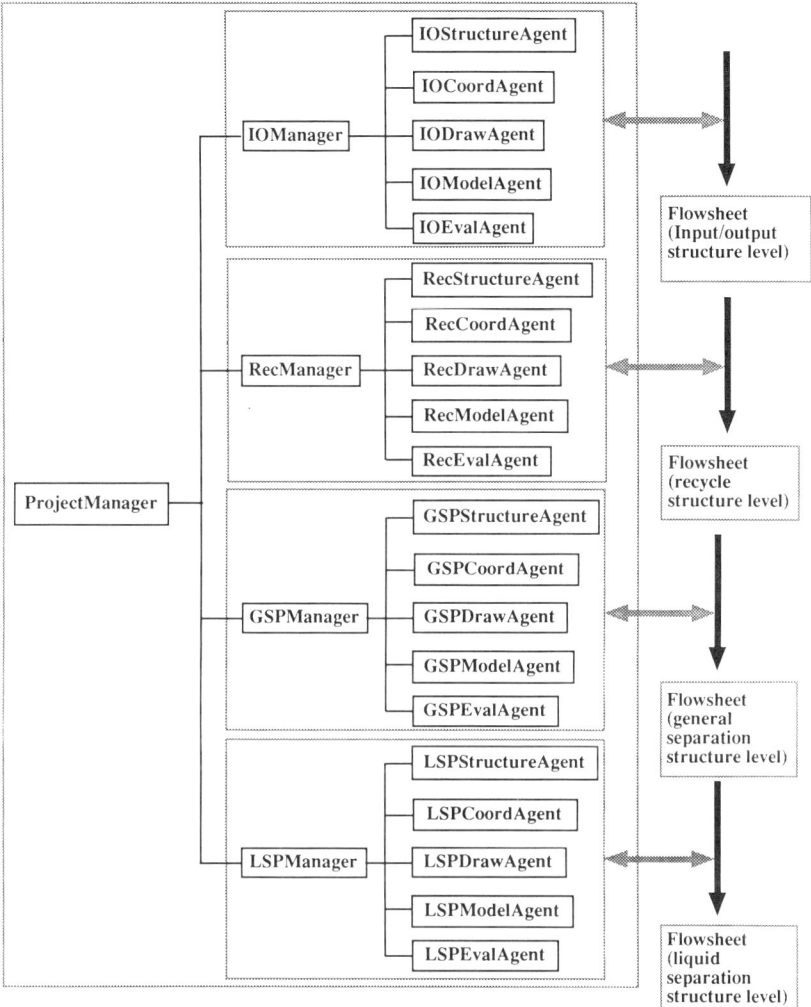

FIG. 17. Object model for design managers and design agents.

also has operations to estimate the properties such as the heat of reaction at a certain temperature, the equilibrium constant at a certain temperature and a pressure.

2. Semantic Relationships among Modeling Objects

Semantic relationships provide links between the various modeling elements, described above, and give rise to a network known as *semantic network.* Then, the description of a process is given by a semantic network. For more information on the nature of the semantic relationships and their role in modeling complex objects, the reader is referred to the more extensive discussion on the subject in the first chapter in the volume. For describing the structural aspects of a process plant, the following semantic relationships from MODEL.LA. are available in HDL. In the descriptions below, semantic relationships are written in boldface and italics. The class name starts with the uppercase letter, e.g., `Reactor`, and the instance name with a definite or indefinite article followed by its class name, e.g., `aReactor` or `theReactor`. A parenthesis represents a set, e.g., (`aReactor, aSeparator`).

> *is-a.* The *is-a* relationship indicates the parent/child relationship of two classes, e.g., `Plant` *is-a* `GenericUnit`.
>
> *is-a-member-of.* The *is-a-member-of* relationship indicates the relationship between a class and its instance, e.g., `aReactor` *is-a-member-of* `Reactor`.
>
> *is-composed-of.* The *is-composed-of* relationship is used to identify the modeling objects that are parts of another object, e.g., `aHeatExchanger` *is-composed-of* (`aTube, aBundle, aShell`).
>
> *is-part-of.* The *is-part-of* relationship is the inverse relationship of the *is-composed-of* relationship. For example, `aShell` *is-part-of* `aHeatExchanger`.
>
> *is-attached-to.* The *is-attached-to* relatinship provides the means for connecting process units and streams. The connection takes place through a linkage object called a port, e.g., `aPort` *is-attached-to* `aStream`.
>
> *is-connected-by.* The *is-connected-by* relationship is the inverse relationship of the *is-attached-to* relationship, e.g., `aGenericUnit` *is-connected-by* `aStream`.

In addition to the above, the following semantic relationships were found to be valuable and were introduced into the HDL.

> *is-a-model-of.* The *is-a-model-of* relationship indicates the relationship between a user interface and a model. An instance of a `GenericUnit` or its descendant classes represents the model of the process unit. For example, `ReactionSystem` has reaction information and has the operations to compute the heat of reaction and equilibrium constant, etc. A graphic unit associated with the reaction system represents the iconic structure of the reaction system, e.g., the number of input ports and the number of output ports. Figure 25 shows this relationship.

is-alternative-of. The *is-alternative-of* relationship relates a flowsheet to another alternative flowsheet. Both objects are of the same class and satisfy the same functions. However, they may have different structures, which result from different design decisions. Alternatives are used to keep track of the evolution of a flowsheet by recording the changes that are made to them.

B. MODELING THE DESIGN TASKS

In the previous section, we presented the modeling classes that allow one to describe the structure and behavior of chemical processing systems. Such a representation provides and structures the declarative knowledge about the current state of the design.

As a design is a transformation process of functional specifications into realizable physical objects, we need constructs that correspond to these transformations. A design task is a high-level construct that takes an action of transforming a design state to more detailed state, which satisfies the function of the design task. As a design task corresponds to the design action, the task has its own design goal and a design plan about how to accomplish its goal. Specific characteristics of design tasks are presented in this section.

1. Basic Task Elements

There are two kind of design tasks:design managers and design agents. A design manager is concerned primarily with organizing the computation, while the details of carrying out the steps are handled by design agents. Figure 17 shows the computational process for the various design tasks contained in HDL. The position of the task in the hierarchy shown in Figure 17 defines the role and the scope of the task during the design process. Figure 18 shows the inheritance relationships among design tasks.

a. Design-Task-Element 1: `GenericManager`. `GenericManager` can be considered a team leader who exercises control over subordinates, each of whom makes decisions within the context of its own bounded sub tasks. When subordinates encounter insurmountable difficulties in their sub tasks, some overall strategy may be introduced by the team leader to improve interactions between the subordinates.

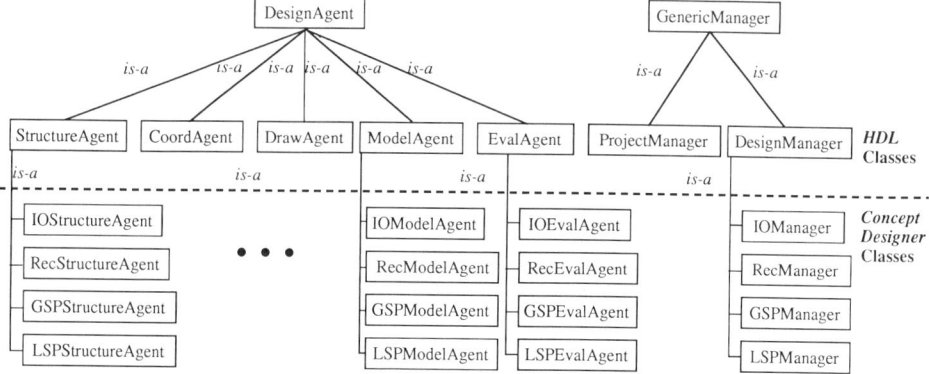

Fig. 18. The class tree of the design managers and the design agents in *ConceptDesigner*. The ancestor classes of DesignAgent and GenericManager exist, but not shown here. HDL classes are shown above the dotted line.

(i) *Design-Task-Element 1.1:* DesignManager. DesignManager is an abstract task that can be used to create a specific design manager. The classes that are descendants of the design manager can refine a process flowsheet from less detailed levels to more detailed levels. Each design manager carries out a functional transformation as shown below:

$$\text{aDesignManager}(\text{aFlowsheet}_i) \Rightarrow \text{aFlowsheet}_{i+1}$$

where subscript *i* denotes the abstraction level. For instance, RecycleManager refines a flowsheet at the input/output structure level into a flowsheet at the recycle structure level as follows:

$$\text{aRecycleManager}(\text{aIOStructure}) \Rightarrow \text{aRecycleStructure}.$$

(ii) *Design-Task-Element 1.2:* ProjectManager. ProjectManager is a task that organizes and coordinates the activities of design managers for a design project. Figure 17 shows ProjectManager at the top of the hierarchy among the design tasks. This hierarchy defines the roles of design tasks, while the class hierarchy defines the inheritance relationships.

b. *Design-Task-Element 2:* DesignAgent. DesignAgent is an abstract task that has a domain-specific knowledge. Specific design knowledge is represented as a design plan for each design agent. The following are subclasses of the DesignAgent class. Design agents 2.1–2.3 are sufficient to synthesize a structure and present it to the user in a very friendly manner. It should be noted that there is a certain order of executing these agents; `StructureAgent` first, `CoordAgent` second, and `DrawAgent`. The structural synthesis is completed by these agents. Agents 2.4 and 2.5 set up the material and energy balance and compute the economic potential of the flowsheet.

(i) *Design-Task-Element 2.1:* `StructureAgent`. `StructureAgent` is an abstract task that decomposes a given system into a set of subsystems and coordinates subsystems so that most of interconnections among sub systems are identified. The specific design knowledge for the decomposition and the coordination should be provided when the subclass is created for specific structure design. For instance, `aIOStructureAgent` shown in Fig. 17 has a knowledge and a design plan on how to synthesize an input/output structure. The `RecStructureAgent` shown in Fig. 17 has a knowledge and a design plan about how to sythesize a recycle structure. The interaction with the user is needed only when the agent does not have heuristic rules about the decision.

(ii) *Design-Task-Element 2.2:* `CoordAgent`. `CoordAgent` is an abstract task that establishes the connections among the `Ports` of the `GenericUnits` that do not have connections. Thus, this agent supplements the `StructureAgent` in terms of coordination.

(iii) *Design-Task-Element 2.3:* `DrawAgent`. `DrawAgent` is an abstract task that transforms a `GenericUnit` into an icon associated with the given `GenericUnit`. It also draws the streams that connect the ports of a generic unit to the other ports of generic unit based on the connection information of each port. It is a precondition that the connection information has been already identified by a `StructureAgent` and a `CoordAgent`.

(iv) *Design-Task-Element 2.4:* `ModelAgent`. `ModelAgent` is an abstract task that is in charge of the analysis. After the structure has been synthesized by the `StructureAgent`, a `ModelAgent` sets up and solves the material and energy balances, based on the synthesized structure. When design specifications are needed to finish the analysis, the agent may ask the user for these specifications.

(v) *Design-Task-Element 2.5:* `EvalAgent`. `EvalAgent` is an abstract task that is in charge of the evaluation. It computes the economic potential of the current process flowsheet.

2. Semantic Relationships among Design Tasks

The following list of semantics establishes the requisite relationships among the various design-task elements described above. The resulting semantic network (nodes are design tasks, edges are semantic relationships) models the overall design methodology.

decompose. The *decompose* relationship is used to decompose a boundary system into a set of subsystems. The syntax for this relationship is

>   aStructureAgent *decompose* aBoundarySystem

For example, aRecStructureAgent *decompose* aPlantSystem.

is-decomposed-into. The *is-decomposed-into* relationship is used to indicate the results of applying *decompose* relationship. The syntax is

>   aGenericUnit *is-decomposed-into* (a list of aGenericUnits)

For example, aPlantSystem *is-decomposed-into* (aReactionSystem, aSeparationSystem).

transform. The *transform* relationship is used to transform a list of generic units into a list of graphic units. Each graphic unit becomes associated with the corresponding generic unit. The syntax is

>   aDrawAgent *transform* (a list of aGenericUnits)

For example, aIODrawAgent *transform* (aPlantSystem).

is-transformed-into. The *is-transformed-into* relationship is used to indicate the results of applying transform relationship. The syntax is

>   (a list of aGenericUnits) *is-transformed-into*
>   (a list of aGraphicUnits)

For example, (aReactionSystem) *is-transformed-into* (aGraphicUnit).

compute. The *compute* relationship is used to compute the material and energy balances for a flowsheet. A ModelAgent computes the material and energy balances for all the process units that exist on the flowsheet. The syntax is

>   aModelAgent *compute* aFlowsheet

For example, aRecModelAgent *compute* aRecycleFlowsheet.

evaluate. The *evaluate* relationship is used to evaluate the economic potential of a flowsheet. An EvalAgent evaluates an economic potential of a Flowsheet by sending the message "computeCost" to all the process units which exist on the flowsheet. The syntax is

   aEvalAgent *evaluate* aFlowsheet

For example, aRecEvalAgent *evaluate* aRecycleFlowsheet.

refine. The *refine* relationship is used by a design manager to refine an instance of aGenericSystem at the current abstraction level into a more detailed flowsheet at the next abstraction level. The syntax is

   aDesignManager *refine* aGenericSystem

For example, aRecManager *refine* aPlantSystem at the input/output structure.

is-refined-into. The *is-refined-into* relationship is used to indicate the results of applying *refine* relationship. The syntax is

   aGenericSystem *is-refined-into* aFlowsheet

For example, aPlantSystem at the input/output structure level *is-refined-into* aFlowsheet at the recycle structure level.

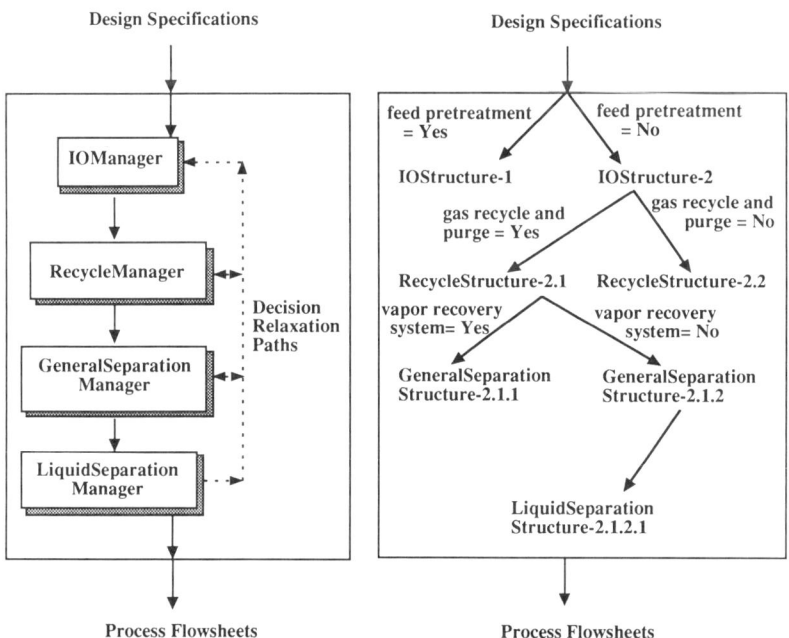

FIG. 19. Generation of design alternatives by relaxing design decisions.

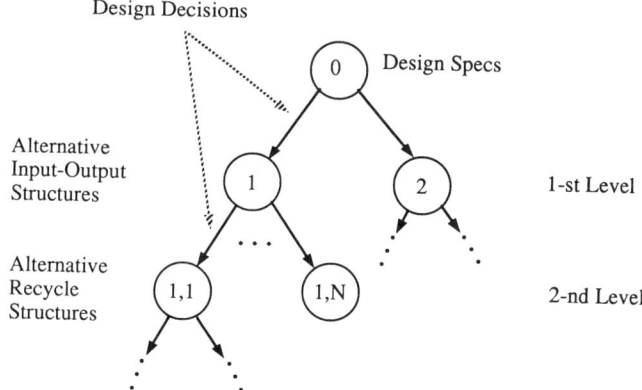

FIG. 20. The tree data structure maintaining design alternatives.

C. ELEMENTS FOR HUMAN–MACHINE INTERACTION

In order to communicate with the user, a computational procedure should possess very informative user interfaces, such as those described in following paragraphs, which are part of the *ConceptDesigner* (see Section IV; see also Figs. 22–25).

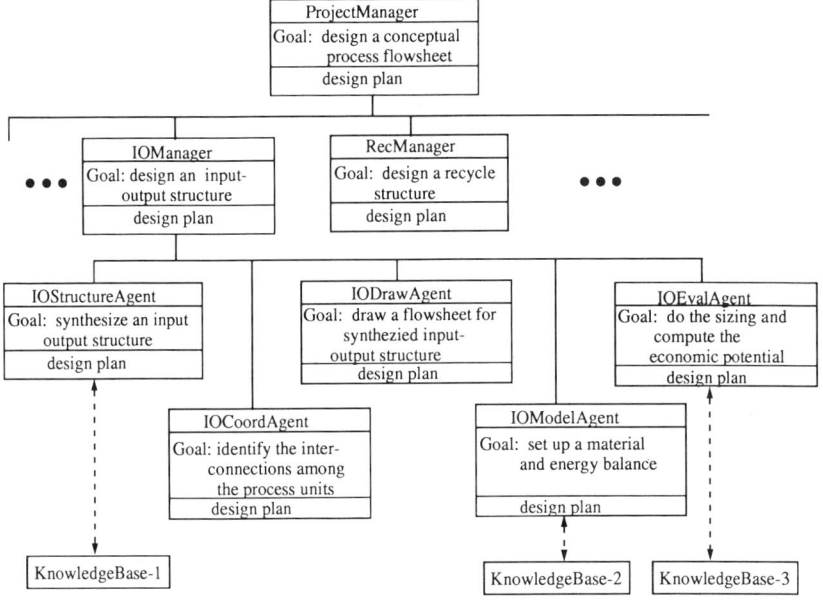

FIG. 21. Design task hierarchy in *ConceptDesigner*.

1. Human Interaction Element 1: `MainWindow`

This is an abstraction for the main interface. It provides several capabilities to define a design project. A user can define new project, load an old project, or save a current project into a file through `MainWindow`. The user can also define a new physical property database, load an existing database, modify the loaded database, and save the loaded database to a file. The user can specify the design mode: an automatic design mode, or an interactive design mode. In the automatic design mode, the design will be done by the system and the system will ask questions when they are needed. In the interactive design mode, the user can control the design process by executing each design task.

2. Human Interaction Element 2: `LayoutWindow`

This is an abstraction for the work space. A process flowsheet is developed and presented at the `LayoutWindow`. All design tasks are applied to an instance of `LayoutWindow` to transform a current process flowsheet. A snapshot of the transition in the process flowsheet is a design state at the moment. Figures 22 and 23 show the snapshots of a `LayoutWindow` during the design of hydrodealkylation-of-toluene process.

The elements described above are places where new modeling elements are designed and their values are displayed. The displayed elements are associated with the modeling elements and provide the user interface through which the user can access all the information generated during the design process. Those elements are presented below.

3. Human Interaction Element 3: `GraphicUnit`

This is an abstraction associated with a `GenericUnit`. `GraphicUnit`, displayed on the display area, is a means through which the user can communicate with the `GenericUnit`. A user can see the values of the design parameters. The user can also modify the graphic elements of the process unit, i.e., move the icon to the different place, rotate, or zoom. In Fig. 23b, `aGraphicUnit` displays the information of the associated distillation unit.

4. Human Interaction Element 4: `GraphicPort`

This is an abstraction associated with a `Port`. A user can modify the `GraphicPort` graphically, i.e., move, rotate, or zoom. The user can also see the values of `aPort` associated with the `GraphicPort`. In Fig. 23a, a `GraphicPort` shows the information of the associated port.

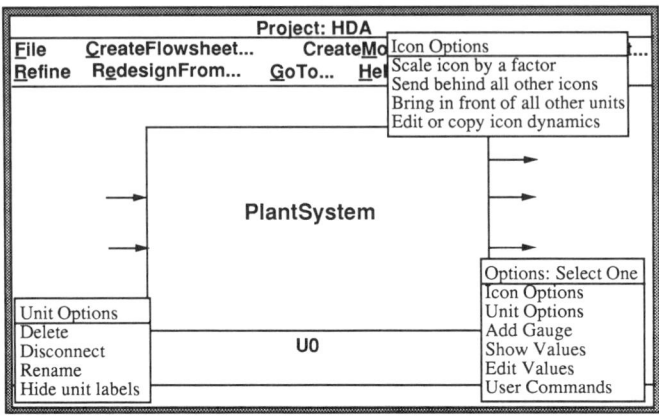

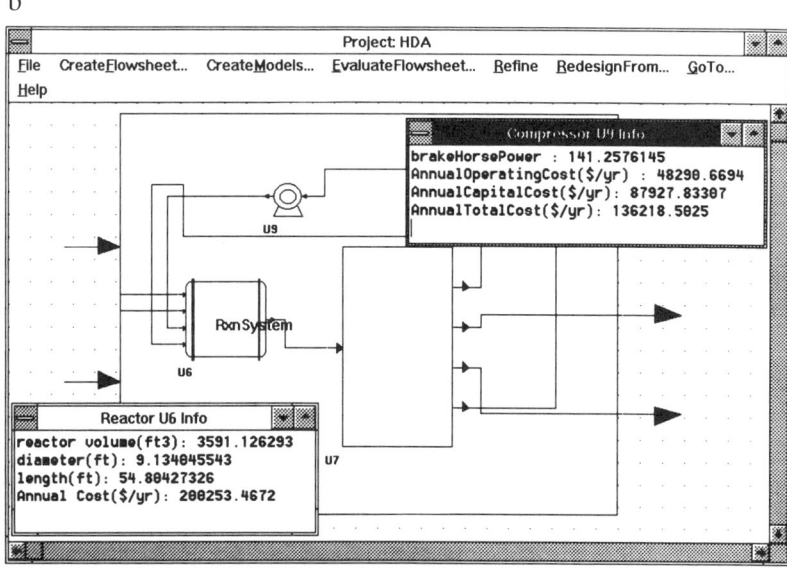

FIG. 22. User interfaces at (a) input/output structure level and (b) recycle structure level.

a

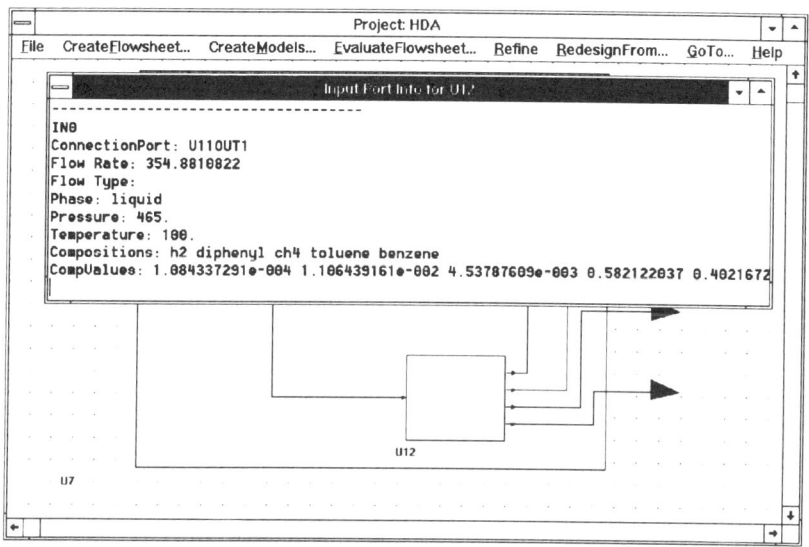

b

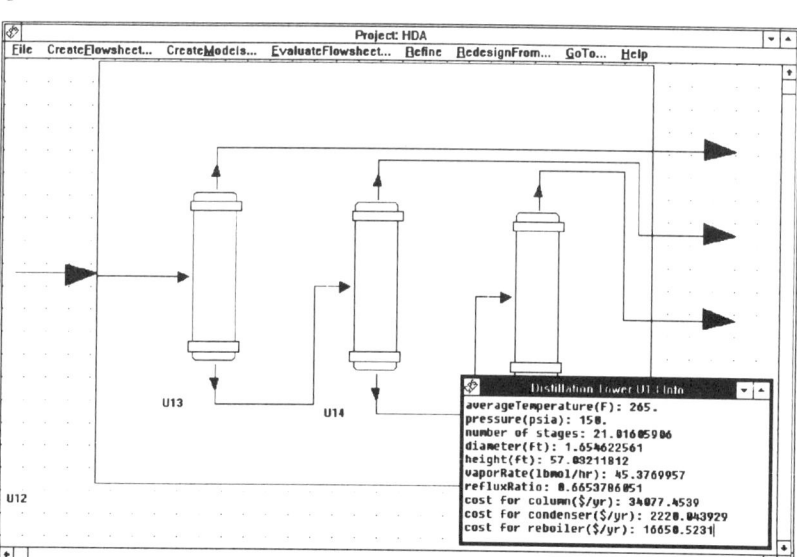

FIG. 23. User interfaces at (a) general separation structure level and (b) liquid separation structure level.

5. *Human Interaction Element 5:* `GraphicStream`

This is the abstraction associated with `Stream`. Like a `GraphicUnit`, the user can modify a `GraphicStream` graphically. The user can also see the values of a `aStream` associated with a `GraphicStream`.

D. OBJECT-ORIENTED FAILURE HANDLING

In producing a design, the software system will design a process flowsheet in a direction specified by its design plan. During the design process, the user may encounter obstacles or commit errors. Examples include violations of design constraints and numerical errors. These errors or failures are likely to occur more frequently in the conceptual design where HDL is intended to be used.

HDL models the way of producing a design as hierarchically structured design tasks. A hierarchically structured model leads to hierarchically structured failure handling. As shown in Fig. 17, each design agent has an explicitly defined role. The explicit structure of the model of the design process makes it possible to know what to do in the even of failure. When a failure occurs, the relevant information becomes immediately accessible due to the explicit localization of the failure. In some case, the system can guide the user to the point where the control should backtrack. Therefore, it is easy to correct the local error, and then to start a redesign from the agent which is responsible for the error.

E. MANAGEMENT OF DESIGN ALTERNATIVES

As the design decisions committed at the conceptual design fix 70–80% of the project cost, it is very important to design the economically optimal process flowsheet. A mechanism for managing alternatives has been developed, based on the characteristics of the conceptual process design itself: evolutionary, iterative decisionmaking process. The mechanism helps the user generate design alternatives by relaxing the previous design decisions.

The user can use the mechanism in the following way. First, he or she develops the base-case design which does not violate any process constraints. Then, the user improves the base-case design in the evolutionary manner by relaxing the decisions that look relatively promising in terms of contribution to the economic potential of the process flowsheet. Suppose we are working on the recycle structure level and we find one design decision about the gas recycle difficult to make because there are no heuristics available for the decision. We must then make a decision to recycle and purge gas components. After we complete the design of the recycle structure by deciding the values for the design variables and computing the economic potential, we can go back to the input/output structure level and restart the design. As all design tasks (including design managers and design agents) work as independent objects, it is a simple matter to go back to any abstraction level and start a redesign. The loaded design managers do the design in the same way except the decision about the gas recycle (e.g., we will choose not to recycle gas components). Figure 19 shows the design alternative generation process by relaxing design decisions. Figure 20 shows the tree structure which is dynamically generated by *ConceptDesigner* and is used to maintain design alternatives. The design decisions are recorded to a file and retrieved from the file along with design alternative. This helps a user identify the design decisions associated with a design alternative and relax them.

IV. ConceptDesigner: The Software Implementation

ConceptDesigner (Han, 1994) is the software system that implements the computational model, described in Section II, of the design process, using the linguistic constructs of HDL. The *ConceptDesigner* was designed to automate large segments of the design process with minimal interaction with the user. Nevertheless, a highly informative user-interface allows the human designer to monitor continuously the progress of the evolving design, and to override the automatic mode with a manual interactive mode of decisionmaking.

A. OVERALL ARCHITECTURE

ConceptDesigner is an object-oriented software system, which was built entirely on the top of the HDL design language. Figure 18 shows the complete set of HDL objects that *ConceptDesigner* used to create its own customized design tasks. In the design of *ConceptDesigner*, there are one-to-one mappings between the design-oriented tasks, identified in Section II, and the modeling elements of HDL, discussed in Section III. In other words, the structure of the software objects in *ConceptDesigner* is in direct and explicit correspondence to the design methodology we have

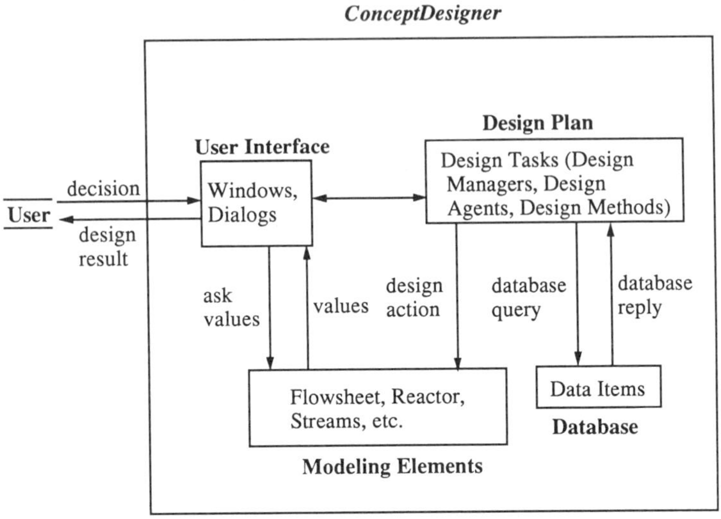

FIG. 24. Functional modules of *ConceptDesigner*.

adopted for the synthesis of process flowsheets. For example, Fig. 21 shows the hierarchical interrelationship among various objects (provided by HDL) in *ConceptDesigner*. Notice the one-to-one correspondence of the objects in Fig. 21, to the design tasks of the goal structure in Fig. 5, which models the design methodology. For example, the object `Project-Manager` in Fig. 21 corresponds to the goal "design a conceptual chemical process" of Fig. 5. Similarly, we can see the following correspondence; objects, `IOManager, RecManager` in Fig. 21 correspond to goals "design an input/output structure" and "design a recycle structure" of Fig. 5. Furthermore, the objects in Fig. 21, `IOStructureAgent, IOModelAgent`, and `IOEvalAgent`, correspond to the goals of Fig. 9, "synthesize an input/output structure," "analyze an input/output structure," and "evaluate an input/output structure."

Figure 24 shows the four major modules of *ConceptDesigner*. Figure 25 illustrates the semantic relationships among major objects in those modules.

1. Design Plan Module

This contains all the design-task modeling elements of HDL, and has structured them, through message passing methods, in such a way that

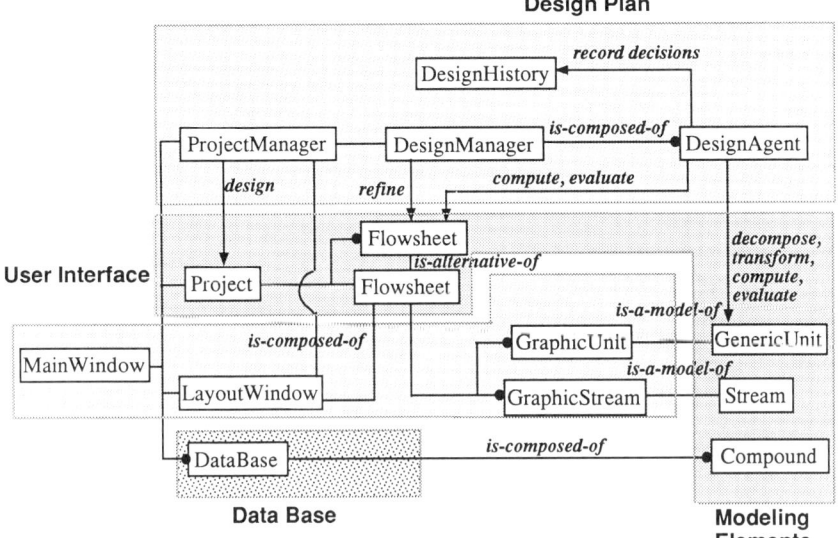

Fig. 25. Object model for *ConceptDesigner*.

they implement the goal structures of Figs. 8, 9, 10, 12, and 13, which model the design methodology. Thus, the `ProjectManager` can access information about the process design at any stage of its development, possess a "design plan" and calls on the `DesignManagers` to execute it. These managers are the softward objects: `IOManager, RecManager, GSPManager`, and `LSPManager` (see Fig. 17). Each of these managers possesses the data and methods to carry out the design of input/output structure, recycle structure, generalized separation structure, and liquid separation subsystem, respectively. They carry out their corresponding design tasks through the activation of five software objects, the `DesignAgents`, with common structure but different implementations. Figure 17 shows the structure of the specific `DesignAgents`, used at various stages of the design processed. The various `DesignAgents` contain specific methods, which are used to define the structure of processing systems, set up mathematical models, solve mathematical models, do sizing and costing of units, call on databases for physical properties, make decisions, etc. These methods are implemented through algorithms that contain symbolic manipulations, numerical procedures, or/and production rule-based systems.

2. Modeling Objects Module

This contains all the modeling elements that HDL uses to provide contextual, hierarchical and multi-view representation of process flowsheets (see Section III, A).

3. User Interface Module

This contains a broad variety of classes that the *ConceptDesigner* acquired from *x-Kit* (a computer-aided environment for the expeditious tailoring of graphic-oriented software modules, developed at MIT-LISPE). These classes provide the following capabilities: construct icons for abstract or detailed processing units; construct iconic flowsheets; provide animation and dynamic views to the process flowsheets; dialog windows, tables, lists, control buttons, and other user interface elements; and graphs, and charts for the presentation of data. *x-Kit* has also provided the following facilities that can be used for linking *ConceptDesigner* with external programs and databases:

(a) *File-Linker*. A high-level editor to configure the mapping of data between the objects of *ConceptDesigner* and external ASCII (American Standard Code for Information Interchange) files. These mappings can be used to transfer data to input files for an external FORTRAN program, or retrieve data from output files that carry the results of the FORTRAN program.
(b) *C-function Linker*. A program that uses dynamic link libraries and allows any object-method to be implemented through an external C function.
(c) *Database-Linker*. A high-level editor to configure the mapping of data between the tables of external relational databases and the *ConceptDesigner*'s objects. Among the many graphic user interfaces, the *LayoutWindow* occupies a central position, since this is the main canvas where the flowsheet is shown. Various instances of the *LayoutWindow* depict the input/output, recycle, generalized separation, liquid separation, and other abstractions of the evolving flowsheet (see Figs. 22 and 23).

Figure 26 provides a different view of the *ConceptDesigner*'s structure. It shows that every project generates a series of distinct process flowsheets, all of which are handled by a file management system, which is organized in a hierarchical tree to reflect the evolution of the design and the various alternative designs. It also depicts the various sources of data

AUTOMATION IN DESIGN: CHEMICAL PROCESSING SCHEMES

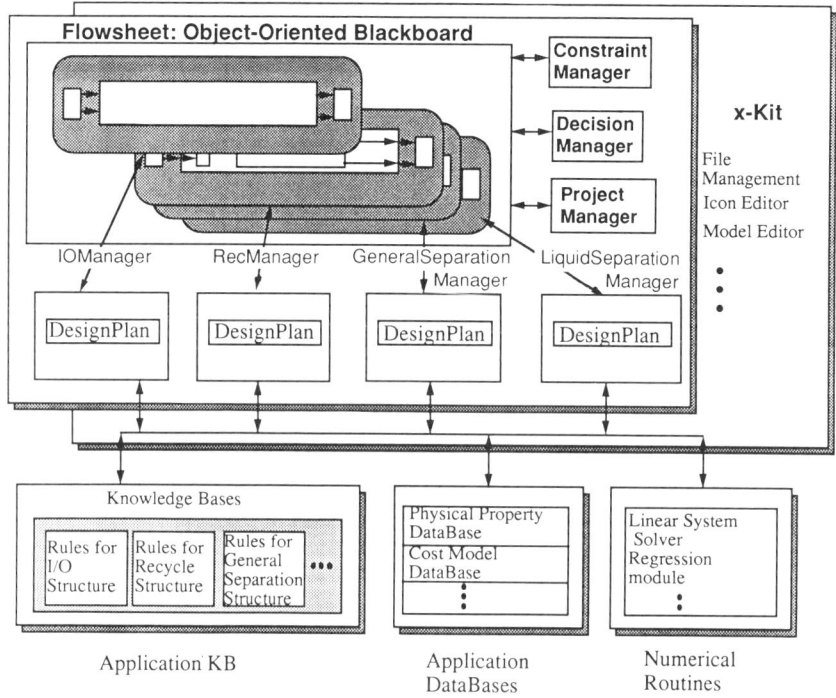

FIG. 26. Overall architecture of *ConceptDesigner*.

used and the types of knowledge bases and numerical procedures employed.

B. IMPLEMENTATION DETAILS

HDL has been implemented in Actor, which is an object-oriented language. The user interfaces have been developed from Microsoft Windows 3.1. *ConceptDesigner* can be run in any PC (personal computer)-compatible computer.

Conceptual process design is a complex process. As the design goes through several abstraction levels and generates many design alternatives at each abstraction level, a huge space is needed to store all information which is not usually available in PC or workstation environment. *ConceptDesigner* uses a dynamic memory allocation and a paging to resolve these memory problems. When an object is temporarily created, a certain amount of memory is allocated to the object. When the object is no longer needed, the allocated memory is immediately recovered by a garbage collection system. Another technique is a paging. After a process flowsheet is designed and the design process moves to the next abstraction level, the current process flowsheet is automatically saved to a file with a unique name. The memory consumed by the flowsheet is automatically recovered. These techniques enable *ConceptDesigner* to deal with large design cases.

V. Summary

In this chapter, we have tried to present a detailed approach for automating a complex design methodology. Three requirements are considered pivotal for the execution of projects, namely:

1. An explicit and formal model of the design methodology must be created. This model is the essence of the computational procedure to be developed.
2. A modeling language is needed to provide multi-faceted representation of the evolving design artifact, and represent the design tasks and their semantic relationships.
3. An object-oriented implementation of the software system is absolutely necessary.

The *ConceptDesigner* is a software system that was developed to automate large segments of Douglas' methodology for the synthesis of conceptual process designs. It satisfies all three of the above requirements. The most important lesson learned can be summarized as follows:

1. Hierarchical planning is an essential ingredient for modeling explicitly a design methodology. It sketches the general specs of the intermediate design milestones, without restricting the design activities too severely. Goal structures based on the concept of abstract refinement offer a very good model for describing the design plan.

2. Goal structures based on the notion of a unified, global transformation represent a very attractive generic model for the representation of the design methodology that transforms the artifact from an intermediate design milestone to the next. Such goal structures can accommodate equally well depth-first evolution of designs (such as the one suggested in Douglas' methodology), or exhaustive searches through implicit enumeration of design alternatives (such as those used in the MILP or MINLP formulation of process synthesis problems).

3. We cannot overstate the significance of modeling languages. A well-defined language captures explicitly, through its modeling elements,

domain-specific knowledge, and structures in a generic manner most of the design tasks.

4. The software that automates segments of a design activity should represent an explicit map of the design methodology. This is a necessary and sufficient condition in order to produce a software system which is (a) maintainable and (b) open to modifications and extensions with new knowledge.

References

Balzer, R., Goldman, N., and Wile, D., On the transformational implementation approach to programming. *Proc. Int. Conf. Software Eng.*, 2nd, p. 337 (976).
Chandrasekaran, B., Generic tasks in knowledge-based reasoning: high-level building blocks for expert system design. *IEEE Expert* **1**, 3 (1986).
Douglas, J. M., As hierarchical decision procedure for process synthesis. *AIChE J.* **31**(3), 353 (1985).
Douglas, J. M., "Conceptual Design of Chemical Processes." McGraw-Hill, New York, 1988.
Douglas, J. M., Synthesis of multistep reaction processes. *In* "Foundations of Computer-Aided Process Design" (J. J. Siirola, I. E. Grossmann, and G. Stephanopoulos, eds.), p. 79. CACHE Corp., Austin, TX, and Elsevier, New York, 1989.
Edgar, T. F., and Himmelblau, D. M., "Optimization of Chemical Processes," pp. 413–422. McGraw-Hill, New York, 1988.
Grossmann, I. E., Mixed-integer programming approach for the synthesis of integrated process flowsheets. *Comput. Chem. Eng.* **9**, 463 (1985).
Grossmann, I. E., MINLP optimization strategies and algorithms for process synthesis. *In* "Foundations of Computer-Aided Process Design" (J. J. Siirola, I. E. Grossmann, and G. Stephanopoulos, eds.), p. 105. CACHE Corp., Austin, TX, and Elsevier, New York, 1989.
Han, C., Human-aided, computer-based design paradigm: The automation of conceptual process design. Ph.D. Thesis, Department of Chemical Engineering, Massachusetts Institute of Technology, Cambridge, MA (1994).
Johnston, J., Synthesis of control structures for complete chemical plants. Ph.D. Thesis, Department of Chemical Engineering, Massachusetts Institute of Technology, Cambridge, MA (1991).
Kirkwood, R. L., PIP-process invention procedure, a prototype expert system for synthesizing chemical process flowsheets. Ph.D. Thesis, Department of Chemical Engineering, University of Massachusetts, Amherst (1987).
Kirkwood, R. L., Locke, M. H., and Douglas, J. M., A prototype expert system for synthesizing chemical process flowsheets. *Comput. Chem. Eng.* **12**, 329 (1988).
Kocis, G. R., and Grossmann, I. E., A modelling/decomposition strategy for MINLP optimization of process flowsheets. *Comput. Chem. Eng.* **13**, 797 (1989).
Kritikos, T., A model for process design automation. Ph.D. Thesis, Department of Chemical Engineering, Massachusetts Institute of Technology, Cambridge, MA (1991).
Mahalec, V., and Motard, R. L., Procedures for the initial design of chemical processing systems. *Comput. Chem. Eng.* **1**, 57 (1977).

Malone, M. F., Glinos, K., Marquez, F. E., and Douglas, J., Simple, analytical criteria for the sequencing of distillation columns. *AIChE J.* **31**, 683 (1985).

McKenna, T. F., and Malone, M. F., Polymer process design. I. Continuous production of chain-growth homopolymers. *Comput. Chem. Eng.* **14**, 1127 (1990).

Mitchell, T., Steinberg, L., Reid, G., Schooley, P., Jacobs, H., and Kelly, V., Representations for reasoning about digital circuits. *Proc. IJCAI*-81 (1981).

Mostow, J., Toward better models of the design process. *AI Mag.* Spring, p. 44 (1985).

Newell, A., The knowledge level. *AI J.* **19**(2), 87 (1982).

Piela, P. C., ASCEND—An object-oriented environment for the development of quantitative models. Ph.D. Thesis, Department of Chemical Engineering, Carnegie Mellon University, Pittsburgh, PA (1989).

Powers, G. J., Heuristic synthesis in process development. *Chem. Eng. Prog.* **68**, 88 (1972).

Quantrille, T. E., Liu, Y. A., "Artificial Intelligence in Chemical Engineering." Academic Press, San Diego, CA, 1991.

Rajagopal, S., Ng, K. M., and Douglas, J. M., A hierarchical procedure for the conceptual design of solids processes. *Comput. Chem. Eng.* **16**, 675 (1992).

Rich, E., Knight, K., "Artificial Intelligence," 2nd ed. McGraw-Hill, New York, 1991.

Rossiter, A. P., and Douglas, J. M., Design and optimization of solids processes. Part 1. A hierarchical decision procedure for process synthesis of solids systems. *Chem. Eng. Res. Des.* **64**, 175 (1986a).

Rossiter, A. P., and Douglas, J. M., Design and optimization of solids processes. Part 2. Optimization of crystallizer, centrifuge and dryer systems. *Chem. Eng. Res. Des.* **64**, 184 (1986b).

Rudd, D. F., Powers, G. J., and Siirola, J. J., "Process Synthesis." Prentice-Hall, Englewood Cliffs, NJ, 1973.

Scherlis, W., and Scott, D., First steps towards inferential programming. *IFIP Congr. '83* (1983).

Siirola, J. J., Powers, G. J., and Rudd, D. F., Synthesis of system design. II. Toward a process concept generator. *AIChE J.* **17**, 677 (1971).

Siletti, C. A., Computer-aided design of protein recovery processes. Ph.D. Thesis, Department of Chemical Engineering, Massachusetts Institute of Technology, Cambridge, MA (1988).

Siletti, C. A., and Stephanopoulos, G., BioSep designer: A knowledge-based process synthesizer for bioseparations. In "Artificial Intelligence Approaches in Engineering Design" (C. Tong and D. Sriram, eds.). Vol. I, p. 295. Academic Press, San Diego, CA, 1992.

Stefik, M., Bobrow, D., Bell, A., Brown, H., Conway, L., and Tong, C., The partitioning of concepts in digital system design. *Proc. Conf. Adv. Res. VLSI*, p. 43 (1982).

Stephanopoulos, G., Artificial intelligence and symbolic computing. In "Foundations of Computer-Aided Process Design" J. J. Siirola, I. E. Grossmann, and G. Stephanopoulos, eds.), p. 21. CACHE Corp., Austin, TX, and Elsevier, New York, 1989.

Stephanopoulos, G., Henning, G., and Leone, H., MODEL.LA.:A modeling language for process engineering: Part I. The formal framework. *Comput. Chem. Eng.* **14**, 813 (1990a).

Stephanopoulos, G., Henning, G., and Leone, H., MODEL.LA.:A modeling language for process engineering: Part II. Multifaceted modeling of processing systems. *Comput. Chem. Eng.* **14**, 847 (1990b).

Stephanopoulos, G., Han, C., Linninger, A., Ali, S., and Stephanopoulos, E., Concept of ZAP (Zero Avoidable Pollution) in the synthesis and evaluation of batch pharmaceutical processes. Paper presented at *Ann. AIChE Meet.*, San Francisco (1994).

Tong, C., Toward an engineering science of knowledge-based design. *AI Eng.* **2**, 133 (1987).

Tong, C., and Sriram, D., "Artificial Intelligence in Engineering Design," Vol. I, p. 9. Academic Press, San Diego, CA, 1992.

SYMBOLIC AND QUANTITATIVE REASONING: DESIGN OF REACTION PATHWAYS THROUGH RECURSIVE SATISFACTION OF CONSTRAINTS

Michael L. Mavrovouniotis

Department of Chemical Engineering
Northwestern University
Evanston, IL 60208.

I. Reaction Systems and Pathways	148
II. Catalytic Reaction Systems	151
A. Basic Concepts, Terminology, and Notation	151
B. Previous Work on the Construction of Mechanisms	154
C. Structure of the Algorithm	155
D. Features of the Algorithm	159
E. Examples	160
III. Biochemical Pathways	169
A. Features of the Pathway Synthesis Problem	173
B. Formulation of Constraints	175
C. Algorithm	176
D. Examples	179
IV. Properties and Extensions of the Synthesis Algorithm	183
V. Summary	185
References	185

Given a fixed, predetermined set of elementary reactions, compose reaction pathways (mechanisms) that satisfy given specifications in the transformation of available raw materials to desired products. This is a problem encountered quite frequently during research and development of chemical and biochemical processes. As in the assembly of a puzzle, the pieces (available reaction steps) must fit with each other (i.e., satisfy a set of constraints imposed by the precursor and successor reactions) and conform with the size and shape of the board (i.e., the specifications on the overall transformation of raw materials to products). This chapter draws from *symbolic and quantitative reasoning* ideas of AI which allow the systematic synthesis of artifacts through a *recursive satisfaction of constraints* imposed on the artifact as a whole and on its components. The artifacts in this chapter are mechanisms of catalytic reactions and pathways of biochemical transformations. The former require the construction of *direct* mechanisms, without cycles or redundancies, to determine the basic legitimate chemical transformations in a reacting system. The latter are the chemical engines of living cells, and they represent legitimate routes for the biochemical conversion of substrates to products either desired from a bioprocess or essential for cell survival. The algorithms discussed in this chapter could be used in one of the following two settings: (1) synthesize alternative pathways of chemical or biochemical reactions as a means to interpret overall transformations which are experimentally observed and (2) synthesize reaction pathways in the course of exploring new, alternative production routes. In this chapter, we will discuss examples in both directions. Although we will be concerned only with constraints on the directionality and stoichiometry of elementary reactions, the ideas can be extended to include other types of constraints arising, for example, from kinetics or thermodynamics.

I. Reaction Systems and Pathways

Complex systems of chemical reactions, and the problems associated with their design and analysis, are omnipresent in process engineering applications. In biochemical processes, complex networks of biochemical reactions, catalyzed by enzymes, accomplish both the growth of cells in the bioreactor and the conversion of the raw materials to the target products. Catalytic processes with solid catalysts and fluid-phase raw materials and products are used for many inorganic chemicals as well as refining and petrochemics. Noncatalytic fluid-phase reaction processes involving complex feeds and products are also common.

1. Hierarchical Structure

In all of these systems, chemical reactions have a hierarchical structure. What we normally think of as one reaction is actually composed from several elementary steps, which are often called, collectively, the *mechanism* of the reaction. For processes with a solid catalyst, these steps describe the interaction of the catalyst's active sites with the fluid-phase reactants. For biochemical reactions, there are association and dissociation steps that involve complexes of the enzyme and the substrates. In all cases, unstable, short-lived intermediates may be formed and destroyed in a rapid succession of steps; the overall reaction we observe might not involve these unstable intermediates because it is a suitable composition of such elementary steps.

The hierarchical character of reaction systems does not end with decomposition of reactions into steps. At the next level, many reactions taken together make up pathways. In bioprocesses, such pathways form long chains, cycles, and branching structures that accomplish biologically identifiable functions. In other processes, a pathway is used to describe the sequence of transformations that are needed to obtain the desired product from the available raw materials. One might even envision, at the next level, pathways combined to describe entire chemical plants or families of processing technologies.

2. Synthesis of Pathways

This chapter concerns itself with the hierarchical view of reaction systems. We are not interested in the decomposition of a reaction into mechanism steps or a pathway into reactions, but rather the synthesis or design of these composite entities from appropriate building blocks: Construction of reactions and mechanisms from elementary steps, or construction of pathways from reactions.

The common feature is the imposition of constraints on the overall chemical transformation, and the assembly of building blocks so that these constraints are met by the construct (reaction mechanism or pathway) as a whole. The constraints refer to the stoichiometry of the transformations, i.e., the reactants and products that are involved, along with their proportions. As we will see, there are different ways to formulate these constraints, and each way is suitable for different applications.

In all applications, the core problem can be stated as follows. We have a set of constraints that specify what may or may not be used as a starting material and final product of the process, and we are given a set of transformations (steps or reactions) that can be used as building blocks. For each transformation, we know the participating compounds and their stoichiometric coefficients. We need to assemble the transformations into a composite transformation, so that the stoichiometry of this composite transformation satisfies the constraints. This composite transformation may utilize any compound it needs along the way, as long as in the end its net stoichiometry satisfies the constraints. We can think of the construction procedure as assembly of a device from components, or we can abstract it mathematically into the formation of suitable linear combinations of the building-block transformations. In the second view, we should keep in mind that these linear combinations must respect the initial direction of reactions, since some reactions may be permitted only in their forward direction.

3. Application Domains

Two types of applications that involve this core problem will be discussed in this chapter. In the synthesis of biochemical pathways, using enzymatic reactions as building blocks, it is useful to examine each chemical species and decide whether it is required or allowed to be a starting material for the pathway, and likewise whether it is required or allowed to be a final product of the pathway.

The second application is the construction of catalytic reaction mechanisms out of elementary steps, involving only on type of catalytic site. A more useful way to formulate the stoichiometric constraints for these systems is to classify every chemical species as either a terminal species or an intermediate. Terminal species represent stable compounds that can be produced or consumed in significant quantities, while intermediates are short-lived unstable species that participate in the mechanism but are neither raw materials nor final products of the process.

4. Interpretation of Pathway Design

There are various ways in which we can interpret the idea of meeting stoichiometric constraints by combining transformations or assembling steps. We alluded to these interpretations in earlier sections, in our informal statement of the problem. As with the formulation of the constraints, each application domain accepts more naturally a different interpretation.

One interpretation, well suited to the synthesis of biochemical pathways, is process design. In the development and design of a new bioprocess, there are specifications on the raw materials that may be used and the desired and permitted products. One must select a pathway to use and a microorganism that possesses this pathway. With the advent of genetic engineering techniques, pathways and microorganisms are not restricted to existing ones: It is possible to modify a pathway of an organism by introducing biotransformations from other organisms. Therefore, the design of pathways takes on a true synthetic character rather than mere selection among enumerated alternatives.

Another interpretation is related to the dynamic behavior of a closed chemical system and the quasi-steady-state assumption. It is common in the kinetics of reaction systems to assume that the concentrations of intermediates reach a quasi-steady state: The rate of production and the rate of consumption of the intermediate approximately balance each other, and their difference (the net rate that leads to accumulation or depletion of the intermediate) is small compared to each of the two. The quasi-steady-state assumption allows simplification of the equations, by setting the accumulation term in the mass balance for an intermediate to zero. The construction of pathways is tantamount to identification of quasi-steady-state patterns, because the pathway makes sure that the intermediates are consumed at the same rate as they are produced. We are specifically interested in the smallest possible pathways, which cannot be shortened through elimination of a step and cannot be reduced to a superposition of smaller subpathways.

A similar interpretation can be proposed for open systems if only some species are permitted to enter and leave the system: The remaining species are the intermediates, and the construction of pathways arranges the stoichiometries to reflect the fact that, at steady state, the production of intermediates must balance their consumption.

II. Catalytic Reaction Systems

Our discussion of pathway design in the context of catalytic reaction mechanisms will summarize the treatment presented by Mavrovouniotis (1992) and Mavrovouniotis and Stephanopoulos (1992). The interested reader may refer to these for mathematical details, analysis of computational complexity, and comparison to other approaches in the context of model reaction systems.

A. Basic Concepts, Terminology, and Notation

We will follow the terminology and symbols used by Mavrovouniotis (1992) and Mavrovouniotis and Stephanopoulos (1992), who in turn followed the nomenclature of Happel and Sellers (1982, 1983, 1989), Happel (1986), Sellers (1984, 1989), and Happel *et al.* (1990). For the rest of this chapter, H&S will denote the latter set of references and the entire approach of Happel and Sellers. In accordance with H&S, we use the term *mechanism* rather than *pathway* for this class of systems.

1. Intermediate and Terminal Species

Consider a given set of elementary reaction steps that are feasible in a system, and the species involved in these steps. Species can be classified as either *intermediates*, which occur in very small amounts, or *terminal* species that can occur in significant amounts and constitute the raw materials and products of the process. The intermediates are precisely those species for which the quasi-steady-state assumption could be made.

A chemical systems consists of A species, designated as $a_1, a_2, \ldots, a_A$, and S mechanism steps, designated as $s_1, s_2, \ldots, s_S$. Each step s_i accomplishes a specific elementary reaction or transformation, $r_i = R(s_i)$. Let α_{ij} represent the stoichiometric coefficient of species a_j in step s_i, with the usual convention that $\alpha_{ij} > 0$ if and only if a_j is a product of s_i, $\alpha_{ij} < 0$ if and only if a_j is a reactant of s_i, and $\alpha_{ij} = 0$ if and only if a_j does not participate in s_i. The stoichiometry of the transformation $R(s_i)$ can then be written as

$$r_i = R(s_i) = \sum_{j=1}^{A} \alpha_{ij} a_j.$$

The coefficients α_{ij} of each step s_i denote only ratios of species. Thus, the meaning of the reaction r_i is not affected if we multiply all corresponding α_{ij} by a positive constant.

The directionality of a step s_i will be denoted by a label, as either $\leftrightarrows s_i$ or $\to s_i$. The sign $\leftrightarrows$ denotes a *reversible* step, whose net rate may be either positive (i.e., in the forward direction) or negative (i.e., in the reverse direction). The sign $\to$ denotes an *irreversible* step that is either thermodynamically irreversible or known to proceed with a net positive rate (i.e., in the forward direction).

The first I species (a_j with $j = 1, \ldots, I$) are assumed to be the intermediates, whereas the remaining $T = A - I$ species (a_j, $j = I + 1, \ldots, I + T$) are terminal species. If intermediates are present in very low concentrations, and their high rate of production is balanced by a high rate of consumption, the quasi-steady-state assumption allows dropping the accumulation term from the mass balance of an intermediate.

2. Overall Mechanisms

An *overall* mechanism (Horiuti and Nakamura, 1967; Horiuti, 1973; Temkin 1973, 1979) is one that obeys the quasi-steady-state assumption: It must consist of steps combined in specific proportions, such that the net transformation involves only terminal species.

Although there is significant net production of some terminal species, and significant net consumption of others, within the quasi-steady-state assumption the concentrations of intermediates are low and the rate of production of each intermediate by some steps is approximately balanced by its rate of consumption by other steps. *Overall* reactions can thus be defined as the set of net transformations permissible under the quasi-steady-state assumption; *overall* mechanisms are the combinations of steps that accomplish this.

A *mechanism*, m_k, is a linear combination of steps:

$$m_k = \sum_{i=1}^{S} \sigma_{ki} s_i,$$

and belongs to an S-dimensional vector space. The coefficients σ_{ki} represent the proportional participation of the steps, and their signs must reflect the proper direction of steps: for irreversible steps, σ_{ki} must be nonnegative.

For clarity, the index i will hereafter be used for steps, the index j for species, and the index k for mechanisms. A reaction vector r_k, associated with each mechanism, describes the net transformation that is accomplished by the mechanism:

$$r_k = R(m_k) = \sum_{i=1}^{S} \sigma_{ki} R(s_i).$$

Through substitution of the expression of $R(s_i)$ the reaction $R(m_k)$ can be written as

$$R(m_k) = \sum_{j=1}^{A} \beta_{kj} a_j \quad \text{with} \quad \beta_{kj} = \sum_{i=1}^{S} \sigma_{ki} \alpha_{ij},$$

where β_{kj} is the stoichiometric coefficient of species a_j in the net reaction accomplished by mechanism m_k.

A reaction $r_k = R(m_k)$ that involves only terminal species (i.e., $\beta_{kj} = 0$ for $j = 1, \ldots, I$) is called an *overall reaction*, and the mechanism m_k is an *overall mechanism*, corresponding to a quasi-steady state.

3. Direct Mechanisms

Clearly, a linear combination of overall mechanisms is also an overall mechanism. We are interested only in the smallest possible overall mechanisms, which describe the most extreme modes of operation of the chemical system. To define these, we examine the set of steps participating in a mechanism—with nonzero coefficients σ_{ki}. We are interested in overall mechanisms whose set of participating steps is not a proper superset of some other overall prime mechanism m'_k. These mechanisms, called *direct* mechanism by H&S, are minimal with respect to inclusion (for the set of participating steps). One might equivalently state that a mechanism m_k is direct if and only if every other overall mechanism includes at least one step not participating in m_k. The definition used here, in contrast to the one given by H&S, makes no particular reference to the *reaction* accomplished by the mechanism, other than the requirement that only overall mechanisms be considered.

Reaction mechanisms in the literature are often direct, because physical intuition encourages avoidance of excess steps. In considering the mechanism of a reaction, one often postulates ahead of time an implicit quasi-steady state, i.e., a combination of steps that explains the observed net reaction and conserves all intermediate species. In our approach, however, we are interested in identifying all possible direct mechanisms that are consistent with the postulated set of steps.

Direct mechanisms are not necessarily linearly independent; it may be possible to express a direct mechanism as a linear combination of other direct mechanisms. However, direct mechanisms are chemically distinct, because each involves a unique combination of steps.

B. Previous Work on the Construction of Mechanisms

A related concept is that of *laminar* mechanisms, introduced by Sinanoğlu (1975), Sinanoğlu and Lee (1978), and Lee and Sinanoğlu (1981). These mechanisms can be constructed a priori (i.e., without regard to a specific set of mechanism steps) given the number of catalysts involved. Sellers (1971, 1972) also presented procedures for construction of a priori mechanisms for certain chemical systems. In both these instances of mechanism construction, the focus was the a priori enumeration of potential pathway shapes, without regard to the specific reactions permitted by the system. These techniques are thus not applicable to the synthesis of pathways or mechanisms from a given set of reaction steps.

Milner (1964) used the term *direct* mechanisms to describe the smallest possible overall mechanisms. Milner (1964) showed that, under certain conditions, a direct mechanism can involve at most $I + 1$ steps, where I is the number of intermediates; this allows the construction of the mechanisms by examining combinations of $I + 1$ steps at a time. Another algorithm for the construction of direct mechanisms was presented by H&S, under the fundamental assumption that all steps are reversible and the net reaction taking place is actually known. This chapter presents an algorithm characterized by much greater generality and efficiency. The advantages of the algorithms presented here over those of H&S will be pointed out in the description of the algorithms and the examples throughout this section.

C. STRUCTURE OF THE ALGORITHM

We will describe here the general operation of the algorithm for the construction of direct mechanisms for chemical systems. We will examine the algorithm only in its simplest form. Mavrovouniotis (1992) presents various enhancements and discusses computational implementation considerations. We will begin with two examples (ammonia and methanol synthesis) that illustrate the step-by-step operation of the algorithm; these should be particularly useful for readers that are not accustomed to the abstract description of algorithms, who may wish to study the examples first, or study the abstract algorithm and the examples simultaneously.

Initially, the algorithm considers each individual reaction step as a partial mechanism. Then, one intermediate after another are examined, and the set of partial mechanisms is modified so that the intermediate does not appear in the net stoichiometry; the modification of mechanisms is carried out in a way that preserves the correct direction of irreversible reaction steps. By processing all intermediates in this way, a set of overall mechanisms is constructed. This final set of mechanisms might include duplicate mechanisms and even indirect ones; these can be easily discarded. Similar action must be taken in the procedure of H&S. Mavrovouniotis (1992) discusses procedures for eliminating such redundant mechanisms in the end, or even preventing their construction.

The algorithm operates on a set of intermediate species N, the set of terminal species N_T, and a set of partial mechanisms M, which is iteratively modified. Other information maintained by the algorithm during its operation is arranged in a convenient format in Fig. 1, which also explains how the entries are initialized. The algorithm proceeds from this setup by successively eliminating each intermediate species from the system. In order to eliminate an intermediate, we consider all the mechanisms whose reactions involve the intermediate species at hand. We can create combinations of two mechanisms at a time, with combination coefficients such that the intermediate vanishes from the reaction. The new row for a combination of mechanisms has elements σ_{ki} (left portion of Fig. 1) and β_{kj} (right portion of Fig. 1) that are simply the linear combinations of the respective elements of the old mechanisms. To permit the best choice of intermediate to be made, the setup of the algorithm (Fig. 1) lists, below

s_1	s_2	.....	s_S		number of combinations ⇒	n_1	n_2	.....	n_l	–	.....	–
ε_1	ε_2	.....	ε_S	mechanism ⇓	origin ⇓	a_1	a_2	.....	a_l	a_{l+1}	.....	a_A
σ_{11}	σ_{12}	.....	σ_{1S}	ε_1 m_1	–	β_{11}	β_{12}	.....	β_{1l}	$\beta_{1(l+1)}$	.....	β_{1A}
σ_{21}	σ_{22}	.....	σ_{2S}	ε_2 m_2	–	β_{21}	β_{22}	.....	β_{2l}	$\beta_{2(l+1)}$	.....	β_{2A}
⋮	⋮	⋮	⋮	⋮	⋮	⋮	⋮	⋮	⋮	⋮	⋮	⋮
σ_{S1}	σ_{S2}	.....	σ_{SS}	ε_S m_S	–	β_{S1}	β_{S2}	.....	β_{Sl}	$\beta_{S(l+1)}$	.....	β_{SA}

FIG. 1. Initialized setup for the application of the algorithm for the construction of mechanisms. Rows correspond to mechanisms, and the operation of the algorithm adds and deletes rows. The left portion of the table contains the coefficients σ_{ki}; initially, each mechanism m_k contains only the corresponding step s_k (i.e., σ is the identity matrix). With each step symbol s_i we list the directionality label ε_i ($\rightarrow$ for an irreversible step, or $\leftrightarrows$ for a reversible step). Directionality labels are also listed for each mechanism m_k. The right portion of the table shows the reactions accomplished by the mechanisms, and groups intermediate species as a_1 to a_I, and terminal species as a_{I+1} to a_A. The algorithm deletes columns corresponding to intermediates as these are processed. Initially $\beta_{kj} = \alpha_{ki}$ when $i = j$, but this will not be the case for new mechanisms that are constructed. Above each intermediate species symbol the number of combination mechanisms that must be created to eliminate the species is listed; this will be explained further in the description of the algorithm. The column marked "origin" in the middle of the table is not essential; it is used for keeping track of how each new mechanism is constructed; one can list the intermediate that was eliminated and the mechanisms that were combined in the construction. [Reprinted with permission from Mavrovouniotis, M. L., and Stephanopoulos, G. "Synthesis of reaction mechanisms consisting of reversible and irreversible steps: I. A synthesis approach in the context of simple examples". *Ind. Eng. Chem. Res.* **31**, 1625–1637. (1992). Copyright 1992 American Chemical Society.]

each species, the number of combination mechanisms that must be created to eliminate the species. The description given below explains the individual operations involved in more detail.

1. Initialization

We initialize N to the set of all intermediates (assuming $I \neq 0$), i.e., $N := \{a_1, a_2, \ldots, a_I\}$. The set M is initialized with one-step mechanisms. Therefore, the initial σ is just the identity matrix, whereas the initial β matrix is equal to the α matrix (Fig. 1):

$$M := \{m_1, m_2, \ldots, m_S\} \quad \text{with} \quad \beta_{kj} := \alpha_{kj} \text{ and } \sigma_{ki} := \begin{cases} 1 \text{ if } k = i, \\ 0 \text{ if } k \neq i. \end{cases}$$

The initial directionalities are the same as those of the corresponding individual steps (Fig.1).

2. Number of Combinations for Each Intermediate

We will now define the sets of partial mechanisms in which each intermediate participates. We divide them into irreversible and reversible ones to take directionality restrictions properly into account. For each species $a_j \in N$, let Y_j be a subset of M, defined as follows:

A mechanism $m_k \in M$ belongs to Y_j if and only if it is irreversible and its net reaction contains a_j as a reactant, i.e., $m_k \in Y_j$ if and only if $m_k \in M$, $\to m_k$, and $\beta_{kj} < 0$.

Let y_j be the cardinality of Y_j, i.e., $y_j = |Y_j|$. Define similarly a subset Z_j of irreversible mechanisms as follows:

A mechanism $m_k \in M$ belongs to Z_j if and only if it is irreversible and its net reaction contains a_j as a product (i.e., $\beta_{kj} > 0$).

Let z_j be the cardinality of this set. For reversible partial mechanisms, it does not matter whether a_j is a net reactant or a net product, as long as it appears in the net transformation. Thus, we define another subset, X_j, as follows:

A mechanism $m_k \in M$ belongs to X_j if and only if it is reversible and its net reaction contains species a_j (i.e., $\beta_{kj} \neq 0$).

Let x_j be the cardinality of X_j, i.e., $x_j = |X_j|$. One can obtain the numbers y_j, z_j, and x_j by scanning the column of a_j in Fig. 1 for nonzero entries. A positive entry corresponding to an irreversible mechanism is counted into z_j. A negative entry corresponding to an irreversible mechanism is counted into y_j. A positive or negative (but not zero) entry corresponding to a reversible mechanism is counted into x_j.

We want to combine partial mechanisms, two at a time, so that a_j is not present in the net stoichiometry of the combination, always respecting directionality restrictions on mechanisms. We can combine a mechanism from X_j with one from Y_j, because the mechanisms in X_j can be reversed if necessary to make a_j a product (to balance with a mechanism from Y_j that uses a_j as a reactant); this gives rise to $x_j y_j$ combinations. Similarly, we can combine a mechanism from X_j with one from Z_j, reversing mechanisms in X_j if necessary to make a_j a reactant; this gives rise to $x_j z_j$ combinations. We can also combine a mechanism from X_j with any other mechanism from X_j in the same way; the number of combinations in this case is

$$\binom{x_j}{2} = \frac{x_j(x_j - 1)}{2}.$$

Finally, a mechanism from Z_j can be combined with any mechanism from Y_j, without any reversal, giving rise to $z_j y_j$ combinations.

The total number of possible combinations, n_j (placed above a_j in the layout of Fig. 1), necessary to eliminate a_j, is thus given by

$$n_j := \frac{x_j(x_j - 1)}{2} + x_j z_j + x_j y_j + z_j y_j.$$

Two limiting cases of this formula will be used later in the ammonia and methanol examples. If all steps are reversible, $n_j := [x_j(x_j - 1)]/2$, while if all the steps are irreversible $n_j := z_j y_j$.

3. Selection and Elimination of an Intermediate

The algorithm ultimately eliminates all intermediates, and one could process them in any order. However, to minimize computational effort in the current iteration, we select to eliminate the a_J that corresponds to the smallest n_J. For the intermediate a_J, the elimination constructs precisely those combinations that were counted into n_J in the previous phase of the algorithm. We put the new mechanisms in a set M_J. Specifically, the construction of M_J is governed by the following rules.

Rule 1. For each distinct combination $m_k \in X_J$, $m_b \in X_J$, the combination mechanism is $m_c := \beta_{bJ} m_k - \beta_{kJ} m_b$ and it is reversible ($\leftrightarrows m_c$).

Rule 2. For each $m_k \in Z_J$ and each $m_b \in X_J$, the mechanism $\to m_c$ must be constructed according to the sign of β_{bJ}, because the sign determines whether we need to reverse the mechanism m_b, and we must ensure that the coefficient that multiplies m_k is positive. Thus
(a) If $\beta_{bJ} > 0$, we form $m_c := \beta_{bJ} m_k - \beta_{kJ} m_b$.
(b) If $\beta_{bj} < 0$, we form $m_c := \beta_{kJ} m_b - \beta_{bJ} m_k$.

Rule 3. Similarly, for each $m_k \in Y_J$ and each $m_b \in X_J$, for $m \to m_c$ as follows:
(a) If $\beta_{bJ} > 0$, we form $m_c := \beta_{bJ} m_k - \beta_{kJ} m_b$.
(b) If $\beta_{bJ} < 0$, we form $m_c := \beta_{kJ} m_b - \beta_{bJ} m_k$.

Rule 4. Finally, for each $m_k \in Y_J$ and each $m_b \in Z_J$, we form the mechanism $\to m_c$ as $m_c := \beta_{bJ} m_k - \beta_{kJ} m_b$. From the definition of the set Z_J it is easy to see that $\beta_{bJ} > 0$ and $-\beta_{kJ} > 0$, which guarantees that the irreversibility of mechanisms m_k and m_b is respected in the construction of m_c.

The combination mechanisms in all of these categories constitute the set M_J, whose cardinality is $n_J = x_J(x_J-1)/2 + x_J z_J + x_J y_J + z_J y_J$. The directionality of each new mechanism is $\leftrightarrows$ (reversible) if and only if both mechanisms used in the combination were reversible. The σ and β coefficients of each combination mechanism are obtained as linear combinations of the respective coefficients of the constituent mechanisms. For $m_c := \beta_{bJ} m_k - \beta_{kJ} m_b$, one obtains $\sigma_{ci} := \beta_{bJ} \sigma_{ki} - \beta_{kJ} \sigma_{bi}$ (for all i) and $\beta_{cj} := \beta_{bJ} \beta_{kj} - \beta_{kJ} \beta_{bj}$ (for all j). From the latter expression the coefficient for a_J can be computed by setting $j := J$, which gives $\beta_{cJ} = \beta_{bJ} \beta_{kJ} - \beta_{kJ} \beta_{bJ} = 0$, verifying that the choice of combination coefficients achieves the elimination of a_J, as intended.

A new mechanism is logged as a new row in Fig. 1, and its σ and β portions are computed as linear combinations of the rows of the mechanisms being considered. The "origin" entry in Fig. 1 is used for bookkeeping. In the examples presented later, we record the name of the intermediate eliminated and a symbolic linear expression of the combination of mechanisms (see Figs. 3–7 and 11–16).

4. Update of Active Sets

In order to eliminate a_J we can now modify the current set of partial mechanisms M by adding the new combination mechanisms M_J, and removing their ancestors X_J, Y_J, and Z_J, i.e., $M := (M \cup M_J) - (X_J \cup Y_J \cup Z_J)$. We can then remove the intermediate a_J, by setting $N := N - \{a_J\}$. Within Fig. 1, in this phase we remove all rows that have nonzero entries in the a_J-column and then remove the column a_J itself.

5. Termination or New Iteration

If Fig. 1 still contains active columns corresponding to intermediates (i.e., $N \neq \varnothing$), we go back to phase 2 of the algorithm for the next iteration. If no intermediate columns remain ($N = \varnothing$) the elimination is complete, and all direct mechanisms are in the set M.

D. FEATURES OF THE ALGORITHM

Our algorithm for the synthesis of direct mechanisms initially considers each reaction step as a partial mechanism. Then, one intermediate after another are examined, and the set of partial mechanisms is modified so that the intermediate does not appear in the net stoichiometry; the modification of mechanisms is carried out in a way that preserves the correct direction of irreversible reaction steps. By processing all intermediates in this way, we can construct a set of overall mechanisms.

The set of mechanisms M resulting from the application of the basic algorithm may contain duplicate mechanisms (as in the ammonia example, see next subsection) or even indirect ones (as in the methanol example, see next subsection). A simple procedure can be applied in the end to remove redundant (duplicate or indirect) mechanisms from the final set of mechanisms. Alternatively, incremental checking can be carried out for each new mechanism or each iteration of the algorithm. Redundancy of mechanisms is one of a number of issues that affect the operation of the algorithm, and variations of the basic algorithm can be defined to enhance either its computational efficiency or its conceptual simplicity. For example, when two or more intermediates entail the same number of combinations (n_J), the basic algorithm makes a completely arbitrary choice; a variation of the algorithm might instead make the selection in a manner that removes as many existing mechanisms as possible from the set M, increasing computational efficiency. In the direction of conceptual simplicity, instead of assembling and using the sets X_j, Y_j, and Z_j separately, one may modify the algorithm by dropping directionality labels from mechanisms, and lumping X_j, Y_j, and Z_j into a single set W_j; the construction of combinations that eliminate a selected intermediate a_J using every possible pair of mechanisms $m_k \in W_J$ and $m_b \in W_J$, and rejecting the ones that contain steps in the wrong direction. These and related issues are discussed by Mavrovouniotis (1992).

The algorithm presented here for the construction of direct mechanisms differs significantly in its operation from the method of H&S. First, the irreversibility of mechanism steps is taken into account by our algorithm as mechanisms are constructed; i.e., it is not necessary to reserve directionality considerations until the end. Second, the mechanisms do not have to be restricted a priori to a particular reaction. Third, direct mechanisms are constructed recursively through combination of steps or partial mechanisms, rather than as linear combinations from a basis set of linearly independent mechanisms.

E. Examples

The mechanism–construction approach is illustrated here through two examples, which have also been treated by H&S: an example involving reversible steps for the synthesis of ammonia, and an example with irreversible steps for the synthesis of methanol. We note, in particular, that in the methanol example H&S assumed that the steps that lead to ethane and dimethylether (which are byproducts) are irrelevant for methanol mechanisms; this assumption was motivated by the need to keep

									mechanism											
$N_2+I\rightleftharpoons N_2I$	$N_2I+H_2\rightleftharpoons N_2H_2I$	$N_2H_2I+I\rightleftharpoons 2NHI$	$N_2+2I\rightleftharpoons 2NI$	$NI+HI\rightleftharpoons NHI+I$	$NHI+HI\rightleftharpoons NH_2I+I$	$NHI+H_2\rightleftharpoons NH_3+I$	$H_2+2I\rightleftharpoons 2HI$	$NH_2I+HI\rightleftharpoons NH_3+I$		species ⇒	N_2I	N_2H_2I	NHI	NI	HI	NH_2I	I	N_2	H_2	NH_3
s_1	s_2	s_3	s_4	s_5	s_6	s_7	s_8	s_9		number of combinations ⇒	1	1	6	1	6	1	28	—	—	—
⇌	⇌	⇌	⇌	⇌	⇌	⇌	⇌	⇌		origin ⇓	a_1	a_2	a_3	a_4	a_5	a_6	a_7	a_8	a_9	a_{10}
1									⇌ m_1	—	1							−1	−1	
	1								⇌ m_2	—	−1	1							−1	
		1							⇌ m_3	—		−1	2				−1			
			1						⇌ m_4	—				2			−2	−1		
				1					⇌ m_5	—			1	−1	−1		1			
					1				⇌ m_6	—			−1		−1	1	1			
						1			⇌ m_7	—			−1				1		−1	1
							1		⇌ m_8	—					2		−2		−1	
								1	⇌ m_9	—					−1	−1	2			1

FIG. 2. Application of the algorithm on the ammonia mechanism. The mechanism steps for the synthesis of ammonia are shown on the top left portion, above each step symbol s_i. The steps are those used by Happel and Sellers (1982, 1983) and Happel et al. (1990), and were originally derived from the work of Horiuti (1973) and Temkin (1973). The active site of the catalyst is denoted as "I," and all species that contain this active site are intermediates; N_2, H_2, and NH_3 are terminal species. The setup follows the description given in Fig. 1, with the entries that are equal to zero left blank. Directionalities are shown, but for this example all steps are considered reversible and the directionality does not matter. We start by taking partial mechanisms $m_1, m_2, \ldots, m_9$, each of which has length 1 and includes only the step with the same index (i.e., $\sigma_{ki} = 1$ if $k = i$; $\sigma_{ki} = 0$ if $k \neq i$). Naturally, these are only partial mechanisms, which will be gradually transformed to overall mechanisms. Since all steps are reversible, the number of combination mechanisms that must be created to eliminate each intermediate (listed below the species) is simply $x(x-1)/2$, where x is the number of mechanisms whose reactions involve the species. The algorithm proceeds by successively eliminating each intermediate from the system. If, for example, we pursue the elimination of NHI, there are four existing partial mechanisms, m_3, m_5, m_6, and m_7, to be considered; there are $4 \times 3/2 = 6$ pairwise combinations of these that must be constructed to eliminate NHI. The choice to eliminate the intermediate a_3 (NHI) first, however, is not computationally efficient; if we choose NH_2I (a_6) instead, we only need to construct one mechanism. In fact, a_6 has the minimum number of combinations and its column has been highlighted in Fig. 2, to show that this is the species which will be eliminated first. Intuitively, one can think of this elimination as follows. There are only two steps that involve NH_2I; if any mechanism uses one of the two steps, it must use the other, with a coefficient that would eliminate the intermediate NH_2I; hence, the two steps can be combined. [Reprinted with permission from Mavrovouniotis, M. L., and Stephanopoulos, G. "Synthesis of reaction mechanisms consisting of reversible and irreversible steps: I. A synthesis approach in the context of simple examples". *Ind. Eng. Chem. Res.* **31**, 1625–1637. (1992). Copyright 1992 American Chemical Society.]

the size of the problem small, since the H&S algorithm is not very efficient computationally. As we will see, our algorithm constructs one mechanism that involves simultaneous production of ethane and methanol (i.e., m_{35} in Figs. 16 and 17), showing that the omission of byproducts is misleading.

1. Synthesis of Ammonia

We first consider an example on ammonia synthesis that was used by Happel and Sellers (1982, 1983) and Happel *et al.* (1990) and contains only

$N_2+I\rightleftharpoons N_2I$	$N_2I+H_2\rightleftharpoons N_2H_2I$	$N_2H_2I+I\rightleftharpoons 2NHI$	$N_2+2I\rightleftharpoons 2NI$	$NI+HI\rightleftharpoons NHI+I$	$NHI+HI\rightleftharpoons NH_2I+I$	$NH_2+H_2\rightleftharpoons NH_3+I$	$H_2+2I\rightleftharpoons 2HI$	$NH_2I+HI\rightleftharpoons NH_3+I$	mechanism	species $\Rightarrow$	N_2I	N_2H_2I	NHI	NI	HI	NH_2I	—	N_2	H_2	NH_3
s_1	s_2	s_3	s_4	s_5	s_6	s_7	s_8	s_9		number of combinations $\Rightarrow$	1	1	6	1	3	##	21	—	—	—
$\rightleftharpoons$	$\rightleftharpoons$	$\rightleftharpoons$	$\rightleftharpoons$	$\rightleftharpoons$	$\rightleftharpoons$	$\rightleftharpoons$	$\rightleftharpoons$	$\rightleftharpoons$		origin $\Downarrow$	a_1	a_2	a_3	a_4	a_5	a_6	a_7	a_8	a_9	a_{10}
1									$\rightleftharpoons m_1$	—	1					##	−1	−1		
	1								$\rightleftharpoons m_2$	—	−1	1				##			−1	
		1							$\rightleftharpoons m_3$	—		−1	2			##	−1			
			1						$\rightleftharpoons m_4$	—				2		##	−2	−1		
				1					$\rightleftharpoons m_5$	—			1	−1	−1	##	1			
##	##	##	##	##	##	##	##	##	$\rightleftharpoons m_6$	—	##	##	##	##	##	##	##	##	##	##
						1			$\rightleftharpoons m_7$	—		−1				##	1		−1	1
							1		$\rightleftharpoons m_8$	—				2		##	−2		−1	
##	##	##	##	##	##	##	##	##	$\rightleftharpoons m_9$	—	##	##	##	##	##	##	##	##	##	##
						1		1	$\rightleftharpoons m_{10}$	$a_6: m_6+m_9$		−1				−2	##	3		1

FIG. 3. Setup for the application of the algorithm on the ammonia mechanism, after the elimination of a_6 (NH_2I). In the "origin" column, the construction of the mechanism m_{10} is documented as resulting from the elimination of a_6 through a specific linear combination of the mechanisms m_6 and m_9. The row of m_{10} has been obtained by carrying out the linear combination of the rows of m_6 and m_9. The rows for m_6 and m_9, as well as the column for a_6 have been crossed out from the table. The numbers of combinations have changed for a_5 and a_7 (but remain unchanged for other intermediates). Any one of the species a_1, a_2, and a_4, each entailing only one new combination, could be eliminated next. Both a_2 (N_2H_2I) and a_4 (NI) have been highlighted; their elimination can be carried out in parallel, because the two sets of mechanisms involved are disjoint; i.e., there is no mechanism whose reaction stoichiometry involves both a_4 and a_2. [Reprinted with permission from Mavrovouniotis, M. L., and Stephanopoulos, G. "Synthesis of reaction mechanisms consisting of reversible and irreversible steps: I. A synthesis approach in the context of simple examples". *Ind. Eng. Chem. Res.* **31**, 1625–1637, (1992). Copyright 1992 American Chemical Society.]

$N_2+I\leftrightarrows N_2I$	$N_2I+H_2\leftrightarrows N_2H_2I$	$N_2H_2I+I\leftrightarrows 2NHI$	$N_2+2I\leftrightarrows 2NI$	$NI+HI\leftrightarrows NH_2I+I$	$NHI+HI\leftrightarrows NH_2I$	$NHI+H_2\leftrightarrows NH_3+I$	$H_2+2I\leftrightarrows 2HI$	$NH_2I+HI\leftrightarrows NH_3+I$	mechanism	species ⇒	N_2I	N_2H_2I	NHI	NI	HI	NH_2I	–	N_2	H_2	NH_3
s_1	s_2	s_3	s_4	s_5	s_6	s_7	s_8	s_9	⇓	number of combinations ⇒	1	##	6	##	3	##	10	–	–	–
⇄	⇄	⇄	⇄	⇄	⇄	⇄	⇄	⇄		origin ⇓	a_1	a_2	a_3	a_4	a_5	a_6	a_7	a_8	a_9	a_{10}
1									⇄ m_1	–	1	##		##		##		–1	–1	
##	##	##	##	##	##	##	##	##	⇄ m_2	–	##	##	##	##	##	##	##	##	##	##
##	##	##	##	##	##	##	##	##	⇄ m_3	–	##	##	##	##	##	##	##	##	##	##
##	##	##	##	##	##	##	##	##	⇄ m_4	–	##	##	##	##	##	##	##	##	##	##
##	##	##	##	##	##	##	##	##	⇄ m_5	–	##	##	##	##	##	##	##	##	##	##
##	##	##	##	##	##	##	##	##	⇄ m_6	–	##	##	##	##	##	##	##	##	##	##
			1						⇄ m_7	–		##	–1	##		##	1		–1	1
					1				⇄ m_8	–		##		##	2	##	–2		–1	
##	##	##	##	##	##	##	##	##	⇄ m_9	–	##	##	##	##	##	##	##	##	##	##
					1			1	⇄ m_{10}	a_6: $m_6 + m_9$		##	–1	##	–2	##	3			1
	1	1							⇄ m_{11}	a_2: $m_2 + m_3$	–1	##	2	##		##	–1		–1	
			1	2					⇄ m_{12}	a_4: $m_4 + 2m_5$		##	2	##	–2	##			–1	

Fig. 4. Application of the algorithm on the ammonia mechanism, after the elimination of a_2 and a_4, which yielded two new partial mechanisms, m_{11} and m_{12}, and eliminated m_2, m_3, m_4, and m_5. The number of combinations has been recomputed for each remaining intermediate species. The intermediate N_2I (a_1) will be eliminated next. [Reprinted with permission from Mavrovouniotis, M. L., and Stephanopoulos, G. "Synthesis of reaction mechanisms consisting of reversible and irreversible steps: I. A synthesis approach in the context of simple examples". *Ind. Eng. Chem. Res.* 31, 1625–1637, (1992). Copyright 1992 American Chemical Society.]

reversible steps. The initial setup for the application of the algorithm is shown in Fig. 2, which also presents the actual elementary steps above the corresponding symbols s_i. The progression of the algorithm is explained in Figs. 2–7. Each figure shows one or more iterations through phases 2–5 of the algorithm; thus each figure eliminates one or more intermediates and flags those intermediates that will be eliminated next. Figure 7 provides the final results and compares them to the results obtained by Happel *et al.* (1990).

2. Synthesis of Methanol

Consider, as a second example, the mechanism for the synthesis of methanol (Fig. 8), analyzed by Happel *et al.* (1990). All the steps are

164   MICHAEL L. MAVROVOUNIOTIS

s_1	s_2	s_3	s_4	s_5	s_6	s_7	s_8	s_9	mechanism ⇩	species ⇒ origin ⇩	a_1 N_2I	a_2 N_2H_2I	a_3 NHI	a_4 NI	a_5 HI	a_6 NH_2I	a_7 —	a_8 N_2	a_9 H_2	a_{10} NH_3
$N_2{+}I{\rightleftarrows}N_2I$	$N_2I{+}H_2{\rightleftarrows}N_2H_2I$	$N_2H_2I{+}I{\rightleftarrows}2NHI$	$N_2{+}2I{\rightleftarrows}2NI$	$NI{+}HI{\rightleftarrows}NHI{+}I$	$NHI{+}HI{\rightleftarrows}NH_2I{+}I$	$NH_2I{+}HI{\rightleftarrows}NH_3I{+}I$	$H_2{+}2I{\rightleftarrows}2HI$	$NH_2I{+}HI{\rightleftarrows}NH_3{+}I$		number of combinations ⇒	##	##	6	##	3	##	6	—	—	—
⇌	⇌	⇌	⇌	⇌	⇌	⇌	⇌	⇌												
##	##	##	##	##	##	##	##	##	⇌ m_1	—	##	##	##	##	##	##	##	##	##	##
##	##	##	##	##	##	##	##	##	⇌ m_2	—	##	##	##	##	##	##	##	##	##	##
##	##	##	##	##	##	##	##	##	⇌ m_3	—	##	##	##	##	##	##	##	##	##	##
##	##	##	##	##	##	##	##	##	⇌ m_4	—	##	##	##	##	##	##	##	##	##	##
##	##	##	##	##	##	##	##	##	⇌ m_5	—	##	##	##	##	##	##	##	##	##	##
##	##	##	##	##	##	##	##	##	⇌ m_6	—	##	##	##	##	##	##	##	##	##	##
						1			⇌ m_7	—	##	##	−1	##		##	1		−1	1
							1		⇌ m_8	—	##	##		##	2	##	−2		−1	
##	##	##	##	##	##	##	##	##	⇌ m_9	—	##	##	##	##	##	##	##	##	##	##
					1			1	⇌ m_{10}	a_6: $m_6 + m_9$	##	##	−1	##	−2	##	3			1
##	##	##	##	##	##	##	##	##	⇌ m_{11}	a_2: $m_2 + m_3$	##	##	##	##	##	##	##	##	##	##
		1	2						⇌ m_{12}	a_4: $m_4 + 2m_5$	##	##	2	##	−2	##		−1		
1	1	1							⇌ m_{13}	a_1: $m_1 + m_{11}$	##	##	2	##		##	−2	−1	−1	

Fig. 5. Application of the algorithm on the ammonia mechanism, after the elimination of a_1, with only five mechanisms (m_7, m_8, m_{10}, m_{12}, and m_{13}) remaining active. This reduction in the number of mechanisms as the first few intermediates are eliminated is quite common; however, the number of mechanisms tends to increase again at the end. The intermediate HI (a_5) will be eliminated next. [Reprinted with permission from Mavrovouniotis, M. L., and Stephanopoulos, G. "Synthesis of reaction mechanisms consisting of reversible and irreversible steps: I. A synthesis approach in the context of simple examples". *Ind. Eng. Chem. Res.* **31**, 1625–1637, (1992). Copyright 1992 American Chemical Society.]

irreversible, providing an opportunity to show how directionality of steps is taken into account as the mechanisms are constructed. In the beginning of their analysis, Happel *et al.* (1990) remove steps s_9 and s_{11}, because they lead to byproducts (ethane and dimethylether) rather than methanol. Here, we will keep all steps; it turns out that the products are not formed independently of each other, and isolation of the main product from the byproducts of the mechanism can be misleading.

The application of the algorithm is shown in Figs. 9–16, with explanation of the differences arising from the directionality of the steps; each figure eliminates one or more intermediates and flags those intermediates

												N_2I	N_2H_2I	NHI	NI	HI	NH_2I	—	N_2	H_2	NH_3
$N_2+I\leftrightarrows N_2I$	$N_2I+H_2\leftrightarrows N_2H_2I$	$N_2H_2I+I\leftrightarrows 2NHI$	$N_2+2I\leftrightarrows 2NI$	$NI+HI\leftrightarrows NH_2I+I$	$NHI+HI\leftrightarrows NH_2I+I$	$NHI+H_2\leftrightarrows NH_3+I$	$H_2+2I\leftrightarrows 2HI$	$NH_2I+HI\leftrightarrows NH_3+I$	mechanism	species ⇒										
s1	s2	s3	s4	s5	s6	s7	s8	s9	⇓	number of combinations ⇒	##	##	10	##	##	##	10	–	–	–
⇆	⇆	⇆	⇆	⇆	⇆	⇆	⇆	⇆		origin ⇓	a_1	a_2	a_3	a_4	a_5	a_6	a_7	a_8	a_9	a_{10}
									⇆ m_7	—	##	##	-1	##	##	##	1		-1	1
##	##	##	##	##	##	##	##	##	⇆ m_8	—	##	##	##	##	##	##	##	##	##	##
##	##	##	##	##	##	##	##	##	⇆ m_{10}	a_6: m_6+m_9	##	##	##	##	##	##	##	##	##	##
##	##	##	##	##	##	##	##	##	⇆ m_{12}	a_4: m_4+2m_5	##	##	##	##	##	##	##	##	##	##
1	1	1							⇆ m_{13}	a_1: m_1+m_{11}	##	##	2	##	##	##	-2	-1	-1	
					1		1	1	⇆ m_{14}	a_5: m_8+m_{10}	##	##	-1	##	##	##	1		-1	1
			1	2			1		⇆ m_{15}	a_5: m_8+m_{12}	##	##	2	##	##	##	-2	-1	-1	
			1	2	-1			-1	⇆ m_{16}	a_5: $m_{12}-m_{10}$	##	##	3	##	##	##	-3	-1		-1

FIG. 6. Application of the algorithm on the ammonia mechanism, after the elimination of a_5. We can generally, at any time, drop inactive (crossed out) mechanisms or eliminated intermediates, but we have so far maintained them so that one can easily keep track of the progress of the algorithm. At this point, in order to reduce the size of the setup, we drop from the figure all those mechanisms that had been crossed out in Fig. 5 or earlier. The two intermediates (a_3 and a_7) remaining in Fig. 6 involve the same number of combinations; in fact elimination of either one results automatically in the elimination of the other (Fig. 7), because the columns for a_3 and a_7 contained precisely opposite numbers; a_3 is arbitrarily chosen. [Reprinted with permission from Mavrovouniotis, M. L., and Stephanopoulos, G. "Synthesis of reaction mechanisms consisting of reversible and irreversible steps: I. A synthesis approach in the context of simple examples". *Ind. Eng. Chem. Res.* **31**, 1625–1637, (1992). Copyright 1992 American Chemical Society.]

that will be eliminated next. Figure 17 analyzes the results and compares them to those of Happel *et al.* (1990). With respect to the mechanisms m_{29} and m_{35}, which lead to byproducts, we observe (Figs. 16 and 17) that m_{29} can be thought of as unrelated to the synthesis of methanol, but omission of m_{35} may be misleading: This mechanism leads to simultaneous production of methanol and ethane, in stoichiometric proportions. Thus, the mechanism m_{35} should properly be viewed as one of the mechanisms that lead to methanol—with the drawback, of course, that it also leads to an equal number of moles of an undesired byproduct. The omission of byproducts is therefore risky simplification. We also note that the inclusion of directionality considerations appears particularly cumbersome in the treatment of Happel *et al.* (1990, p. 1061): The direct submechanisms for one of the basis reactions violate directionality restrictions; thus, an

$N_2+I\leftrightarrows N_2I$	$N_2I+H_2\leftrightarrows N_2H_2I$	$N_2H_2I+I\leftrightarrows 2NHI$	$N_2+2I\leftrightarrows 2NI$	$NI+HI\leftrightarrows NHI+I$	$NHI+HI\leftrightarrows NH_2I+I$	$NH+H_2\leftrightarrows NH_2I+I$	$H_2+2I\leftrightarrows 2HI$	$NH_2I+HI\leftrightarrows NH_3I+I$	mechanism	corresponding mechanism from Happel et al. (1990)	species ⇒	N_2I	N_2H_2I	NHI	NI	HI	NH_2I		N_2	H_2	NH_3
s_1	s_2	s_3	s_4	s_5	s_6	s_7	s_8	s_9	⇓	⇓	number of combinations ⇒	##	##	##	##	##	##	0	–	–	–
⇆	⇆	⇆	⇆	⇆	⇆	⇆	⇆	⇆			origin ⇓	a_1	a_2	a_3	a_4	a_5	a_6	a_7	a_8	a_9	a_{10}
1	1	1			2				⇆ m_{17}	m_4,Table V	a_3: $m_{13}+2m_7$	##	##	##	##	##	##		–1	–3	2
				1	–1	1	1		⇆ m_{18}	null,Table IV	a_3: $m_{14}-m_7$	##	##	##	##	##	##		0	0	0
			1	2		2	1		⇆ m_{19}	m_1,Table V	a_3: $m_{15}+2m_7$	##	##	##	##	##	##		–1	–3	2
			1	2	–1			–1	⇆ m_{20}	m_3,Table V	a_3: $m_{16}+3m_7$	##	##	##	##	##	##		–1	–3	2
1	1	1			2		2	2	⇆ m_{21}	m_5,Table V	a_3: $m_{13}+2m_{14}$	##	##	##	##	##	##		–1	–3	2
–1	–1	–1	1	2			1		⇆ m_{22}	null,Table IV	a_3: $m_{15}-m_{13}$	##	##	##	##	##	##		0	0	0
3	3	3	–2	–4	2			2	⇆ m_{23}	m_6,Table V	a_3: $3m_{13}-2m_{16}$	##	##	##	##	##	##		–1	–3	2
			1	2	2		3	2	⇆ m_{24}	m_2,Table V	a_3: $m_{15}+2m_{14}$	##	##	##	##	##	##		–1	–3	2
			1	2	2		3	2	⇆ m_{25}	m_2,Table V	a_3: $m_{16}+3m_{14}$	##	##	##	##	##	##		–1	–3	2
			1	2	2		3	2	⇆ m_{26}	m_2,Table V	a_3: $3m_{15}-2m_{16}$	##	##	##	##	##	##		–1	–3	2

FIG. 7. The final set of mechanisms produced here and their correspondence to the mechanisms constructed by Happel et al. (1990). As was explained earlier, the intermediate a_7 does not need to be eliminated because its column contains only zeros. All the direct mechanisms identified by Happel et al. (1990) have been produced. However, the last three mechanisms in Figure 7 are actually identical. Hence, the simple procedure discussed here does not preclude multiple occurrences of the same mechanism, or, as the next example will show, the occurrence of mechanisms that are not direct. For small studies this is not a significant drawback, because one can easily eliminate the redundancies in the end. The potential duplication of mechanisms and construction of indirect mechanisms are addressed by Mavrovouniotis (1992). [Reprinted with permission from Mavrovouniotis, M. L., and Stephanopoulos, G. "Synthesis of reaction mechanisms consisting of reversible and irreversible steps: I. A synthesis approach in the context of simple examples". *Ind. Eng. Chem. Res.* **31**, 1625–1637, (1992). Copyright 1992 American Chemical Society.]

alternative reaction is manually constructed which gives acceptable direct submechanisms.

3. Importance of Reaction Directionality

We provide here an abstract example (Fig. 18) to discuss the complexity characteristics of the method of H&S, and in particular that method's assumption at the outset that all reactions are reversible. This assumption might be insignificant for very simple reaction mechanisms or those that in fact do not contain irreversible steps. However, the assumption constitutes

s_1: $CH_4 + O_2 \rightarrow CH_3 + HO_2$

s_2: $CH_3 + O_2 \rightarrow CH_3O_2$

s_3: $CH_3O_2 \rightarrow CH_2O + OH$

s_4: $CH_3O_2 + CH_4 \rightarrow CH_3O_2H + CH_3$

s_5: $CH_3O_2H \rightarrow CH_3O + OH$

s_6: $CH_3O \rightarrow CH_2O + H$

s_7: $CH_3O + CH_4 \rightarrow CH_3OH + CH_3$

s_8: $OH + CH_4 \rightarrow CH_3 + H_2O$

s_9: $CH_3 + CH_3 \rightarrow C_2H_6$

s_{10}: $CH_3 + OH \rightarrow CH_3OH$

s_{11}: $CH_3 + CH_3O \rightarrow CH_3OCH_3$

s_{12}: $CH_2O + CH_3 \rightarrow CH_4 + CHO$

s_{13}: $CHO + O_2 \rightarrow CO + HO_2$

s_{14}: $CH_2O + CH_3O \rightarrow CH_3OH + CHO$

s_{15}: $CHO + CH_3 \rightarrow CO + CH_4$

FIG. 8. Mechanism steps for the synthesis of methanol, as used in an example by Happel *et al.* (1990). The mechanism was proposed by Yarlagadda *et al.* (1988) and assumes that all steps have a net rate in the indicated direction. The species CH_4, O_2, CH_3OH, CO, H_2O, C_2H_6, and CH_3OCH_3 are terminal and all others are intermediates; formaldehyde (CH_2O) is an intermediate because it is present in small amounts (Happel *et al.*, 1990). [Reprinted with permission from Mavrovouniotis, M. L., and Stephanopoulos, G. "Synthesis of reaction mechanisms consisting of reversible and irreversible steps: I. A synthesis approach in the context of simple examples". *Ind. Eng. Chem. Res.* **31**, 1625–1637, (1992). Copyright 1992 American Chemical Society.]

a significant drawback if many steps are irreversible and one starts with a very large reaction system, a large portion of which eventually does not participate in the final solution. In particular, the presence of many parallel irreversible routes leading from intermediates to other intermediates can have devastating consequences on efficiency. Figure 18 shows an idealized schematic of these features; there is actually only one mechanism that converts A to E, but a method that postpones consideration of reaction directionality until the final mechanisms have been constructed will produce an enormous number of possibilities. Thus, the procedure of H&S, as well as any method that ignores directionality, faces great difficulties for large systems of mostly irreversible reactions.

s_1	s_2	s_3	s_4	s_5	s_6	s_7	s_8	s_9	s_{10}	s_{11}	s_{12}	s_{13}	s_{14}	s_{15}	mechanism ⇓	species ⇒ combinations⇒ origin ⇓	CH_3 24 a_1	HO_2 0 a_2	CH_3O_2 2 a_3	OH 4 a_4	CH_3O_2H 1 a_5	CH_3O 4 a_6	H 0 a_7	CHO 4 a_8	CH_2O 4 a_9
→	→	→	→	→	→	→	→	→	→	→	→	→	→	→											
1															→m_1	—	1	1							
	1														→m_2	—	-1		1						
		1													→m_3	—			-1	1					1
			1												→m_4	—	1		-1		1				
				1											→m_5	—				1	-1	1			
					1										→m_6	—						-1	1		1
						1									→m_7	—	1					-1			
							1								→m_8	—	1		-1						
								1							→m_9	—	-2								
									1						→m_{10}	—	-1		-1						
										1					→m_{11}	—	-1					-1			
											1				→m_{12}	—	-1							1	-1
												1			→m_{13}	—		1						-1	
													1		→m_{14}	—						-1		1	-1
														1	→m_{15}	—	-1							-1	

FIG. 9. Initial setup for the application of the algorithm on the methanol mechanism. Terminal species are not shown in this figure; they will be incorporated in Fig. 10. The initial arrangement shows the mechanisms m_1 to m_{15}, each using only the step with the same index. The arrows below each step and to the left of each mechanism serve as indicators of directionality; here, all steps are irreversible (→), and the algorithm that is proposed in this chapter can take directionality into account. In the formation of combinations of partial mechanisms to eliminate an intermediate species, it is no longer possible to take *any* combination of two mechanisms whose net reactions involve the species. If, for example, we attempt to combine m_1 and m_{13} to eliminate a_2 (HO_2), then the combination expression $\beta_{b2}m_k - \beta_{k2}m_b$ would lead to either $m_1 - m_{13}$ (for $k = 1$ and $b = 13$) or $m_{13} - m_1$ (for $k = 13$ and $b = 1$), violating the directionality of one of the mechanisms m_1 and m_{13}. This happens because a_2 participates as a net product in both mechanisms; it is simply not possible to eliminate a_2 if we insist on using both mechanisms in the forward direction. Thus, for irreversible steps and mechanisms, a legitimate combination that eliminates an intermediate must include one mechanism in which the species is a net product and one mechanism in which the species is a net reactant. This makes each intermediate's number of combinations equal to yz, where z is the number of mechanisms for which the species is a net product, and y is the number of mechanisms for which the species in question is a net reactant. The somewhat more complicated procedure for chemical systems that include both irreversible and reversible steps was given in the algorithm. As in the previous example, the intermediates with the smallest numbers of combinations are highlighted. The species a_2 (HO_2) and a_7 (H) are chosen for elimination, because they give rise to zero combinations (they occur only as products). [Reprinted with permission from Mavrovouniotis, M. L., and Stephanopoulos, G. "Synthesis of reaction mechanisms consisting of reversible and irreversible steps: I. A synthesis approach in the context of simple examples". *Ind. Eng. Chem. Res.* **31**, 1625-1637. (1992), Copyright 1992 American Chemical Society.]

																mechanism	species ⇒	CH_3	CH_3O_2	OH	CH_3O_2H	CH_3O	CHO	CH_2O	CH_4	O_2	CH_3OH	CO	H_2O	C_2H_6	CH_3OCH_3
s_1	s_2	s_3	s_4	s_5	s_6	s_7	s_8	s_9	s_{10}	s_{11}	s_{12}	s_{13}	s_{14}	s_{15}		⇓	combinations⇒	18	2	4	1	3	2	2							
→	→	→	→	→	→	→	→	→	→	→	→	→	→	→	→		origin ⇓	a_1	a_3	a_4	a_5	a_6	a_8	a_9	a_{10}	a_{11}	a_{12}	a_{13}	a_{14}	a_{15}	a_{16}
	1															→m_2	—	-1	1								-1				
		1														→m_3	—	-1	1		1										
			1													→m_4	—	1	-1		1				-1						
				1												→m_5	—			1	-1	1									
					1											→m_7	—	1			-1			-1	1						
						1										→m_8	—	1	-1					-1		1					
							1									→m_9	—	-2												1	
								1								→m_{10}	—	-1	-1					1							
									1							→m_{11}	—	-1			-1										1
										1						→m_{12}	—	-1				1	-1	1							
												1				→m_{14}	—				-1	1	-1		1						
														1		→m_{15}	—	-1				-1		1		1					

FIG. 10. Setup for the methanol mechanism, after the elimination of a_2 (HO_2) and a_7 (H). Mechanisms that have been abolished and intermediates that have been eliminated are immediately removed from the figures; thus, the rows of m_1, m_6, and m_{13}, and the columns of a_2 and a_7 are not present in Fig. 10 and subsequent figures. The terminal species, not shown in Fig. 9, have been included in the setup. The number of combinations has been recalculated for each species. The intermediates a_5 (CH_3O_2H) and a_8 (CHO), each of which gives rise to only two combinations, are next eliminated in parallel, since their sets of mechanisms (m_4 and m_5 for a_5; m_{12}, m_{14}, and m_{15} for a_8) are disjoint. [Reprinted with permission from Mavrovouniotis, M. L., and Stephanopoulos, G. "Synthesis of reaction mechanisms consisting of reversible and irreversible steps: I. A synthesis approach in the context of simple examples". *Ind. Eng. Chem. Res.* **31**, 1625-1637. (1992). Copyright 1992 American Chemical Society.]

III. Biochemical Pathways

The algorithm described for direct reaction mechanisms can also be used for the synthesis of biochemical pathways from basic bioreactions in the pursuit of quasi-steady-state behaviors of bioprocesses. More importantly, the algorithm identifies pathways leading from available raw materials to desired target products, enabling more informed decisions in the early stages of the design of a bioprocess. In translating the algorithm to biochemical systems, mechanism steps would map to individual bioreactions (usually catalyzed by enzymes), whereas overall reaction mechanisms correspond to acceptable biochemical pathways for a bioprocess.

Here, we will discuss a different formulation of the problem, which is better suited to the preliminary-design interpretation of the synthesis of biochemical pathways. This section will be based on the algorithm and

Figure 11 Table

s1	s2	s3	s4	s5	s6	s7	s8	s9	s10	s11	s12	s13	s14	s15	mechanism ⇓	species ⇒ combinations⇒ origin ⇓		CH_3 18 a_1	CH_3O_2 2 a_3	OH 4 a_4	CH_3O 3 a_6	CH_2O 2 a_9	CH_4 a_{10}	O_2 a_{11}	CH_3OH a_{12}	CO a_{13}	H_2O a_{14}	C_2H_6 a_{15}	CH_3OCH_3 a_{16}
1															→m_2	—		−1	1								−1		
	1														→m_3	—			−1	1			1						
				1											→m_7	—		1		−1		−1	1						
					1										→m_8	—		1		−1		−1					1		
						1									→m_9	—		−2										1	
							1								→m_{10}	—		−1	−1				1						
								1							→m_{11}	—		−1		−1									1
		1	1												→m_{16}	a_5: m_4+m_5		1	−1	1	1		−1						
										1				1	→m_{17}	a_8: $m_{12}+m_{15}$		−2			−1		2			1			
													1	1	→m_{18}	a_8: $m_{14}+m_{15}$		−1			−1	−1	1			1	1		

FIG. 11. Setup for the application of the algorithm on the methanol mechanism, after the elimination of a_5 (CH_3O_2H) and a_8 (CHO). Elimination of the intermediate a_9 (CH_2O), which involves two combination mechanisms, is carried out next. [Reprinted with permission from Mavrovouniotis, M. L., and Stephanopoulos, G. "Synthesis of reaction mechanisms consisting of reversible and irreversible steps: I. A synthesis approach in the context of simple examples". Ind. Eng. Chem. Res. **31**, 1625–1637. (1992). Copyright 1992 American Chemical Society.]

Figure 12 Table

s1	s2	s3	s4	s5	s6	s7	s8	s9	s10	s11	s12	s13	s14	s15	mechanism ⇓	species ⇒ combinations⇒ origin ⇓		CH_3 18 a_1	CH_3O_2 3 a_3	OH 6 a_4	CH_3O 3 a_6	CH_4 a_{10}	O_2 a_{11}	CH_3OH a_{12}	CO a_{13}	H_2O a_{14}	C_2H_6 a_{15}	CH_3OCH_3 a_{16}
1															→m_2	—		−1	1					−1				
			1												→m_7	—		1		−1	−1	1						
				1											→m_8	—		1		−1	−1					1		
					1										→m_9	—		−2									1	
						1									→m_{10}	—		−1	−1			1						
							1								→m_{11}	—		−1		−1								1
		1	1												→m_{16}	a_5: m_4+m_5		1	−1	1	1	−1						
	1									1				1	→m_{19}	a_9: m_3+m_{17}		−2	−1	1		2			1			
	1												1	1	→m_{20}	a_9: m_3+m_{18}		−1		1	−1	1			1	1		

FIG. 12. State of the algorithm, for the mechanism, after the elimination of a_9 (CH_2O). Similarly, elimination of the intermediate a_3 (CH_3O_2) leads to Fig. 13. [Reprinted with permission from Mavrovouniotis, M. L., and Stephanopoulos, G. "Synthesis of reaction mechanisms consisting of reversible and irreversible steps: I. A synthesis approach in the context of simple examples". Ind. Eng. Chem. Res. **31**, 1625–1637. (1992). Copyright 1992 American Chemical Society.]

SYMBOLIC AND QUANTITATIVE REASONING

S1	S2	S3	S4	S5	S6	S7	S8	S9	S10	S11	S12	S13	S14	S15	mechanism ⇓	species ⇒ combinations⇒ origin ⇓	CH$_3$ 10 a$_1$	OH 6 a$_4$	CH$_3$O 3 a$_6$	CH$_4$ a$_{10}$	O$_2$ a$_{11}$	CH$_3$OH a$_{12}$	CO a$_{13}$	H$_2$O a$_{14}$	C$_2$H$_6$ a$_{15}$	CH$_3$OCH$_3$ a$_{16}$
						1									→m$_7$	—	1		−1	−1	1					
							1								→m$_8$	—	1	−1		−1				1		
								1							→m$_9$	—	−2								1	
									1						→m$_{10}$	—	−1	−1			1					
										1					→m$_{11}$	—	−1		−1							1
1			1	1											→m$_{21}$	a$_3$: m$_2$+m$_{16}$	1	1	−1	−1						
1	1											1		1	→m$_{22}$	a$_3$: m$_2$+m$_{19}$	−3	1		2	−1		1			
1	1												1	1	→m$_{23}$	a$_3$: m$_2$+m$_{20}$	−2	1	−1	1	−1	1	1			

FIG. 13. State of the algorithm, for the methanol mechanism, after the elimination of a_3 (CH$_3$O$_2$). Next, elimination of the intermediate a_6 (CH$_3$O) yields Fig. 14. [Reprinted with permission from Mavrovouniotis, M. L., and Stephanopoulos, G. "Synthesis of reaction mechanisms consisting of reversible and irreversible steps: I. A synthesis approach in the context of simple examples". *Ind. Eng. Chem. Res.* **31**, 1625–1637. (1992). Copyright 1992 American Chemical Society.]

S1	S2	S3	S4	S5	S6	S7	S8	S9	S10	S11	S12	S13	S14	S15	mechanism ⇓	species ⇒ combinations⇒ origin ⇓	CH$_3$ 10 a$_1$	OH 8 a$_4$	CH$_4$ a$_{10}$	O$_2$ a$_{11}$	CH$_3$OH a$_{12}$	CO a$_{13}$	H$_2$O a$_{14}$	C$_2$H$_6$ a$_{15}$	CH$_3$OCH$_3$ a$_{16}$
						1									→m$_8$	—	1	−1	−1				1		
							1								→m$_9$	—	−2							1	
								1							→m$_{10}$	—	−1	−1		1					
1	1											1		1	→m$_{22}$	a$_3$: m$_2$+m$_{19}$	−3	1	2	−1		1			
1		1	1	1											→m$_{24}$	a$_6$: m$_{21}$+m$_7$	1	1	−2	−1	1				
1		1	1							1					→m$_{25}$	a$_6$: m$_{21}$+m$_{11}$	−1	1	−1	−1					1
2	1	1	1										1	1	→m$_{26}$	a$_6$: m$_{21}$+m$_{23}$	−2	2		−2	1	1			

FIG. 14. Setup of the methanol mechanism, after the elimination of a_6 (CH$_3$O). The intermediate a_4 (OH) will be eliminated next. [Reprinted with permission from Mavrovouniotis, M. L., and Stephanopoulos, G. "Synthesis of reaction mechanisms consisting of reversible and irreversible steps: I. A synthesis approach in the context of simple examples". *Ind. Eng. Chem. Res.* **31**, 1625–1637. (1992). Copyright 1992 American Chemical Society.]

s1	s2	s3	s4	s5	s6	s7	s8	s9	s10	s11	s12	s13	s14	s15	mechanism	species ⇒ origin ⇓	CH$_3$ combinations⇒ 5 a_1	CH$_4$ a_{10}	O$_2$ a_{11}	CH$_3$OH a_{12}	CO a_{13}	H$_2$O a_{14}	C$_2$H$_6$ a_{15}	CH$_3$OCH$_3$ a_{16}
→	→	→	→	→	→	→	→	→	→	→	→	→	→	→										
						1									→m$_9$	—	−2						1	
1	1				1									1	→m$_{27}$	a_4: m$_8$+m$_{22}$	−2	1	−1	.	1	1		
1		1	1		1	1									→m$_{28}$	a_4: m$_8$+m$_{24}$	2	−3	−1	1		1		
1		1	1		1		1								→m$_{29}$	a_4: m$_8$+m$_{25}$		−2	−1			1	1	
2	1	1	1			2							1	1	→m$_{30}$	a_4: 2m$_8$+m$_{26}$		−2	−2	1	1	2		
1	1							1	1					1	→m$_{31}$	a_4: m$_{10}$+m$_{22}$	−4	2	−1	1	1			
1		1	1		1			1							→m$_{32}$	a_4: m$_{10}$+m$_{24}$		−2	−1	2				
1		1	1					1	1						→m$_{33}$	a_4: m$_{10}$+m$_{25}$	−2	−1	−1	1				1
2	1	1	1					2					1	1	→m$_{34}$	a_4: 2m$_{10}$+m$_{26}$	−4		−2	3	1			

FIG. 15. Setup of the methanol mechanism, after the elimination of a_4 (OH). Elimination of the only remaining intermediate, a_1 (CH$_3$), is carried out next. [Reprinted with permission from Mavrovouniotis, M. L., and Stephanopoulos, G. "Synthesis of reaction mechanisms consisting of reversible and irreversible steps: I. A synthesis approach in the context of simple examples". *Ind. Eng. Chem. Res.* **31**, 1625–1637. (1992). Copyright 1992 American Chemical Society.]

s1	s2	s3	s4	s5	s6	s7	s8	s9	s10	s11	s12	s13	s14	s15	mechanism	species ⇒ origin ⇓	CH$_4$ a_{10}	O$_2$ a_{11}	CH$_3$OH a_{12}	CO a_{13}	H$_2$O a_{14}	C$_2$H$_6$ a_{15}	CH$_3$OCH$_3$ a_{16}
→	→	→	→	→	→	→	→	→	→	→	→	→	→	→									
1		1	1		1		1								→m$_{29}$	a_4: m$_8$+m$_{25}$	−2	−1			1		1
2	1	1	1			2							1	1	→m$_{30}$	a_4: 2m$_8$+m$_{26}$	−2	−2	1	1	2		
1		1	1		1			1							→m$_{32}$	a_4: m$_{10}$+m$_{24}$	−2	−1	2				
1		1	1			1	1	1							→m$_{35}$	a_1: m$_{28}$+m$_9$	−3	−1	1		1	1	
2	1	1	1			1	2					1		1	→m$_{36}$	a_1: m$_{28}$+m$_{27}$	−2	−2	1	1	2		
3	1	2	2			2	2	1				1		1	→m$_{37}$	a_1: 2m$_{28}$+m$_{31}$	−4	−3	3	1	2		
2		2	2			1	1		1	1					→m$_{38}$	a_1: m$_{28}$+m$_{33}$	−4	−2	2		1		1
4	1	3	3			2	2		2				1	1	→m$_{39}$	a_1: 2m$_{28}$+m$_{34}$	−6	−4	5	1	2		

FIG. 16. The final results for the methanol mechanism. [Reprinted with permission from Mavrovouniotis, M. L., and Stephanopoulos, G. "Synthesis of reaction mechanisms consisting of reversible and irreversible steps: I. A synthesis approach in the context of simple examples". *Ind. Eng. Chem. Res.* **31**, 1625–1637. (1992). Copyright 1992 American Chemical Society.]

s_1	s_2	s_3	s_4	s_5	s_6	s_7	s_8	s_9	s_{10}	s_{11}	s_{12}	s_{13}	s_{14}	s_{15}	mechanism	direct?	mechanism in Table X of Happel et al. (1990)
1		1	1			1		1							m_{29}	√	not found, because s_{11} was omitted
2	1	1	1			2					1			1	m_{30}	√	m_2/r_0; m_3/r_0 (lines 5 and 7)
1		1	1		1			1							m_{32}	√	m_1/d_4; m_3/d_4; m_4/d_4; m_6/d_4 (lines 2,6,8,10)
1		1	1		1	1	1								m_{35}	√	not found, because s_9 was omitted
2	1	1	1		1	2				1				1	m_{36}	√	m_4/r_0 (line 7)
3	1	2	2		2	2		1		1				1	m_{37}	$=m_{36}+m_{32}$	—
2		2	2		1	1			1	1					m_{38}	$=m_{29}+m_{32}$	—
4	1	3	3		2	2		2					1	1	m_{39}	$=m_{30}+2m_{32}$	—

FIG. 17. Analysis of the results for the methanol mechanism. The last three mechanisms that were produced (m_{37}, m_{38}, and m_{39}) are not direct; they can be formed from the direct ones. The last column shows the correspondence of the direct mechanisms to those of Happel et al. (1990). It is important to note that the same mechanism can be constructed many times in the procedure of Happel and Sellers (1983, p. 290)—as well as the simple algorithm presented here. It should also be noted that, in Table X of Happel et al. (1990), the mechanisms m_1/r_0 (line 1), m_2/d_4 (line 4), m_5/r_0 (line 9), m_5/d_4 (line 10), and m_6/r_0 (line 11) are all infeasible because they use either s_1 or s_5 in the wrong direction. The algorithm presented in this chapter never constructs mechanisms that violate the directionality of irreversible steps. [Reprinted with permission from Mavrovouniotis, M. L., and Stephanopoulos, G. "Synthesis of reaction mechanisms consisting of reversible and irreversible steps: I. A synthesis approach in the context of simple examples". Ind. Eng. Chem. Res. 31, 1625–1637. (1992). Copyright 1992 American Chemical Society.]

examples presented by Mavrovouniotis (1989) and Mavrovouniotis et al. (1990, 1992). There are two differences between this view and the one adopted in the previous section. First, the classification of compounds as either intermediates or terminal species, introduced by H&S and used in Section II, does not distinguish between appearance of a compound as a reactant (raw material) and appearance as a product; in this section, we will permit separate specifications on net reactants and net products. Second, in Section II reversibility considerations were taken into account in the construction of mechanisms. Since most bioreactions are irreversible, we will treat reversible bioreactions as an exception, by splitting them into forward and reverse portions.

A. Features of the Pathway Synthesis Problem

Biochemical pathway synthesis is the construction of consistent sets of enzyme-catalyzed bioreactions meeting certain specifications. One seeks to construct pathways which produce certain target bioproducts, under partial constraints on the available substrates (reactants), allowed byproducts, desired yield, productivity, etc. The pathway must include all reactions needed to convert initial substrates supplied to the bioprocess into final

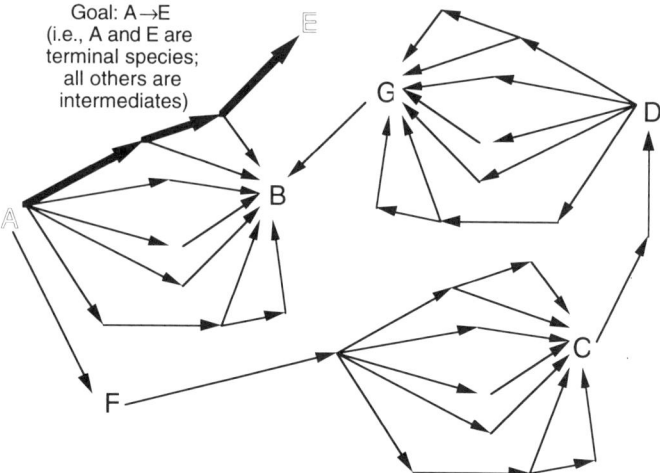

FIG. 18. The consequences of disregarding reaction irreversibility, in an abstract example. The species A and E are terminal and all others are intermediates. There are many parallel irreversible mechanisms that transform various intermediates to others (e.g., from A to B, from F to C, and from D to G). The direction of these internal mechanisms, however, is such that they do not participate in the final solution—which requires mechanisms only between A and E. There is actually only one mechanism (shown in thick arrows) that converts A to E in three steps. A method like that of H&S, which disregards directionality of steps and treats all steps as reversible initially, will come up with a very large number of mechanisms, going through the intermediates F, C, D, G, B (in that order). Our algorithm would *immediately* prune out the irrelevant portions of the network (starting with B) and identify the one feasible mechanism.

products that will be recovered from the process. Even if a computationally synthesized pathway is not present in any single known microorganism, it is of interest because of the possibilities offered by genetic engineering, which achieves the transfer of reactions and pathways from one microorganism to another. Through genetic engineering, a missing step can be inserted in a microorganism so that the synthesized new biochemical pathway (and hence a new bioprocess) is realizable.

Just like a direct reaction mechanism discussed in Section II, a pathway is not merely a set of reactions, because many distinct pathways can be constructed to include the same reactions but achieve different transformations. A pathway must include a *coefficient* (denoted by σ_{ki} in Section II) for each reaction, indicating the proportions in which the stoichiometries are combined. In this section, we will call these coefficients the *reaction stoichiometry* of the pathway. The overall transformation of net reactants to net products will be called the *molecular stoichiometry* of the pathway, and was denoted by β_{kj} in Section II.

Whereas in Section II we assumed that the participating steps will be given for a particular application, here it is useful to have a *database* with the most common enzymatic reactions and compounds, so that one only needs to add the specialized reactions for a given investigation.

B. FORMULATION OF CONSTRAINTS

A whole class of specifications in the synthesis of biochemical pathways can be formulated by classifying each compound and each reaction from the database according to the role it is required or allowed to play in the synthesized pathways' stoichiometries.

A given compound may occur in a pathway in any of three capacities: (1) as a net reactant or substrate of the pathway, (2) as a net product of the pathway, or (3) as an intermediate in the pathway. Constraints state whether a compound is allowed or possibly required to participate in the synthesized pathways in each of these three capacities.

For compounds as net reactants in the pathways, these conditions are inequality constraints on the coefficients β_{kj} (see Section II) that describe the participation of each species a_j in a pathway m_k. The specifications take the following forms:

(a) The species a_j is called a required reactant if it *must be consumed* by the pathway; this corresponds to a requirement that the coefficient β_{kj} be negative, i.e., $\beta_{kj} < 0$.
(b) A species is designated as an excluded reactant if it *cannot be consumed* by the pathway, and the corresponding restriction on the coefficient is $\beta_{kj} \geq 0$.
(c) Finally, the classification of a_j as an *allowed reactant* introduces no restriction on the coefficient β_{kj}.

In a realistic synthesis problem, the default constraint for compounds is that of excluded reactants, since most compounds in the database would not be available as economical raw materials.

The constraints for potential products of the pathways involve similar inequalities. A species a_j can be specified as (1) a required product, $\beta_{kj} > 0$, (2) an excluded product, $\beta_{kj} \leq 0$, which is the default, or (3) an allowed product, which involves no restriction on the sign of β_{kj}. Each compound a_j thus acquires one constraint from the reactant specifications and one from the product specifications. One of these combinations, however, the designation of the same compound as a required reactant ($\beta_{kj} < 0$) and a required product ($\beta_{kj} > 0$), is inconsistent. Of the remaining combinations, the designation of a compound as an allowed reactant and allowed product corresponds to the absence of any constraint on β_{kj}. The reader may compare these sets of constraints to the specification of intermediate and terminal species in Section II. There, we classified each species a_j using only two categories, as follows:

(a) If a species a_j is an intermediate, its coefficient in m_k is restricted to $\beta_{kj} = 0$. Clearly, this corresponds to the combination (conjuction) of an excluded reactant constraint ($\beta_{kj} \geq 0$) and an excluded product constraint ($\beta_{kj} \leq 0$).

(b) If a_j is a terminal species, its coefficient in m_k is not restricted at all. This corresponds to the combination of an allowed reactant and an allowed product.

Thus, in the case of biochemical pathways we have allowed a richer vocabulary for stoichiometric constraints but we can still reproduce the earlier specifications.

Notice the relation between the specification of a required reactant (negative stoichiometric coefficient—a strict inequality) and an excluded product (negative or zero stoichiometric coefficient—a loose inequality). In the operation of the algorithm, the first type of constraint is initially not fully satisfied; instead, the corresponding loose inequality constraint is satisfied in its place. Thus, for most of the phases in the algorithm, required reactants are treated merely as excluded products. As we will see, the distinction is eventually enforced, in the last phase of the algorithm. A similar observation holds for the strict inequality arising from required products, in relation to the loose inequality arising from excluded reactants.

A reaction can participate in pathways in either its forward or its reverse direction, giving rise to additional possible specifications: Reactions may be required, allowed, or prohibited to participate in the synthesized pathways in each of the two directions. Most reactions are likely to be *excluded* in the backward direction, because of prior knowledge about the (thermodynamic or mechanistic) irreversiblity of the reaction.

C. Algorithm

The algorithm for the synthesis of biochemical pathways follows closely the logic of the algorithm for the synthesis of catalytic reactions, i.e., it synthesizes biochemical pathways from a set of enzymatic reactions through an iterative satisfaction of constraints. A few minor differences reflect the richer vocabulary of specifications, given in the preceding section, and the biochemical context of compounds and enzymatic reactions.

Given a set of stoichiometric constraints and a database of biochemical reactions, the algorithm carries out iterative satisfaction of constraints, just like the algorithm in Section II. The algorithm proceeds as follows.

1. Initialization and Reaction Processing

The set M contains a one-step pathway for each individual enzymatic reaction available in the database, which has been compiled from known biochemistry. The set N contains the metabolites present in the enzymatic reactions of the set M. Naturally, these include the substrates (raw materials) that can be used, the desired products, and a large number of other compounds that occur in the bioreactions but will not serve as raw materials or desired products. This last set of compounds will carry the restrictions of excluded reactants and excluded products.

In order to account for the reversibility of some reactions, the algorithm decomposes reversible reactions into a forward and a backward step; the two directions are thereafter prohibited from participating together in the same pathway. This approach, unlike Section II, treats reversible reactions as the exception.

2. Number of Combinations for Each Intermediate

We note that the iteration through phases 2–4 was called compound-processing in the original description of this algorithm (Mavrovouniotis *et al.*, 1990). In these phases, partial pathways are gradually combined to satisfy the constraints (just as in Section II). This takes place as the algorithm tackles one compound at a time.

The objective of phase 2, in particular, is the selection of the most suitable compound for constraint satisfaction. We want to given priority to compounds that participate (as reactants or products) in only a small number of pathways that are active in the current state of the problem.

The computation of combinations (n_j as defined in Section II) facilitates this selection. Since all pathways are irreversible in our biochemical formulation, we only need to assemble (for each species a_j) the set Y_j, which contains all the partial pathways that include a_j as a net reactant, and the set Z_j, which contains all the partial pathways that include a_j as a net product. The cardinalities of these sets are y_j and z_j, respectively. It is clear that, in order to eliminate the metabolite a_j, we need to combine any pathway from Y_j, giving rise to $n_j = x_j y_j$ combinations.

3. Elimination of an Intermediate Metabolite

Partial pathways (which initially are just individual enzymatic reactions) are gradually combined in an effort to satisfy the constraints on the role of each metabolite. To minimize the computational effort, we choose in each iteration that metabolite that has the smallest number of combinations n_j. Let a_J be this metabolite. For the satisfaction of the constraints, a set M_J of new pathways is constructed through combinations of exactly one partial pathway from Z_J and exactly one pathway from Y_J. These pathways have the form

$$m_c = \beta_{bJ} m_k - \beta_{kJ} m_b,$$

where m_k are m_b are taken from the sets Y_J and Z_J, respectively (i.e., $m_k \in Y_J$ and $m_b \in Z_J$) and β_{bJ}, β_{kJ} are the net coefficients with which a_J participates in m_b and m_k, respectively. The net coefficient of a_J in the new pathway, m_c, is $\beta_{bJ}\beta_{kJ} - \beta_{kJ}\beta_{bJ} = 0$, verifying that, in the resulting new pathway, a_J has been eliminated.

A subtle precaution must be taken in the formation of the combinations, because of the way in which reversible reactions were decomposed into forward and reverse portions. Specifically, if m_k and m_p involve the same reaction (*any* reaction) *in different directions*, then the combination is rejected (or rather not formed at all).

4. Update of Active Sets

In order to modify the set of pathways such that the constraints on the compound a_J are met, we make use of the sets M_J, Y_J, and Z_J. The set of active pathways is modified by selecting the first applicable modification from the following list:

1. If a_J is an excluded product and a required or allowed reactant, $M := M \cup M_J - Z_J$.
2. If a_J is an excluded reactant and a required or allowed product, $M := M \cup M_J - Y_J$.
3. If a_J is an excluded reactant and an excluded product, $M := M \cup M_J - Y_J - Z_J$.
4. If none of the previous conditions describe the constraints on a_J, then M remains unchanged.

Then, we can remove the metabolite a_J and set $N := N - \{a_J\}$.

5. Termination or New Iteration

At this point, if no metabolite remains in N (i.e., if $N \neq \varnothing$), we can proceed to the final phase. If metabolites remain, then we return to phase 2, with a new computation for the selection of the next metabolite to process. In effect, the procedure in phases 3–4 is eventually carried out for all compounds.

6. Pathway Processing

The necessity for this phase, which did not exist in the algorithm of Section II, is a consequence of the richer vocabulary of stoichiometric constraints used for biochemical systems. The set of active pathways resulting from the processing of all metabolites as above satisfies all the requirements, except the constraints designating required reactants, products, or reactions. For these, only the corresponding loose-inequality constraints are satisfied; for example, instead of a required reactant and prohibited product, a compound will have been treated, up to this point, as an allowed reactant and prohibited product. The pathways satisfying the original stoichiometric constraints are those combinations of pathways from the final set that use at least one constituent pathway satisfying each of the strict constraints. The combination must thus include at least one pathway consuming each required reactant, at least one pathway producing each required product, etc.

D. EXAMPLES

We will discuss briefly some examples, to illustrate the features of the pathway synthesis algorithm and the results it has produced.

1. Computational Efficiency

The structured character of biochemical reaction networks is exploited by the synthesis algorithm in early pruning and abstraction, with significant gains in computational efficiency. This happens because the algorithm processes first those compounds and reactions that lead to few or no new combinations. This is shown schematically in Fig. 19, where some irrelevant portions of the network are pruned and some linear chains of reactions are compacted. We should note that this kind of reaction sequences are very common in biochemical reaction systems.

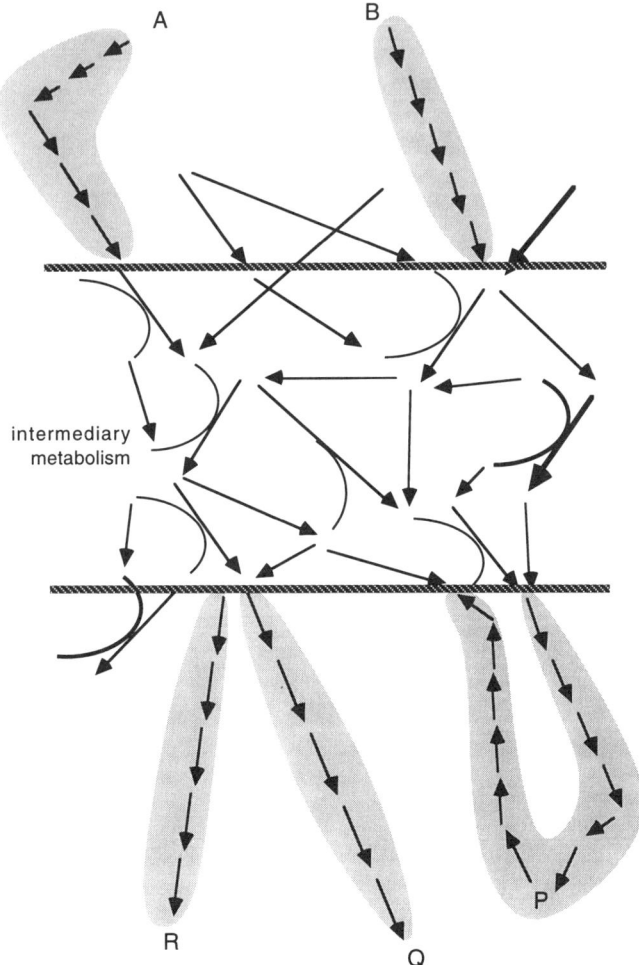

FIG. 19. An instance of efficient processing by the algorithm for the synthesis of biochemical pathways. Suppose that P, A, and Q are excluded reactants and excluded products; B is an allowed reactant; R is an allowed product; and all other compounds are excluded reactants and excluded products. In the early stages of the compound-processing phase, the algorithm will discard, one after the other, all reactions involved in the metabolism of B and R; it will construct the whole pathway that synthesizes and degrades P, discarding the individual reactions, and thus reducing the number of reactions/pathways that are being considered; it will also construct pathways for the consumption of A and the production of Q, discarding the individual reactions for further reduction of the pathway space.

2. Synthesis of Alanine

The synthesis of alanine from glucose will be discussed here. In the database of 250 reactions used by Mavrovouniotis *et al.* (1992), six reactions involve glucose, and four involve alanine (*alanine dehydrogenase, methylserine hydroxymethyltransferase, alanine aminotransferase,* and *β-alanine aminotransferase*).

The initial formulation of the problem has glucose as a required reactant, alanine as a required product, NH_3 as an allowed reactant, and CO_2 as an allowed product. Additionally, a set of compounds that serve as internal currencies in the cell are designated as allowed reactants and allowed products. These are likely to occur in the problem specification, and they include NAD, NADH, NADP, NADPH, ATP, ADP, AMP, coenzyme-A (CoA), phosphate, and pyrophosphate. This initial formulation is actually too tight to generate a manageable number of pathways. By examining the constraints that are difficult to satisfy, we can modify the formulation of the problem to permit a solution. Here, the problem is resolved by designating the compounds malate and acetyl–CoA as allowed reactants and allowed products, reaching a formulation that constructs 1446 useful pathways.

Figures 20 and 21 show two pathways, suppressing many of the side reactants and side products for simplicity. Figure 20 depicts the normal

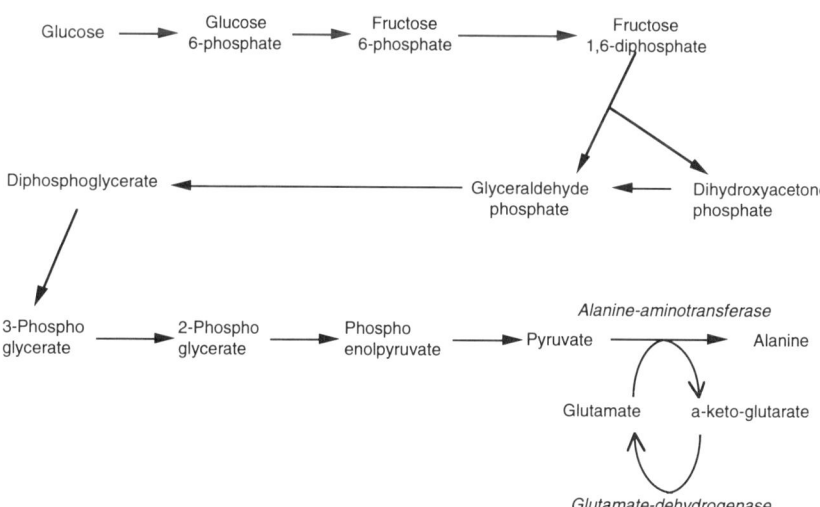

FIG. 20. Synthesis of alanine from glucose that follows the glycolysis pathway to pyruvate, which is in turn converted to alanine by *Alanine aminotransferase*. The glutamate required by this reaction is produced by *glutamate-dehydrogenase*.

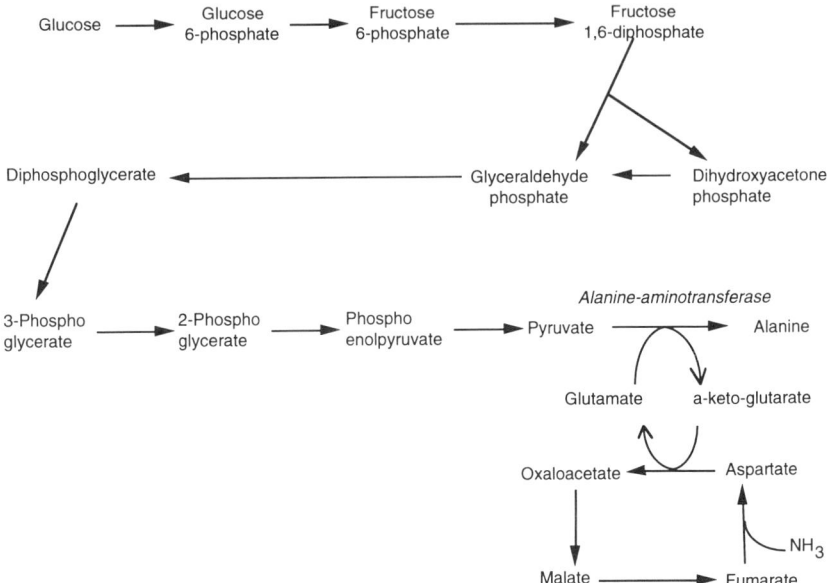

FIG. 21. Synthesis of alanine from glucose, with recovery of glutamate through a loop involving oxaloacetate, aspartate, fumarate, and malate.

pathway for the synthesis of alanine, with glucose as the main reactant. The pathway of Fig. 21 shows an alternative recovery of glutamate through a set of four reactions. The second pathway represents an interesting way for the cell to attach ammonia to an organic backbone; this is a variation that may be relevant for many aminoacid processes.

3. Synthesis of Lysine

In another study (Mavrovouniotis et al., 1990), the biosynthesis of the aminoacid lysine was extensively investigated. In that case study, several reactions (shown in gray in Fig. 22) were assumed nonfunctional, and alternatives that avoid these steps were sought.

Figure 22 shows one of the resulting pathways, which involves the use of glyoxylate as an intermediate to bypass one of the nonfunctional steps. The pathway also includes conversion of malate to fumarate (which is the direction opposite to that normally used by microorganisms) and subsequent conversion of fumarate into aspartate and on to lysine. The use of the latter step is the key to the existence of this innovative pathway.

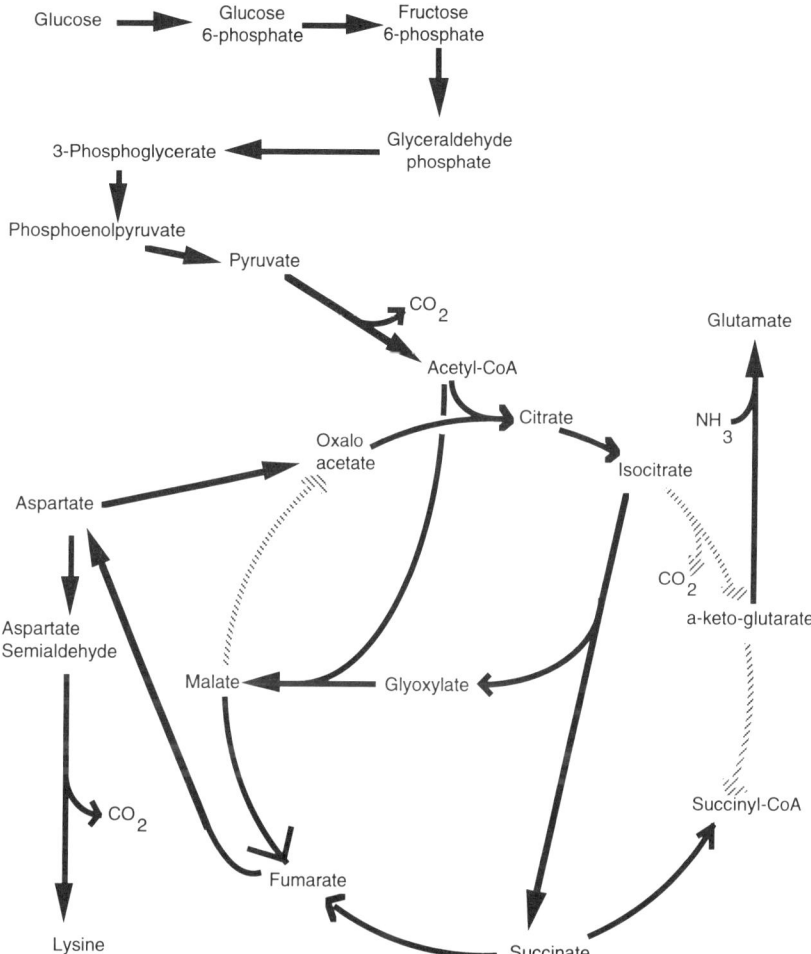

FIG. 22. A pathway for the production of lysine from glucose, drawn in a simplified form (many side reactants and side products are not shown, and many reaction sequences are lumped together).

IV. Properties and Extensions of the Synthesis Algorithm

Both forms of the algorithm (i.e., the basic version of Section II and the alternative of Section III) guarantee that each combination of pathways from the final set (derived by the synthesis algorithm) satisfies all the initial stoichiometric requirements. The correctness of the algorithm is based on the fact that each of the original requirements is satisfied in one of the iterations, and after a constraint is satisfied it is guaranteed to remain satisfied. As was pointed out earlier, constraints on required reactants or required products are satisfied in two phases. Initially, only the corresponding loose-inequality form is satisfied; the satisfaction of the strict-inequality version is ensured in the last phase (6) of the algorithm.

The second important property of the algorithm is its completeness. The synthesis algorithm creates a final set of pathways such that any pathway satisfying the original stoichiometry constraints is present in the final set. The basis of this property is that the elimination procedure is guaranteed to preserve legitimate pathways. A more detailed discussion of these properties is given by Mavrovouniotis et al. (1992).

We note that the algorithm may give rise to duplicate or indirect mechanisms (or pathways). Procedures can be applied in the end to remove such redundant mechanisms from the final set of mechanisms, or even to remove them incrementally during each iteration of the algorithm. These and other variations and implementational choices in the algorithm can be defined to enhance either its computational efficiency or its conceptual simplicity (Mavrovouniotis, 1992).

With respect to its computational complexity, the algorithm would require time, in the worst case, exponential in the number of reactions (Mavrovouniotis et al., 1990). This worst-case complexity, however, is not encountered in practice, because the algorithm exploits the structure of biochemical networks, as was shown in Fig. 19. Another example, in Fig. 18, shows how the careful inclusion of directionality (irreversibility) considerations in the algorithm also serves to curb computational complexity.

A last feature of the algorithm that is worth discussing is its extendibility to other types of constraints. Additional specifications for the pathways or mechanisms under construction might arise from thermodynamics, kinetics, yield or productivity restrictions, biological regulation, etc. The question is whether such additional constraints can easily be incorporated into the synthesis procedure.

Some of these constraints are linear in nature. For example, a restriction on the endothermic or exothermic character of a pathway is a linear constraint on the enthalpy of reaction. If we take a linear combination of reactions (or pathways) the enthalpy of the resulting pathway is the corresponding linear combination of the enthalpies of the constituent reactions or pathways. For these linear situations, the quantity being constrained can be considered as merely another species, whose stoichiometry is subject to the type of constraints already tackled by the algorithm. Thus, the synthesis algorithm can deal with these specifications with little or no modification.

Other types of constraints, not corresponding to quantitative linear relations, can also be included, if they fulfill two requirements. The first requirement is that the constraint must possess the following convexity-like property: If the constraint is satisfied for two individual pathways, m_k and m_b, then it should be satisfied for any positive linear combination of these, $m_c = \mu_k m_k + \mu_b m_b$ (where $\mu_k > 0$, $\mu_b > 0$). The second requirement is that, given a set of partial mechanisms that do not satisfy the constraint, there must be a clear way to identify combination mechanisms that do satisfy it. These two requirements allow a constraint to be processed in much the same way that the stoichiometric constraints were, except that the details of construction of combinations would depend on the nature of the constraint. In one of the iterations, the constraint would be processed (through a procedure dictated by the second requirement), and it would thereafter remain satisfied (because of the first requirement).

V. Summary

In a chemical system we often discriminate between intermediate species and terminal species, and the latter are the only ones that appear in appreciable amounts in the stoichiometry of the overall transformation. Alternatively, we designate compounds as allowed or required to appear as reactants or products in the overall stoichiometry. These formulations are significant in the identification of quasi-steady-state behaviors or the synthesis of pathways to accomplish a desired transformation in a process.

We presented here a conceptual framework and algorithms for the synthesis of pathways or mechanisms given a set of steps. The algorithms have been applied to catalytic reaction systems and to biochemical pathways. The basic approach is based on successive processing and elimination of reaction intermediates that should not appear in the net stoichiometry of the overall reactions accomplished.

References

Happel, J., "Isotopic Assessment of Heterogeneous Catalysis." Academic Press, Orlando, FL, 1986.
Happel, J., and Sellers, P. H., Multiple reaction mechanisms in catalysis. *Ind. Eng. Chem. Fundam.*, **21**, 67–76 (1982).

Happel, J., and Sellers, P. H., Analysis of the possible mechanisms for a catalytic reaction system. *Adv. Catal.* **32**, 273–323 (1983).

Happel, J., and Sellers, P. H., The characterization of complex systems of chemical reactions. *Chem. Eng. Commun.*, **83**, 221–240 (1989).

Happel, J., Sellers, P. H., and Otarod, M., Mechanistic study of chemical reaction systems. *Ind. Eng. Chem. Res.* **29**, 1057–1064 (1990).

Horiuti, J., Theory of reaction rates as based on the stoichiometric number concept. *Ann. N.Y. Acad. Sci.* **213**, 5–30 (1973).

Horiuti, J., and Nakamura, T., On the theory of heterogeneous catalysis. *Adv. Catal.* **17**, 1–74 (1967).

Lee, L.-S., and Sinanoğlu, O., Reaction mechanisms and chemical networks—Types of elementary steps and generation of laminar mechanisms. *Z. Phys. Chem.* **124**, 129–160 (1981).

Mavrovouniotis, M. L., Computer-aided design of biochemical pathways. Ph.D. Thesis, Massachusetts Institute of Technology (1989).

Mavrovouniotis, M. L., Synthesis of reaction mechanisms consisting of reversible and irreversible steps: II. Formalization and analysis of the synthesis algorithm. *Ind. Eng. Chem. Res.* **31**, 1637–1653 (1992).

Mavrovouniotis, M. L., and Stephanopoulos, G., Synthesis of reaction mechanisms consisting of reversible and irreversible steps: I. A synthesis approach in the context of simple examples. *Ind. Eng. Chem. Res.* **31**, 1625–1637 (1992).

Mavrovouniotis, M. L., Stephanopoulos, G., and Stephanopoulos, G., Computer-aided synthesis of biochemical pathways. *Biotechnol. Bioeng.* **36**, 1119–1132 (1990).

Mavrovouniotis, M. L., Stephanopoulos, G., and Stephanopoulos, G., Synthesis of biochemical production routes. *Comput. Chem. Eng.* **16**, 605–619 (1992).

Milner, P. C., The possible mechanisms of complex reactions involving consecutive steps. *J. Electrochem. Soc.* **111**, 228–232 (1964).

Sellers, P. H., An introduction to a mathematical theory of chemical reaction networks I. *Archive for Rational Mech. Anal.* **44**, 23–40 (1971).

Sellers, P. H., An introduction to a mathematical theory of chemical reaction networks II. *Archive for Rational Mech. Anal.* **44**, 376–386 (1972).

Sellers, P. H., Combinatorial classification of chemical mechanisms. *SIAM J. Appl. Math.* **44**, 784–792 (1984).

Sellers, P. H., Combinatorial aspects of enzyme kinetics. In "Applications of Combinatorics and Graph Theory in the Biological and Social Sciences" (F. Roberts, ed.). Springer-Verlag, Berlin, 1989.

Sinanoğlu, O., Theory of chemical reaction networks. All possible mechanisms or synthetic pathways with given number of reaction steps or species. *J. Am. Chem. Soc.* **97**, 2309–2320 (1975).

Sinanoğlu, O., and Lee, L.-S., Finding all possible *a priori* mechanisms for a given type of overall reaction. *Theor. Chim. Acta* **48**, 287–299 (1978).

Temkin, M. I., The kinetics of transfer of labeled atoms by reaction. *Ann. N.Y. Acad. Sci.* **213**, 79–89 (1973).

Temkin, M. I., The kinetics of some industrial heterogeneous catalytic reactions. *Adv. Catal.* **26**, 173–291 (1979).

Yarlagadda, P. S., Morton, L. A., Hunter, N. R., and Gesser, H. D., Direct conversion of methane to methanol in a flow reactor. *Ind. Eng. Chem. Res.* **27**, 252–256 (1988).

INDUCTIVE AND DEDUCTIVE REASONING: THE CASE OF IDENTIFYING POTENTIAL HAZARDS IN CHEMICAL PROCESSES

Christopher Nagel[1] and George Stephanopoulos

Laboratory for Intelligent Systems in Process Engineering
Department of Chemical Engineering
Massachusetts Institute of Technology
Cambridge, Massachusetts 02139

I. Introduction	188
A. Predictive Hazard Analysis	190
B. Incompleteness of Conventional Hazard Analysis Methodologies	192
C. Premises of Traditional Approaches	193
D. Overview of Proposed Methodology	194
II. Reaction-Based Hazards Identification	195
A. System Foundations	196
B. Modeling Languages and Their Role in Hazards Identification	198
C. Generation of Reactions and Evaluation of Thermodynamic States	205
III. Inductive Identification of Reaction-Based Hazards	209
A. Hazards Identification Algorithm	211
B. Properties of Reaction-Based Hazards Identification	214
C. An Example in Reaction-Based Hazard Identification: Aniline Production	217
IV. Deductive Determination of the Causes of Hazards	221
A. Methodological Framework	222
B. Variables as "Causes" or "Effects"	225
C. Construction of Variable-Influence Diagrams	227
D. Characterizing of Variable-Influence Pathways	232
E. Assessment of Hazards-Preventive Mechanisms	235
F. Fault-Tree Construction	238
G. An Example of Reaction-Based Hazard Identification: Reaction Quench	241
V. Conclusion	253
References	254

All reasoning carried out by computers is *deductive*; i.e., any software system has all the necessary data, stored in various forms in a database,

---

[1] Present address: Molten Metal Technology, Inc., Waltham, Massachusetts.

and possesses all the necessary algorithms to operate on the set of data and *deduce* some results. Many researchers in the area of cognitive psychology make similar claims on the reasoning mechanisms of the human beings. The fact, though, remains that both humans and machines can use very simple "algorithms" on a small set of data and produce results, which could not have been visible by the "naked eye" of direct reasoning. In such cases, we tend to talk about the *inductive* capabilities of either of the two. These ideas are nowhere more prominent than in the area of *hazards identification and analysis*. One often hears, "if I knew that the conversion of A to B could be catalyzed by the presence of C then I would have foreseen the last disaster, and have done something about it," with the speaker converting a problem of *inductive* identification (i.e., induce the possibility of a hazard from the list of chemicals) into an issue of deductive statement. In this chapter we try to demonstrate that the identification of hazards is essentially an interplay between inductive and deductive reasoning. Through inductive reasoning we attempt to generate all potential hazardous top-level events, which can be justified by the presence of a set of chemicals. We call the reasoning inductive because it has the potential to generate specific knowledge that was not "visible" ahead of time. Once the potentially harmful top-level events have been identified, deductive reasoning attempts to "walk" through the processing scheme and its unit operations and their design or operating characteristics (assumptions, or decisions), and generate the preconditions, which would enable the occurrence of a specific top-level event. The inductive reasoning procedures operate on a set of chemicals and create in an *exhaustive, bottom–up* manner many alternative reaction-pathways, some of which could lead to a hazard, e.g., release of large amounts of energy over a short period of time. On the other hand, the deductive reasoning procedures are *goal-directed* and operate in a *top–down* manner. In this chapter we will develop the detailed framework for the implementation of these ideas, which among other benefits offer the following advantages: (1) formalize the hazards identification problem and unify the methodological approaches at any stage of the design activities and (2) systematize the generation and evaluation of mechanisms for the prevention of hazards, or containment of their effects.

I. Introduction

When a process is transferred from the laboratory to the pilot or/ and commercial scale, a variety of hazards may appear that had earlier been well controlled under the relatively small scale of the discovery effort. Throughout the period of a process's development, two factors that may introduce new hazards and must be examined continuously are *change* and *scale-up* (Brannegan, 1985). Change complicates hazards evaluation by introducing new components that are associated with hazards that may be unknown. Scale-up, particularly initial scale-up, can generate significant potential hazards by escalating the magnitude of effects, initially thought to be benign. Since changes often occur throughout the life of the plant, the need to identify hazards as early as possible in the development stages does not imply that hazard identification ends when the design specifications have been approved. In fact, approval of a design means only "At the time of the study, the study team believes that, provided that the plant is constructed and operated in accordance with their recommendations, the plant will be acceptably safe" (Lowe and Solomon, 1983). The first uncontrolled change during construction, or the first unapproved modification during operation, negates this approval. Consequently, hazard identification is a continuing concern and a permanent ingredient of safe operations and should be applied, sometimes in a very simple form, to control any changes from the original intentions of the designers. Several approaches have been presented in the past decade to systematize hazard identification and hazard analysis, but procedural robustness is often constrained by the quality of information available and the expertise of the individuals involved.

No one doubts the importance of hazard identification, in advance of an unwanted event. However, the quality of the risk analysis results can be no better than the extent to which hazards are recognized in the first place. Furthermore, the analysis is no better than the analyst's understanding of the plant's design and its operations. Decisions about safety are made continuously. These decisions are made in light of all the uncertainties and are based on the understanding of the characteristics of the facility and the substances involved. Formal analytical processes may or may not be involved in the decision process. Studies recently indicated that design errors—a design error is deemed to have been committed when the design is changed after an incident—were rarely revealed before startup and accounted for 25% of all accidents (Haastrup, 1983). India's experience with design error is closer to 40%, and it has been suggested that if the definition is broadened to include management and organizational aspects of process design and engineering, design errors would account for nearly 90% of the recorded incidents (Batstone, 1987). Moreover, the percentage of precursors (leading to a hazardous incident), perceived to be known at the time of the incident, varies. It depends on the perceptions that an individual formed in his or/her particular capacity. Our survey suggests that plant personnel believe that 90-100% of the precursors leading to a hazard are known at the time of the incident, hazard specialists believe that 40-60% of those precursors are known at the time of the incident, and insurance analysts believe that only 20-30% of the precursors initiating or propagating the preconditions to a hazard are known at the time of the incident. Such data are exacerbating and they suggest, to some, that improved hazard identification methods are unnecessary, perhaps even unwanted. Yet, incidents continue to occur.

A. PREDICTIVE HAZARD ANALYSIS

The basic objective of hazard analysis is to identify and assess potentially hazardous situations, and their possible consequences and associated risk, in order to provide a rational basis for determining where risk reduction measures are needed. Hazard identification always has been an integral part of design and operational practice. However, it is to a large degree still an informal process depending on the experience of those directly involved.

Structured hazard identification methods can be classified roughly in two groups: (1) *comparative* methods that rely on systematic comparison of the process design against some recognized code or standard and (2) *fundamental* methods that can be applied in almost any situation (Boykin and Kazarians, 1987; Ozog and Bendixen, 1987). Individual experience is the essential ingredient of hazard identification for the first group of methods. However, such an approach requires that individual experience be collected, organized, recorded and standardized and become accessible information to those designing the equipment (e.g., through national and international codes and standards). Such codes and practices provide minimum standards against which deviations from safe practice can be identified and appraised. An important feature of these methods is that the experience gained through many years is incorporated in the company's practices and therefore available for use at all stages of design and construction. For new processes, the hazard identification procedure is strongly dependent upon information obtained a priori, and derived from the efforts of the research and development engineers and the hazard identification team members. This requires that hazard identification methods must be directed toward stimulating the team members to apply their own experience of safe and unsafe as the standard by which to appraise the design, mainly by raising a series of "what if" questions. Fundamental methods of hazard identification are aimed at two outcomes: to identify serious incidents that may result in personal injury or financial loss [known as "top-level events" or (TLES)]; and to identify the underlying root causes leading to top-level events.

In general, methods that identify actions that eliminate, avoid, or reduce the potential hazard of a particular design are referred to as *intrinsic*. This approach evolves the design technology toward inherently safer configurations through the use of codes, guidelines, and checklists. Intrinsic methodologies can be effective at the stages where the process scheme is conceived and the process flow diagram is developed. At these stages equipment changes are easily made without adversely affecting construction costs and schedules (see Fig. 1), but they lack formality and afford no means for complete and consistent hazard identification. Alternatively, methods that identify actions that reduce the likelihood of a hazard through control and safety devices are referred to as *extrinsic*. These methods tend to control an identified hazard, or the conditions leading to the hazardous state and often generate solutions that increase plant complexity and operational costs. The more formal methods, such as HazOp, FMECA (Failure Modes, Effects, and Criticality Analysis), fault trees, event trees, and cause–consequence analysis, require information that is often sparse in the early design stages (see Fig. 1) and the certainty of which may be indeterminate.

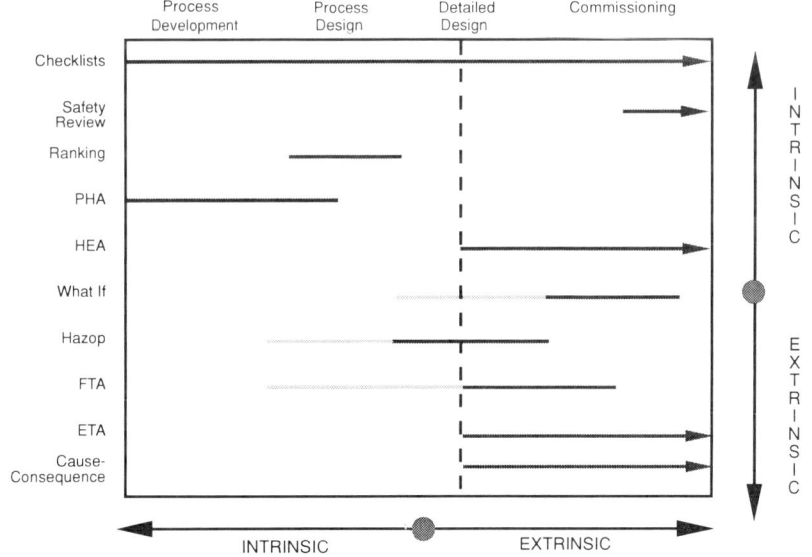

FIG. 1. Applicability of various hazards identification and analysis methodologies during various phases of the design process.

Lees (1980) stated that, "The safety of the plant is determined primarily by the quality of the basic design concept rather than by the addition of special safety features." This point cannot be overemphasized. The degree to which it is economic to eliminate, as opposed to control, a hazard is very dependent on when the hazard is recognized. By the time the design has reached the stage of sufficient documentation to allow a detailed hazard identification, the flexibility to eliminate hazards entirely is very reduced. If hazard/error detection is to be shifted to predesign review, a less encumbered approach must be developed and integrated into the design process.

B. Incompleteness of Conventional Hazard Analysis Methodologies

The identification of potential hazards and the evaluation of their effects should be a continuing process from the conception of the processing scheme to plant shutdown and decommissioning. However, because conventional methods are limited by their scope and useful life span, it is difficult to integrate hazard analysis and evaluation into the design process. More importantly, all conventional methods are incomplete. Their ineffectiveness is inherent in their methodological approach and their limitations in capturing and utilizing all forms of available knowledge. These weaknesses result in two principal deficiencies: (1) the strength of analysis is dependent on the a priori identification of hazards and/or events leading to these hazards, and (2) they are unable to offer a reliable estimate of the quality of the design technology. The same weaknesses prevent the available methods from reaching an adequate balance among the following non-commensurable objectives: (a) early identification of hazards to avoid costly redesign or construction modifications, (b) postponement of evaluation to await more detail, and (c) avoidance of costly duplication of effort. Currently, there is no single solution to the problem. Multiple methods are used over the extended time period of the design process (development, construction, operation). Each of these methods suffers from particular deficiencies, which are born out from the specific design context that they were to service and the character of the approach that they have adopted.

Intrinsic methods, for example, although they tend to increase the quality of the process design, must generally be employed in the early engineering stages. The window of opportunity for their application is very brief since the incorporation of intrinsic safety features at a late stage in the design will usually require major design changes with adverse consequences on the cost and the schedule for the commissioning of the plant's operation. Moreover, they do not provide a creative search for new hazards when experience is lacking, nor can they provide quantification as to the quality of a particular design. Heuristics have been proposed to fortify intrinsic methods (Dale, 1987; Kletz, 1985). They are ad hoc pieces of knowledge, offering no metric of sound origin to discriminate among alternative design concepts. However, they can be suggestive as to a design's quality provided they are intelligently applied; blind usage often leads to unforeseen difficulties.

Extrinsic methods attempt to quantify the implications of a hazard's occurrence by beginning with a detailed design. Unfortunately, by the time a detailed design is available, it is often too late to avoid hazards. The control of the hazards through external, safety devices is the only economic alternative. In principle, controlling the effect of hazards leads to an acceptable solution, although there is no assurance that the envisioned control scheme will effectively mitigate all eventual outcomes. In fact, the fundamental flaw in any method that is based on controlling hazards lies in the assumption that they have the ability to both identify accurately and pinpoint precisely the location of a future hazard. Virtually all experience suggests that the contrary is true. Indeed, wherever we suspect the possibility for the initiation of a hazard, we take added precautions to add control and safety devices in order to reinforce the assurance that the hazard will not occur. Consequently, conventional methods are approximations to hazard analysis and assessment. They are incomplete because the set of axioms they use and the deductive methodologies they employ are both incomplete. Furthermore, their performance and effectiveness is fundamentally limited due to their lack of expressive power. Conventional methods avoid the use of models and seek solutions from techniques that do not have firm chemical engineering foundations. What is needed is a formal, unified approach for systematically, automatically, and completely identifying hazards and pathways leading to hazards in design alternatives —at all stages of the design process. But how does one design such a methodology?

C. PREMISES OF TRADITIONAL APPROACHES

The weaknesses of the traditional approaches, as discussed above, are all due to their inherent *lack of representational expressiveness*. This shortcoming can manifest itself in the following ways: (1) a methodology exhibits strength only in a particular design phase, where it captures and uses most of the knowledge available during that phase of the design process; (2) a methodology is incapable of transferring information derived at one phase to another phase of the design process, nor can it reason about the chemical process and its surrounding environment; (3) discrimination among design alternatives varies depending on the technique employed; and (4) complete identification of hazards cannot be guaranteed and therefore the quality of the analysis cannot be assessed. The fundamental premise of traditional approaches is that *"in hazards analysis there is no unifying theme"*; all hazards are different and that every unreliable design creates hazardous situations in its own way. The stand-alone nature of previous approaches is evidence of this statement.

We have found that the truth is somewhat different: *representation limitations of past efforts have prevented the exploitation of unifying themes.* Moreover, conventional methodologies do not include for use, or produce *explicit* information, such as (1) *underlying assumptions* on operational modes, phase equilibria, fluid mechanical aspects, reaction rates, mechanisms of mass and heat transfer, and selected approaches in estimating the value of thermodynamic properties; (2) *simplifications* made by the analyst to limit the model's validity over a given range of conditions or to underscore the relative importance of various physicochemical phenomena; (3) *missing relationships* including qualitative relationships, order-of-magnitude reasoning, and inequalities; and (4) *scope of the task*, that is, what was the process intended for. Difficulties are encountered in the identification and elucidation of root causes leading to the top-level event because their basis is neither concise nor explicit.

Efforts in engineering science reinforce these observations (Stephanopoulos, 1987). Many tasks are not achievable if representational expressivity isn't sufficiently rich to allow the description of the necessary concepts (Brachman and Levesque, 1985). Concepts must be manipulated directly if powerful reasoning is to be achieved. The success of any advanced computer-aided tool for enhancing the identification of hazards requires: (1) the development of a representational language sufficiently rich to embody advanced scientific concepts and (2) a means for manipulating these concepts and reasoning about them, directly.

D. OVERVIEW OF PROPOSED METHODOLOGY

In this chapter we will attempt to describe a concise framework for the development of hazards identification and analysis techniques. It is based on the following premises:

1. Every top-level hazardous event has been initiated by a physicochemical reaction.
2. A physicochemical reaction initiating a top-level event is determined solely by the type, chemical reactivity, and physical properties of the materials involved.

3. Every top-level event has been activated by the logical satisfaction of all its physicochemical precursor conditions.
4. The satisfaction of the precursor conditions in premise 3 depends on the structure and the design characteristics of the particular process, as well as its operating conditions.

The implications of these four stipulations are very important and determine the logic, character, and implementation of the hazards' identification and analysis approach discussed in this chapter. Specifically, we can conclude the following:

- Premises 1 and 2 (above) imply that an *inductive* generation of *all* physical and chemical reactions leading to potential releases or generations of uncontainable amounts of energy or mass, or both, per unit time is the essential foundation for the complete identification of all potential hazards. Such task *depends entirely on the chemical behavior of the materials/species involved and not on the structure of a processing scheme, or the operating conditions of the associated plant.*
- Premises 3 and 4 (above) imply that the design or operating modifications of a process leading to the elimination or containment of hazards (identified by the inductive approach) can be generated *deductively* from the knowledge of the plant and its operating conditions.

Observing the above logic, we have organized the material of the subsequent sections in this chapter as follows: Section II provides the essential foundation for the reaction-based identification of hazardous top-level events, while Section III describes the inductive reasoning used to identify these hazardous events from the set of available materials. Both of these two sections draw heavily from the material of the first chapter in this volume and specifically from the LCR, the modeling language developed by the authors to describe chemicals and their chemical reactivity. Section IV outlines the methodological framework for the deductive determination of the causes of hazards, using the available knowledge for the description of the plant's design and operating conditions. A fairly detailed example provides an illustration of the ideas and techniques discussed in Section IV.

II. Reaction-Based Hazards Identification

We present a new approach to hazard analysis and assessment, one that begins with the inductive determination of hazardous chemical or physical reactions and proceeds deductively through the network of processing steps to identify completely all causes initiating these hazardous reactions. This strategy is *more efficient* in the identification of reaction-based hazards than conventional methodologies. The approach ensures *completeness* and resolves uncertainty of design quality through first-principles-based quantification of the design risk. Furthermore, it enables a computer-aided automation. The methodology is based on two fundamental postulates that formalize and extend Bretherick's observation: "With the exception of releases of toxic or corrosive material, all accidents in storage, handling, or processing of chemicals involve the release of energy at rates too high to be dissipated in the immediate environment without damage." (Bretherick, 1990). These two postulates establish the theoretical framework that enables the design of a system for automatic identification and analysis of hazards in a decidable manner. By unifying the analysis of potential hazards, the framework can utilize the maximum knowledge available at every point in the design process. Regardless of what stage the design is in, a designer's attention can be focused at (1) earlier design stages and their vulnerable areas so that the associated hazards can be eliminated or (2) later stages, where the a priori sequence of events that leads to hazards can be used as an early warning structure (Lees, 1983).

The methodology employs domain-specific modeling languages (see first chapter in this volume) to describe

(a) Chemicals and their reactivity during the *inductive* identification of potential reaction-based hazards (see the modeling language LCR in first chapter in this volume).
(b) Processes during the *deductive* identification of their process-based causes (see MODEL.LA. in first chapter).

A. SYSTEM FOUNDATIONS

We now establish the *criterion for completeness* for any hazard identification methodology. [Herein, completeness refers to the ability of a methodology to identify all possible top-level events (Nagel, 1991.)] The methodology presented establishes how we use this criterion to completely and systematically identify hazards in an *efficient* manner.

Theorem. *Any hazard identification methodology must cover all subsets of sources present in a process network in order to guarantee completeness.*

In the discussion that follows we restrict our attention to hazards whose enabling path involves reactions among the chemical species, present in the process. This restriction excludes, for example, hazards (i.e., top-level events) created by falls, electrical shock, or impact with stationary objects, since there is no finite set of root causes leading to these hazards, thus making impossible the guaranteeing of completeness for hazards of this type. But, the preceding restriction does not exclude hazards initiated by these processes, provided they cause interactions among the chemical species present in the plant. For example, a static charge initiating a chemical reaction or the overheating of a tank's contents leading to its volatilization and subsequent over-pressuring, would be covered. Our approach is based on two fundamental postulates.

Postulate 1. *Hazards can only be created by the interaction of a system through its boundaries, or the altering of internal restraints of the system, such that the system proceeds toward a new equilibrium state.*

This postulate follows naturally from the First and Second Laws of Thermodynamics. Since states at equilibrium have no irreversible interactions, a state change is required to proceed toward equilibrium; these changes can only be brought about by interactions through a system's boundary or the altering of internal constraints. Therefore, a state change must accompany a hazard or the system would be at equilibrium. Although many nonequilibrium states may be transgressed in the development of a hazardous system, we need only consider the equilibrium states to identify potentially hazardous systems. This is possible because state changes can only be brought about by energy and entropy changes. Since these are state functions, an upper limit can be established on the potential of a hazardous state through the analysis of the equilibrium states leading to it. Our goal in inductive identification of hazards focuses on the identification of these states. Since any equilibrium state can be characterized completely by $(n + 2)$ variables, where n represents the masses of the particular chemical species initially charged and 2 represents two independent variable properties (e.g., temperature and pressure), our task is divided into

- The identification of all chemical species sets, which can potentially be present in the process.
- The identification of the independent variable properties that specify the environment of the chemical species.

The latter requires the identification of the processing environment, whereas the former establishes the need to identify the occurrence of all potential reactions. The inductive generation of all potential reactions will be discussed in detail in Section III. Whereas postulate 1 establishes the framework for inductively identifying hazardous states, postulate 2 advances the mechanism by which the pathway of events leading to a hazardous state can be identified deductively.

Postulate 2. *The degree of completeness of the set of internal restraints and exogenous factors specifying a hazardous system and its transformation, determines the degree of completeness with which the pathways leading to a hazardous state can be identified.*

Moreover, postulate 2 suggests that the deductive identification process is restricted by our understanding of mechanisms that allow states comprising the hazardous system to interact. Since any interaction results in an energy or entropy change of these states, the opportunities for preventing a hazard are limited by the incompleteness of our knowledge to determine the restraints and the external variables that specify these states. These in turn are more limited by the quality of the available details in describing the particular processing system (Battelle, 1985). However, there may not be a finite set of root causes leading to a hazardous state, or the identification of a finite set of causes may not be possible. As a consequence, the pathway of precursor events leading to a hazardous state is incomplete by definition. Despite this limitation, the constitutive equations that promote a hazardous state can be used to optimize the inherent safety of a particular design technology. The combined approach presented in this chapter allows the following:

1. Establishment of a systematic and formal methodology for increasing the inherent safety of a design technology.
2. A means for quantitatively assessing the inherent safety of design alternatives.
3. Establishment of a systematic and formal strategy for optimizing the inherent safety of a design technology through the selection of appropriate control points.
4. Guarantees on the correctness and completeness of the pathways leading to top-level events.

B. Modeling Languages and Their Role in Hazards Identification

Any methodology, whether it is applied in an automatic or manual manner, requires a rich representation of the process if hazards are to be effectively identified. This representation must have sufficiently expressive power to allow chains of precursor states or events to be easily identified. To satisfy these requirements, we have chosen to map the process description into a representational form that allows a multi-level, multi-faceted process description (see the modeling language MODEL.LA. in first chapter in this volume).

Research efforts outside the domain of hazards analysis have established that expressive, fully declarative, domain-rich modeling languages are indispensable, if complex integrated systems capable of synthetic tasks are to developed. Furthermore, they have shown that computer-aided systems with advanced reasoning capabilities require the satisfaction of the following three conditions:

- All declarative models should be fully articulated.
- Declarative knowledge should be completely decoupled from the procedural knowledge.
- A modeling language should be rich with domain-specific knowledge and thus allow the user to think about the task at hand in terms that are familiar (Stephanopoulos *et al.*, 1990a, b; Kritikos, 1991).
- For a computer-aided chemical reasoning system, we add a fourth condition. The logical structure of chemistry should be exploited by imbedding natural constraints into the declarative models describing chemicals and chemical reactivity.

The system described herein for the automatic identification of hazards and the pathways leading to them utilizes two modeling languages which satisfy the above requirements. Specifically, we have used

- LCR (language for chemical reasoning; see first chapter in this volume) to describe chemicals and their structure and atomic and molecular properties, as well as chemical reactions and their structure, directionality and contextual character.
- MODEL.LA. (modeling language; see first chapter in this volume) to describe processing systems and their unit operations and behavior, and to encapsulate the design decisions and operating conditions associated with any specific plant.

These languages are not ad hoc constructs but are based on clear formalisms and satisfy the essential premises of grammar, vocabulary, and semantics on which programming languages are founded. Both modeling languages allow multilevel description of processes, reactions, and materials with internal consistency (logical and quantitative) among the various models defining the corresponding objects at various levels of abstraction. Moreover, they support the representation of materials, reactions, and processes at multiple, coexisting contexts, an explicit comparison of the contextually alternative representations, and a systematic backtracking of decisions and assumptions that led the specific descriptions. Thus, different perspectives of the process, reactions, and materials—such as structural, topological, and physicochemical relationships—can be investigated independently. Both of these languages allow the following operations on processes, reactions, and chemicals:

- Multiple viewing in terms of structure, topology, and behavior
- Disaggregation of abstract descriptions to more detailed ones and aggregation of detailed descriptions to more abstract
- Contextual description of alternative models for the same process
- Controlled flow of information among the models at various levels or in various contexts, and detection of modeling conflicts
- Propagation of qualitative and quantitative knowledge through the defining models of behavior

Utilizing the specialized modeling language, MODEL.LA. (Stephanopoulos et al., 1990a, b; see also first chapter in this volume), we can construct a process description that is suitable for the tasks of hazards analysis or/ and identification. The process description we have employed is an abstraction of the conventional process representation (centered around the topology of a specific network of processing units). It is built on top of the conventional representation and thus provides complete access to the functionality contained in the base representation. This functionality comes with a set of tools that allow us to solve heat and mass balances, identify work interactions, evaluate phase partitioning, etc., and permits the propagation of qualitative and quantitative knowledge through the defining network. By focusing our representation on the thermodynamic state description of the process, we can facilitate the identification of pathways leading to potential hazards in the most efficient manner. The equipment state space (i.e., the process flowsheet representation) can be mapped to a thermodynamic state space, if we know the trajectory of thermodynamic states, comprising a process, combined with the transformations that connect these states. The mapping process that allows us to construct a thermodynamic state-based graph representation of a process flowsheet focuses on the state description of the process. The representation is composed of *streams* and *nodes* in accordance with the following definitions:

- *Streams* are idealized flows that connect nodes, and they have no accumulation or energy losses.
- *Nodes* are identified as points where discrete thermodynamic transformations occur, the result of changes in intensive or extensive variable values.

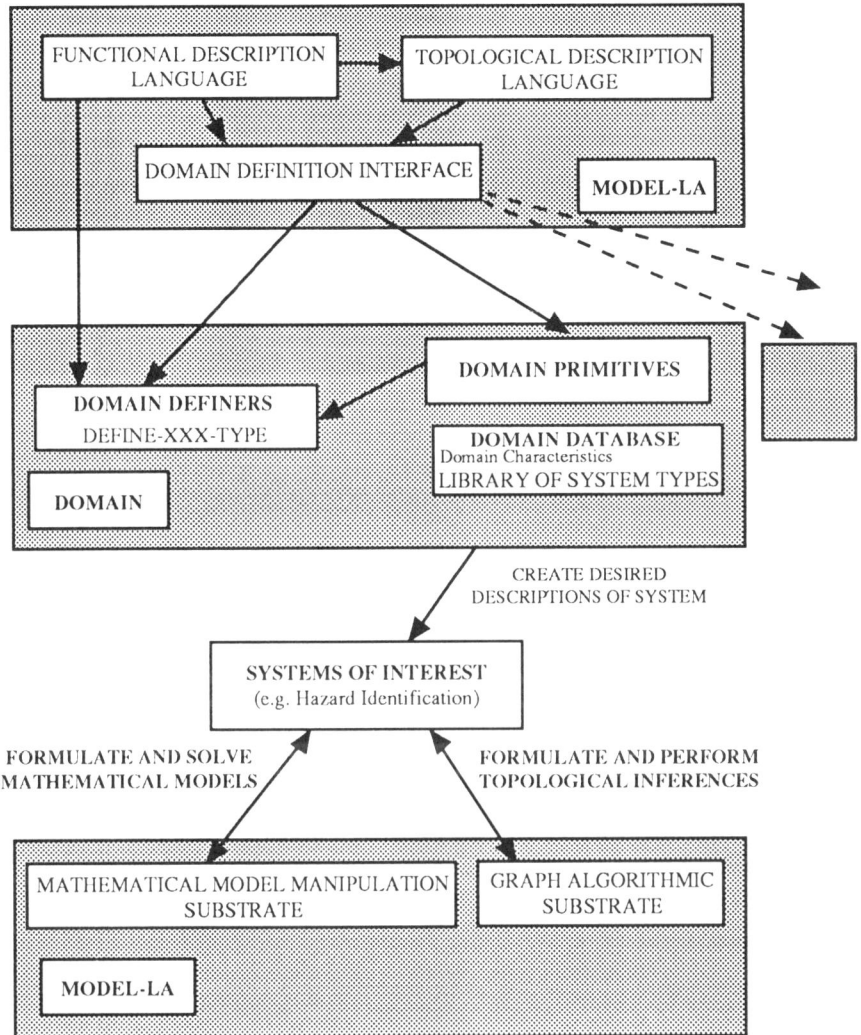

FIG. 2. Overall organization and use of the modeling language MODEL.LA.

These definitions are used to abstract the underlying process representation. As shown in Fig. 2, MODEL.LA. allows us to construct an abstract representation utilizing these definitions for our specific application from a conventional process description. This abstract representation has access to any utility, tool, or component that constitutes the base representation,

```
identifier:            unbound
element-name:          #<state-1>
element-type           composite-state
chemical-species-set:  unbound
operating-conditions:  #<op-cond-1>
boundary-elements:     (#<flow-port-3> ... #<heat-transfer-port-5>)
connecting-states:     (#<state-2>  #<state-3>)
system-description:    #<node-1>
system-volume:         unbound
```

FIG. 3. The attributes of the modeling object, *state*.

e.g.,

- The topology of the network is accessed through the graphical construction of the base representation.
- Mathematical relationships such as phase relations, energy balances, mass balances, and reaction or transport rates can be accessed through the content of the instances of objects of type ***relationships*** (see first chapter in this volume).

Knowledge about the abstract representation and the mapping process are contained in the modeling element, *state*. Figure 3 identifies several of the attributes that describe a state. Notice that the state is a composite object, i.e., an object comprised of objects. (Herein objects bound to attributes are denoted by #⟨NAME⟩ or will appear in bold.) This allows a multilevel, hierarchical representation of a process to be constructed. For example, the operating conditions that specify a state are accessed through the object ***op-cond-1***, which is the attribute value of *operating-conditions*. Attributes of ***op-cond-1*** include the unit with which it is associated as well as the temperature, pressure, flowrate, and composition associated with that unit and their respective interval ranges.

The chemical species set of a state is the set of chemical constituents that are associated with the system description of that state. The values of the attributes *chemical-species-set*, *operating-conditions*, and *system-volume* provide the $(n + 2)$ independent variable quantities that are necessary to define a thermodynamic state. An important feature of this representation is that each state is described by a vector of *intensive* and *extensive* variables. The intensive vector defines the operational state of the process, while the extensive vector defines the maximum accumulation of mass and energy that can occur. This is bounded by flowrate, reaction rate and physical size of the process equipment. The values of these variables are accessed through the attributes: interval flowrate vector, interval accumulation, and system description. The values bounding an interval are dynamically set. (Intervals are defined as the minimum and maximum allowable value that the variable being described can achieve.)

Boundary elements and connecting states are also associated with the state description. Since a state is described as a system and a system has a boundary, boundary elements identify the vehicle for transforming the current state to a new state. The trajectory of connecting states is contained in the value of the attribute, *connecting-states*. This attribute allows a thermodynamic state representation of a process to be constructed from the individual states that compose it.

Alternatively, an equilibrium state in which there is no net effect on system boundaries can still be transformed to a new state when the internal restraints that specify the system are altered. This information, as well as the mapping that takes from an equipment based representation to a state-based representation, is contained in the value of the attribute *system-description*. An expansion of this state attribute value provides access to the base representation through the description of the modeling object, **node**. Figure 4 identifies several of the attributes describing the modeling element, **node**. This detailed description allows us to identify system boundary partitions: flow openings, heat transfer openings, work exchange openings, and mass transfer openings. Additionally, the system boundary type is specified as a thermodynamic boundary. The topological system type and topological connector type provide the links needed to connect the various nodes. Each of these elements has a description that allows us to evaluate its functionality. In addition, application specific handlers are also provided by the underlying modeling language. For example, the boundary element in the attribute, *flow-opening*, may have a description that requires an understanding of vapor-phase, liquid-phase, solid-phase, or multiphase transport. Each of these handlers in turn may access additional methods or handlers to facilitate the evaluation process.

The value of the attribute *system-interior-type* specifies the interior system description of the **node**. The value of this attribute allows us to map the node representation to the process equipment representation. Notice that it is the attributes of the **node** in combination with the features of the underlying specialized modeling language that afford a multilevel, multifaceted view of the process. Hence, we are able to move independently from the node representation to the equipment representation and pursue, as needed, analysis of momentum, mass, and energy interactions of the defining network. It is clear that using the elements of the modeling language, LCR, we also have access to the representation specifying the chemical constituents that are associated with the state. In the sections

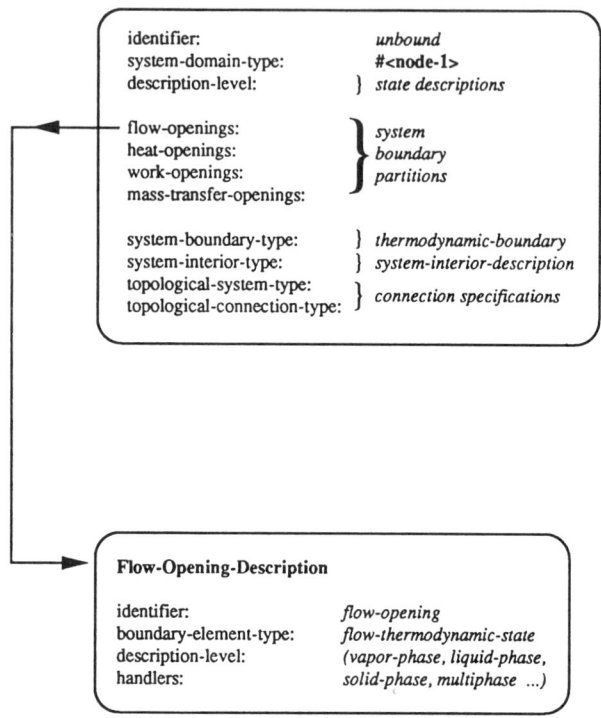

FIG. 4. Select attributes of the modeling objects, *node*, and *flow-opening-description*.

that follow, we will show how the information contained in the equipment-based description of a process through the use of MODEL.LA., is accessed by LCR and the chemical species in order to assist in the generation and evaluation of potential reaction alternatives.

An example of a mapping from the equipment representation to the thermodynamic state representation is shown in Fig. 5. It represents an isothermal vertical packed-bed catalytic reactor equipped with temperature and pressure sensors, an explosion vent, and a distributor plate. Notice that the equipment and sensors are not associated with the state representation. They are contained in the base representation and reside in the process description at the equipment level. As discussed earlier, flow, work, heat, and mass interactions are all modeled independently. This allows us to evaluate independently the effect of these processes. Independent evaluation assists in the identification, evaluation, and assessment of event pathways leading to hazardous states.

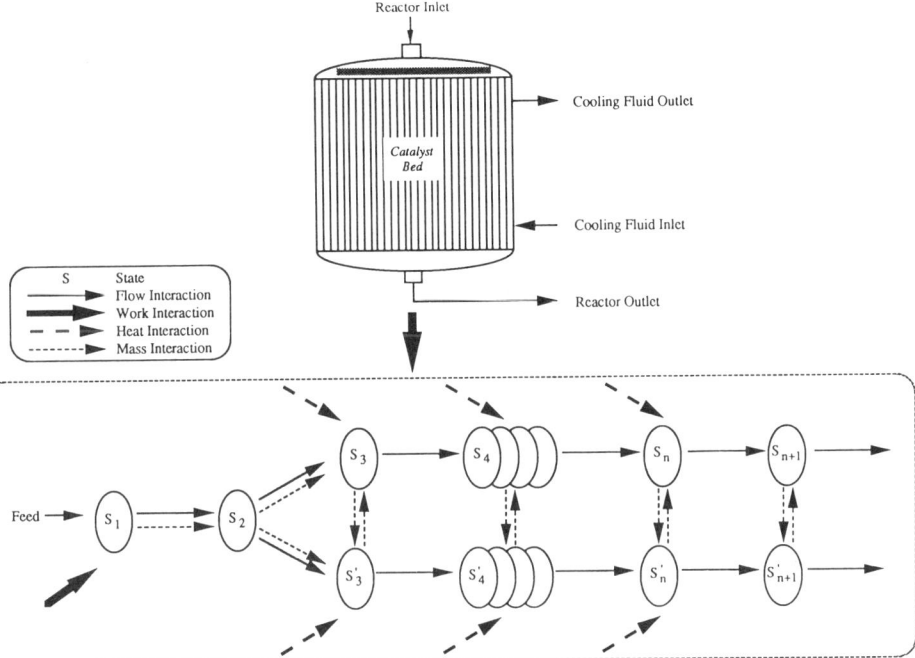

FIG. 5. Mapping of equipment into a thermodynamic state space.

C. Generation of Reactions and Evaluation of Thermodynamic States

Since new thermodynamic states are created by reactions, it is required that we have a strategy for generating potential reactions between species resident in a particular node of the process description. To describe the reacting species and the reactions they undergo as well as the changes in their molecular structures, we use the language, LCR, which is presented in the first chapter in this volume (for a detailed description of the language, see Nagel, 1991). The language is used extensively in the inductive identification of reaction-based hazards. LCR provides a utility for generating reactions and identifying the resulting pathways, given a set of substrates and a reaction environment. The procedure that provides this utility and the keyword arguments (i.e., arguments with a colon prefix) accepted by this method are given below:

(FIND-ALL-PATHWAYS   :substrates   :operators
                      :override-environment :initiator-p)

Let us examine the character of the keywords:

- *:substrates*, is the argument that contains the list of chemicals, which are available at a particular process node.
- *:operators*, is the keyword argument that allows the user to specify the types of transformations to be used in order to focus the generation of reaction pathways. To investigate all theoretically feasible pathways subject to encoded preferences the user may supply the keyword argument *:operators* with the value, **K***, a modified composite operator, and allow the generation and evaluation of all pathways having prespecified features ($\Delta G < 10$ kcal/mol, stereocenters, etc.). Similarly, if there are no preferences, all theoretically feasible reactions can be generated by calling directly FIND-ALL-PATHWAYS with the argument *:operators* having the value $\mathbf{K}_{\text{ab-initio}}$ (Nagel, 1991).
- *:override-environment*, is the argument that lists the operating conditions in which the alternative reaction pathways will be generated. It allows the user (or an automatic procedure) to set alternative operating environments and generate the corresponding alternative reaction pathways.
- *initiator-p*, determines whether an initiator of specific chemical reactions is present or not. It allows the user (or an automatic procedure) to investigate various reaction trajectories without knowing the specifics concerning the mechanisms for initiating reaction pathways.

Using this representation, we can now focus on the identification of the associated thermodynamic states. We assess the likelihood of reactions as well as the inherent instability of each species associated with a node using LCR. The generation of infeasible species is limitated by the following values (i.e., knowledge) embedded in the corresponding attributes;

:override-environment = **reaction-environment**
:operators = **K**, **K***, or $\mathbf{K}_{\text{ab initio}}$.

Let us examine the generation of alternative reactions within the scope of the operating conditions associated with a particular node (or, its corresponding state). For example, suppose that the procedure FIND-ALL-PATHWAYS is applied to a chemical species set (CSS) (i.e., the set of chemicals bound to a specific **state**) composed of three chemicals, **A**, **B**, and **C**, each

Reaction Object
   identifier: *unbound*
   name: *initiation-1*
   reactants: (# <Cl_2>)
   products: (#<Cl> #<Cl>)
   stoichiometry: ((#<Cl_2> . -1)
                                       (#<Cl> . 2))
   reaction-environment: #<reaction-environment-1>
   enabling-conditions: K_f
   composing-transformations: K_t
   composing-reactions: *unbound*
   rate-expression: #<rate-expression-1>
   equilibrium-constant: #<equilibrium-constant-1>
   context: *unbound*

FIG. 6. Description of the modeling object, *reaction*.

of which is described by an instance of the modeling object, **atom-bond-configuration** (see first chapter). The procedure generates the following set of potentially reactive mixtures:

$$A \rightarrow ? \qquad (I)$$

$$B \rightarrow ? \qquad (II)$$

$$C \rightarrow ? \qquad (III)$$

$$A + B \rightarrow ? \qquad (IV)$$

$$A + C \rightarrow ? \qquad (V)$$

$$B + C \rightarrow ? \qquad (VI)$$

$$A + B + C \rightarrow ? \qquad (VII)$$

Unique products generated by these reactions are added to the CSS

Pathway Object
   identifier: *unbound*
   name: *pathway-1*
   reactants: (# <C_2H_6O> #<Cl_2>)
   products: (#<$C_2H_5C_{10}$>)
   stoichiometry: *unbound*
   competing-pathways: (<pathway-2>
                                       #<pathway-3>)
   composing-reactions: (#<initiation-1>
                                       #<abstraction-1>
                                       #<combination-1> ...)
   global-rate-expression: #<composite-rate-exp-1>
   global-equilibrium-constant: *unbound*

FIG. 7. Description of the modeling object, *pathway*.

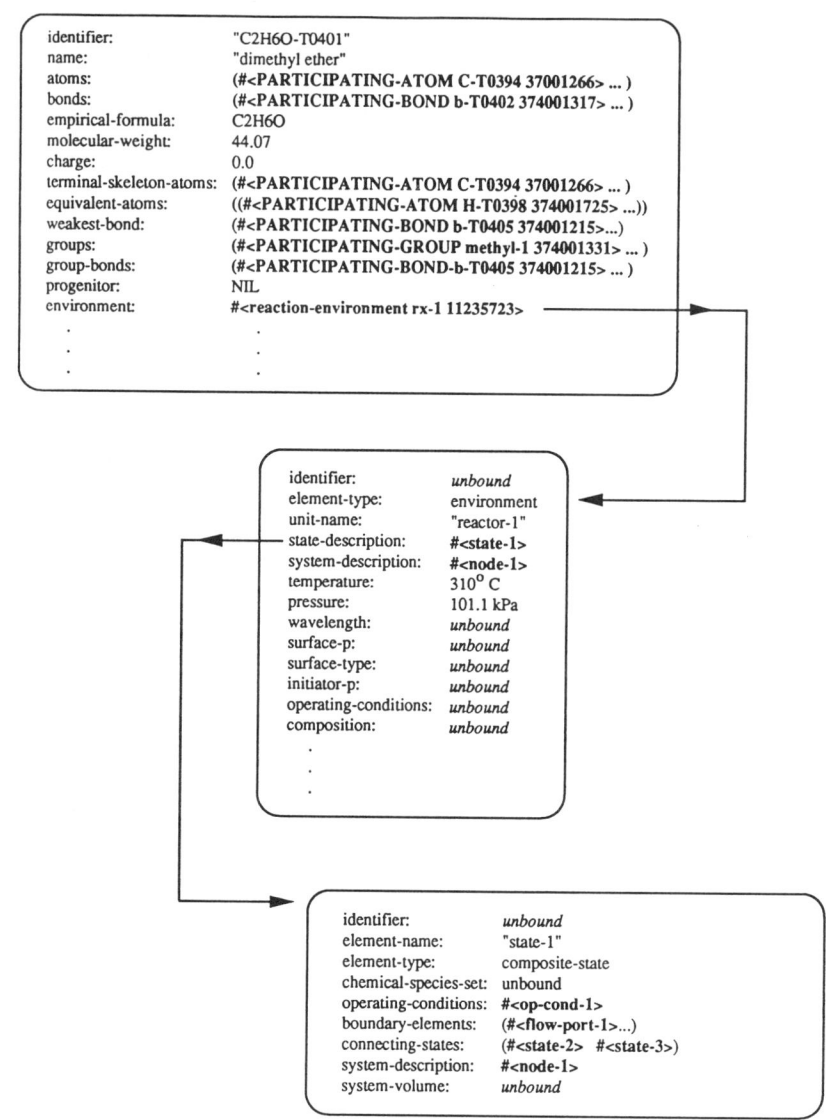

FIG. 8. Accessing process representation from chemical representation.

and the process is repeated. A single call to FIND-ALL-PATHWAYS

(FIND-ALL-PATHWAYS :*substrates* (**ABC**) :*operators* **K***)

generates all possible pathways.

LCR provides modeling objects to contain the information generated by these chemical transformations. For example, the attributes describing the modeling object **reaction** are shown in Fig. 6. These attributes describe not only the reactants and products of the reaction but also contain information on the reaction environment, enabling conditions, composing transformations, reaction stoichiometry, equilibrium constant, rate expression, and identification of competing reactions. Similarly, Fig. 7 shows the attributes describing the object **pathway**, which is made of a network of one-step reactions. Whether any of the above potential reactions, e.g., (I) through (VII), will proceed and develop a hazardous event or not, depends on the value of the operating conditions that characterize the **state** of the corresponding **node**. Therefore, it is imperative that a link be established between the description of the process operations and the **atom-bond-configuration** encoding the information about the available chemicals.

The **atom-bond-configuration** object provides direct access to the **state** description of a **node** through the attribute, :environment. Figure 8 shows how the value of the "state-1," describing the conditions in "reactor-1," is transferred to the instance of the **atom-bond-configuration** describing the chemical reactant, "dimethyl ether." Such links allow the transfer of information among different modeling objects in LCR; a mechanism driven by the semantic relationships of LCR (see First chapter).

In a similar manner, the semantic relationships among the various modeling objects of MODEL.LA. allow the transfer of information among these objects. When used together, LCR and MODEL.LA. *allow functional and topological information derived at one point of the process to be accessible from any other point in the process.*

III. Inductive Identification of Reaction-Based Hazards

Using these tools and the representations that are built with them, potential hazards are inductively identified by combining various chemical and physical environments of the process in an attempt to generate new process states. The types and numbers of process states are determined by both the process configuration and the operating conditions. Enabling

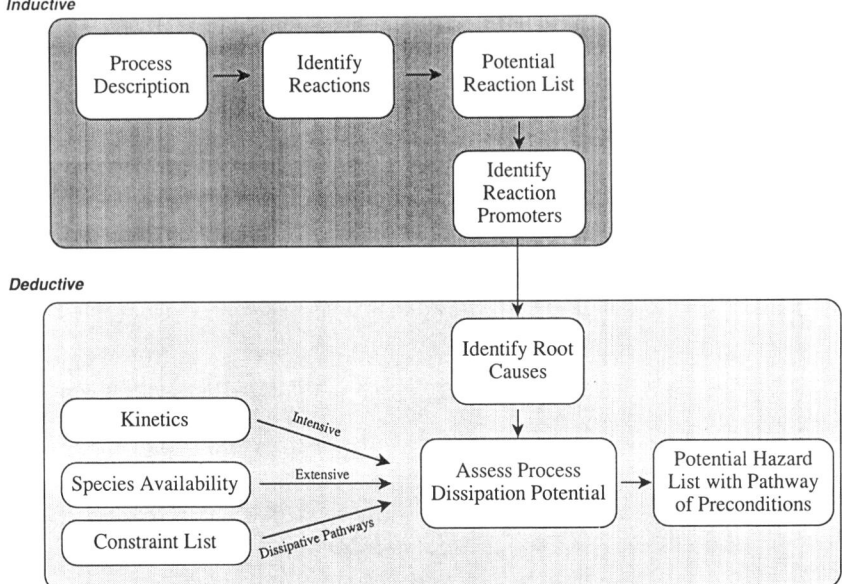

Fig. 9. Overview and implementational elements of the methodology for the identification and analysis of hazards.

conditions of a potential hazard are identified by establishing the conditions specifying the thermodynamic states that precede it. These conditions can be either intended or induced, resulting from changing boundary elements, system descriptions, or reacting states. Normally, they manifest themselves as procedural, material, or boundary changes.

A potential hazard is said to exist when the environment or a sequence of environments cannot be dissipated or prevented by the process. Using this environment as a goal state, we can then deductively identify the sequence of process and/or operational changes that led to its creation. Thus, the root causes and their temporal preconditions (whether they are equipment and/or operational failures) can be identified and associated with the corresponding hazard. Evaluation of the pathways leading to the specific hazard allows the quantification of risk and permits a concrete assessment of the potential hazard in the context of a given design technology.

Information flow during the execution of the hazards identification procedure is shown in Fig. 9. It is composed of two phases:

Phase 1. Inductive identification of reactive states and their assessment as potential hazards.

Phase 2. Deductive identification of pathways leading to those states.

The inductive phase begins with a process description from which reactions are identified using LCR, and the tools described earlier, which allow the identification of potential chemical reacting states. Once all the potential chemical transformations have been identified and the entire chemical species set has been elucidated, a **node** can be fully described and the **states** disseminating from it can be established (i.e., the $n + 2$ independent variable quantities are known). With this description potential physical reactions (e.g., overpressurizing, overheating, overcooling) can be elucidated using classical thermodynamic (open- and closed-system treatment) and transport phenomena analysis around the specific **node**.

Potential physical reactions can only be identified with any assurance of completeness *after* the complete specification of the chemical species set, CSS. This is a result of Postulate 1, $(n + 2)$ *defining variables must be known before a thermodynamic state can be specified*. Since incomplete specification of "n" leads to incomplete specification of potential states, a complete node description is required before physical reactions can be identified completely.

Once potential reactions are identified, they are posted on a potential reaction list. The promoters of each reaction are then identified by tracing the variable values that led to their creation. This is achieved through the links provided in the thermodynamic representation of the **state** and the **atom-bond-configuration** modeling objects. Similarly, by accessing the information contained in K, K^* or $K_{ab\ initio}$, we can also identify the conditions that promoted a particular chemical reaction. These conditions may be chemically induced, as in the case of nucleophilicity, or physically induced, as in the case of high-energy environments enabling radical or photochemistry, or both. These promoters are then used to establish the underlying pathways that enable the top-level event.

In this combined approach, the inductive identification of top-level events and the deductive identification of enabling pathways allow potential hazards and the sequence of events leading to them to be identified completely within the scope of the modeling effort. Moreover, since the top-level event is generated from its underlying states, the pathway of preconditions and the temporal ordering of those preconditions are explicitly spanning only the minimum cut set (see Nagel, 1991).

A. Hazards Identification Algorithm

The algorithms used in this approach are described below using the pseudocode conventions found in Cormen *et al.* (1990). Therein, a *terminal node* is defined as the "first node a feed enters or the last node a product or byproduct exits" in the process flowsheet. The routine calls for two loops, which is necessary to simulate multiple simultaneously occurring boundary failures.

Algorithm 1 ⟨GLOBAL-HAZARD-IDENTIFICATION⟩

  input: process-flowsheet
  initialize
    process-node-representation ← **apply** MAKE-PROCESS-TRANSFORMATION
        to process-flowsheet
    terminal-nodes ← **apply** FIND-ALL-TERMINAL-NODES to process-node-
        representation
    potential-hazard-list ← nil
  for each node in process-node-representation
    potential-hazard-list ← **append** (**apply** IDENTIFY-POTENTIAL-HAZARD
        to (node process-flowsheet))
        to potential-hazard-list
  return
  for each node in terminal-nodes
    expand node-scope
      until UNIQUE-STATE-IDENTIFIED-P   ;predicate test to identify
        unique states
      or EXPANSION-NOT-POSSIBLE   ;empty node-scope
        (i.e., terminal node
        is reached)
    then
      potential-hazard-list ← **append** (**apply** IDENTIFY-POTENTIAL-HAZARD
        to (node process-flowsheet))
        to potential-hazard-list
    return
  end
  return

The internal routine, IDENTIFY-POTENTIAL-HAZARD, called from GLOBAL-HAZARD-IDENTIFICATION, returns a potential hazard and the list of the enabling conditions. The algorithm for IDENTIFY-POTENTIAL-HAZARD, expressed in pseudocode, is given below:

Algorithm 2 ⟨IDENTIFY-POTENTIAL-HAZARD⟩

  input: node   ;starting point-hazards are associated
                          with nodes
      process-flowsheet ;context of the node
  initialize

```
potential-reaction-list ← nil
associated-sates ← nil
extended-CSS ← nil
potential-reaction-list ← apply FIND-ALL-PATHWAYS
                          TO CSS of node; identify
                                          potential
                                          chemical
                                          reactions
extended-chemical-species-set ← collect UNIQUE-CHEMICAL-
                                          CONSTITUENTS from
                                          potential-reaction-list
associated-states ← apply EVALUATE-STATES
   extended-CSS      to node and ;identify states
                                          associated with each
                                          node due to reaction
for each state in associated-states
   when UNIQUE-STATE-DESCRIPTION-P
      potential-reaction-list ← apply EVALUATE-PHYSICAL-
                                          REACTIONS to state
      return
   return
for each reaction in potential-reaction-list
   enabling-criteria ← collect FIND-ENABLING-CRITERIA
   classified-influence-paths ← apply CONSTRUCT-VARIABLE-INFLUENCE-
                                          PATHWAYS to (enabling-criteria process-
                                          flowsheet)
   root-causes ← apply IDENTIFY-NONDISSIPATIVE-PATHWAYS to classified-
                          influence-paths
   when root-causes
      potential-hazard ← list reaction enabling-criteria root-causes
      return
   return
end
return
```

Notice the interplay between the methodology and the specialized modeling languages and the importance of that interplay. These languages enable the methodological approach. For example, FIND-ALL-PATHWAYS applied to a chemical species set (CSS) utilizes the semantic relationships of LCR to construct the *potential-reaction-list*. Similarly, the representation utilized by LCR allows enabling-criteria to be explicitly associated with each reaction; these criteria can come from the **state** representation or from the inherent physicochemical properties associated with the chemical constituents themselves. Likewise, EVALUATE-PHYSICAL-REACTION utilizes functionality of the base representation to assess possible effects and to identify enabling criteria. These criteria come from the operating environment of the process equipment.

By interweaving the semantic relationships of the two modeling languages throughout the methodology, we are able to interplay the respective representations to satisfy the representational needs of the task at hand. For example, root causes are identified by propagating (i.e., solving) the enabling criteria through the network of equations lying in the base representation and defining the behavior of the overall process flowsheet. These criteria, however, are located in different representations, at different abstraction levels, to afford the resolution necessary to identify potentially hazardous conditions. The specialized modeling languages provide the tools to map between different representational forms and identify potential hazards with their enabling conditions and underlying root causes in an efficient manner.

Two additional features of Algorithm 1 are worth discussions:

- Interval values, describing ranges of operating conditions, are *dynamically* redefined whenever a condition is found to exceed the existing value range (values may be exceeded in the positive or negative direction).
- The description of a **node** (i.e., the scope of the defining system) can be expanded to simulate common boundary failure or process equipment malfunction.

B. Properties of Reaction-Based Hazards Identification

Let us now consider the relative efficiency of different methods for the identification of hazards (for details and proofs, see Nagel, 1991). Since it is meaningless to compare the efficiency of methods that are incomplete versus those that are complete, for the purposes of this section we consider only complete methodologies. But before beginning, we should point out one additional result that can be derived from the completeness analysis presented in Section II.B.

Corollary. *Equipment-based methodologies for hazards identification can be complete.*

An equipment-based methodology, e.g., HazOp Analysis, begins the identification of hazards with the postulation of specific process variables

INDUCTIVE AND DEDUCTIVE REASONING: IDENTIFYING HAZARDS

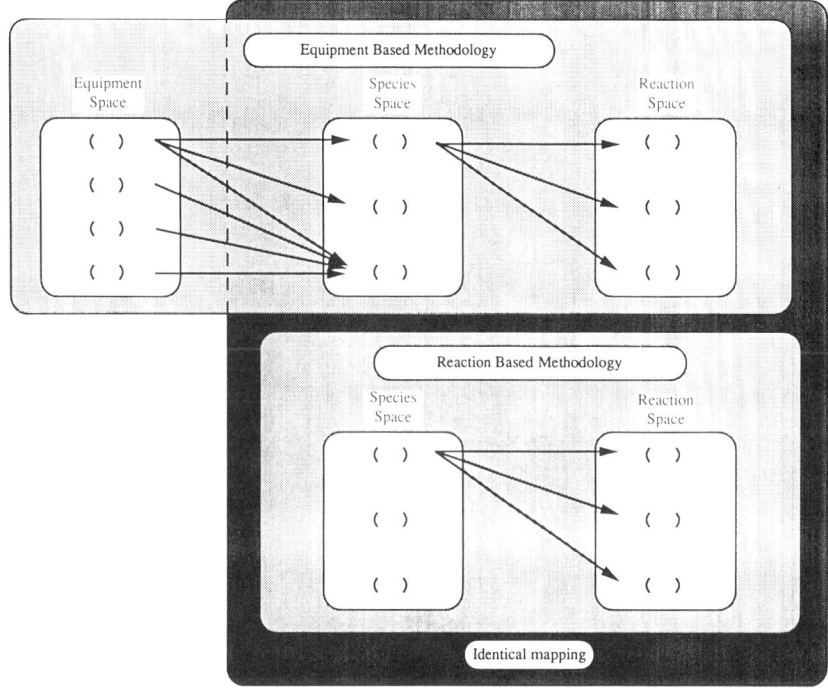

FIG. 10. Mapping of equipment space to species space.

deviations and/or equipment failures. From the point at which the complete chemical species set is generated, an equipment-based methodology is identical to a reaction-based methodology (see Fig. 10), such as the one described in this chapter. The implication of this statement is that the only information generated by the equipment analysis that is used in the identification of top-level events is the state space, which collapses to the species set since temperature and pressure are local variables.

Theorem. *No hazards analysis method exists that is both complete and whose running time is bounded by a polynomial in the number of initial species present in the plant.*

Theorem. *A reaction-based method is more efficient than any equipment-based method.*

The only difference between the equipment and reaction-based methods is in the generation of the chemical species set. Equipment-based methods generate the species set *indirectly* via the equipment state space; a reaction-based method generates it *directly*. Since both approaches must generate every element in order to supply a complete set of chemical species, the difference in efficiency between an equipment-based method and a reaction-based method must result from the fact that the former must generate the members of the chemical species set more than once, while the latter as we know does not. But, this assertion is trivially satisfied if the size of the equipment space is larger than the size of the species set space (see Nagel, 1991).

However, the equipment-based approach is also engineered to perform a second task, namely, *identification of root causes*. It can be shown (see Nagel, 1991) that if the number of chemical species present in the CSS is S_0, the possible number of subsets of species and thus the maximum number of possible reacting mixtures is 2^{S_0}. Consequently, when the complexity of an equipment-based methodology becomes larger than 2^{S_0}, the added complexity is an effect of the method's intention to also identify the root causes of hazards. Assuming that c_1 represents the number of additional guidewords in the selection of hazards, and c_2 the number of additional process parameters invoked by an equipment-based hazards identification method, then the resulting complexity of such method is given by

$$(2+c_1)^{(S_0+c_2)^E},$$

where E represents the pieces of equipment to be searched for the identification of hazards. For a mixture of 2 chemical species the corresponding search space of a method based on the chemical species (as the one presented in this chapter) is $2^2 = 4$. On the other hand, for a simple process with 2 pieces of equipment, $c_1 = 1$ and $c_2 = 2$, an equipment-based hazards identification technique must search a space of $3^{4^2} = 6561$ alternatives.

However, as shown in Section IV, completeness cannot be guaranteed in the identification of root causes by equipment-based techniques, since they cannot guarantee that all reacting mixtures can be identified, and thus not all possible thermodynamic states can be a priori identified. This is because guidewords and variable parameters (e.g., more, less) are designed to trace out *causes*, not generate thermodynamic states. This guarantee only comes when the resolution of the tracing mechanism completely defines the enabling state of the potential hazard.

For example, *to identify a potential hazard that is caused by changing the catalyst shape an equipment-based approach must include surface area as a process parameter to be searched; a piece of information embedded in the phenomenological kinetic rate expression and not explicitly available*. Thus, as a consequence of its formulation and the uncertainty of conditional operators, the equipment-based approach may not guarantee complete identification of potential hazards and cannot guarantee the complete identification of root causes leading to those hazards.

C. AN EXAMPLE IN REACTION-BASED HAZARD IDENTIFICATION: ANILINE PRODUCTION

In this section, we will demonstrate how the methodology can be used to identify hazards inductively from the set of possible chemical reactions. We will focus on the detection of potential hazards arising from changing operating conditions rather than equipment malfunction or failure; the latter is discussed in Section IV.

The ability to elucidate competing chemical pathways when changes in control strategies occur is of particular importance in hazard identification. An understanding of the relationship between the reaction pathway topography and design and operating characteristics allows us to mitigate or constrain the offending reaction trajectory. Consider, for example, the catalytic production of aniline from nitrobenzene and hydrogen. Limiting our attention to the reactor, named **reactor-1**, we see that the initialization routine of the procedure GLOBAL-HAZARD-IDENTIFICATION transforms the reactor into a single terminal process node. The identification of hazards within this **node** is carried out by applying the procedure IDENTIFY-POTENTIAL-HAZARD, to this **node**. The procedure, IDENTIFY-POTENTIAL-HAZARD, begins by applying the procedure FIND-ALL-PATHWAYS to the chemical species set. This set, which is bound to the **node**, contains the known chemical species, i.e., CSS = {*nitrobenzene, hydrogen, Raney nickel,* and *phenol*}. Also, let **reactor-1** be bound to the **node**. The call to the procedure, FIND-ALL-PATHWAYS, and its application to the initial chemical species set, using composite chemical operators involving hydrogenation, is shown below:

(FIND-ALL-PATHWAYS   :*substrates* CSS   :*operators* $K_{hydrogenation}$)

Key reactions identified by the procedure, FIND-ALL-PATHWAYS, are

$$\varphi\text{-}NO_2 + H_2 \rightarrow \varphi\text{-}NO + H_2O, \qquad (VIII)$$

$$\varphi\text{-}NO + H_2 \rightarrow \varphi\text{-}NHOH, \qquad (IX)$$

$$\varphi\text{-}NHOH + H_2 \rightarrow \varphi\text{-}NH_2 + H_2O. \qquad (X)$$

$$C_6H_5NO_2 + H_2 \longrightarrow C_6H_5NO + H_2O$$
$$C_6H_5NO + H_2 \longrightarrow C_6H_5NHOH$$
$$C_6H_5NHOH + H_2 \longrightarrow C_6H_5NH_2 + H_2O$$
$$C_6H_5NHOH + C_6H_5NO \longrightarrow C_6H_5NH_2 + C_6H_5NO_2$$
$$C_6H_5NHOH \longrightarrow C_6H_5N_2$$
$$\longrightarrow (C_6H_5)_2N_2$$
$$C_6H_5NO_2 \longrightarrow \text{decomposition products}$$

FIG. 11. Potential reactions of nitrobenzene.

These reactions combine to form the following overall reaction:

$$\varphi\text{-}NO_2 + 3H_2 \rightarrow \varphi\text{-}NH_2 + 2H_2O. \tag{XI}$$

A less restrictive implementation of the procedure, FIND-ALL-PATHWAYS, on the initial chemical species set, would employ the *:operators* **K** or **K*** and would generate several additional reactions of interest. Among these additional reactions of particular interest are, the decomposition reaction of nitrobenzene and the disproportionation of nitrobenzene with N-phenyl hydroxylamine forming aniline and nitrobenzene:

$$\varphi\text{-}NO_2 \rightarrow \text{products}, \tag{XIII}$$

$$AI\varphi\text{-}NO + \varphi NHOH \rightarrow \varphi NH_2 + \varphi - NO_2. \tag{XIV}$$

Several of the pathways that emanate from these reactions and are constructed by a recursive application of the procedure, FIND-ALL-PATHWAYS, are shown in Fig. 11. Figure 12 shows how this information is managed by LCR using its modeling elements. This information is unavailable in conventional chemical synthesis programs (e.g., SECS and CYCLOPS) and provides an explanation (using the semantic relationships advanced by LCR) as to why the various reactions were activated and the specific intermediates were formed. For example, **pathway-1**, representing the transformation of nitrobenzene into aniline, "knows" that it *is-disaggregated-in* three separate individual reactions, denoted by **hydrogenation-1, hydrogenation-2,** and **hydrogenation-3**. Similarly, it "knows" that it *is-abstracting* the more general global reaction, **reaction-pathways**. It also "understands" which pathways are in competition; these in turn can be abstracted or disaggregated using the semantic relationships of LCR. In this way, the chemical sequence of events that led to a specific reaction can be traced all the way back to the reaction conditions that enabled it.

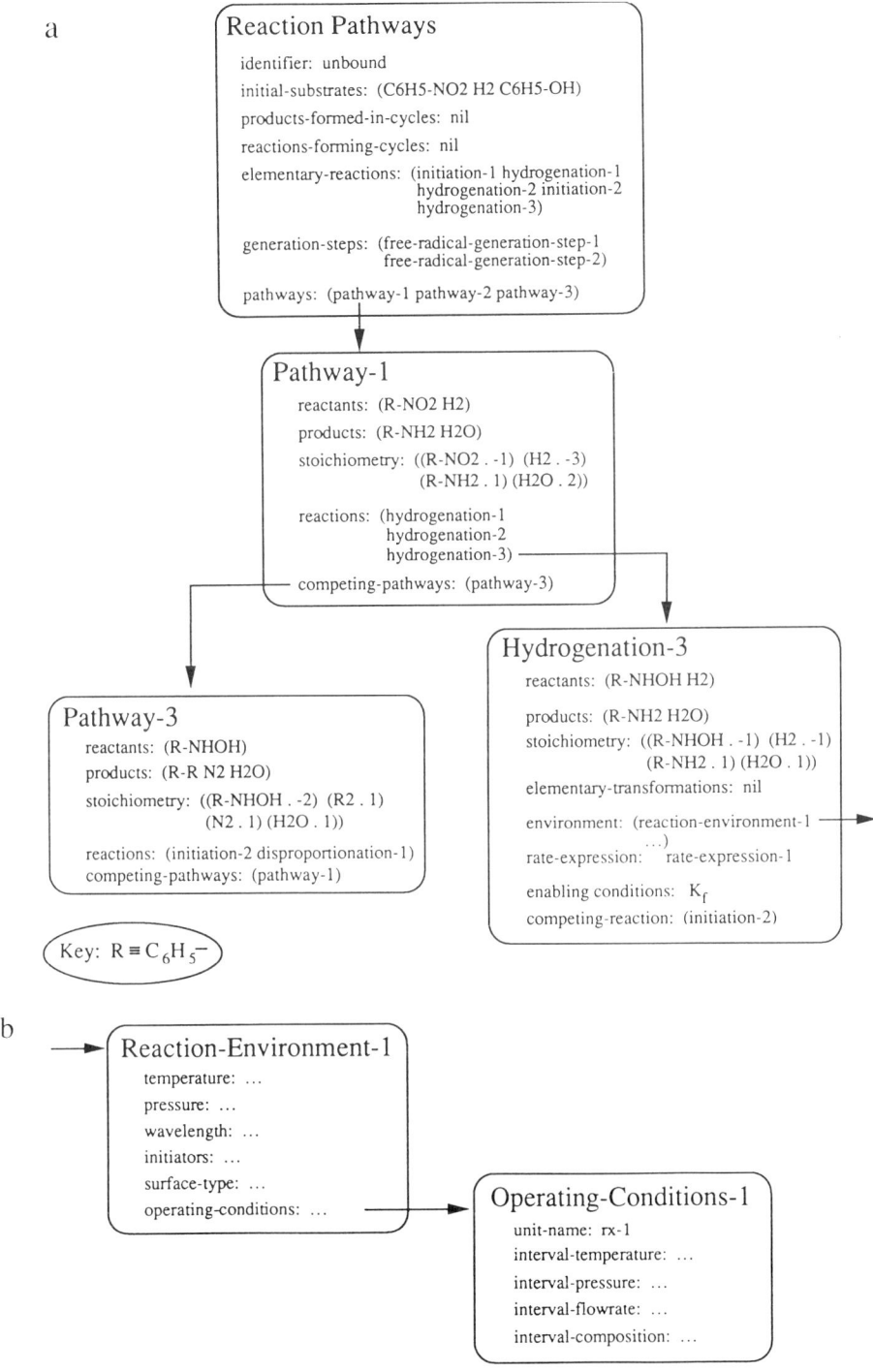

FIG. 12. Linkages between (a) reactions and pathways, (b) reactions and their operating conditions.

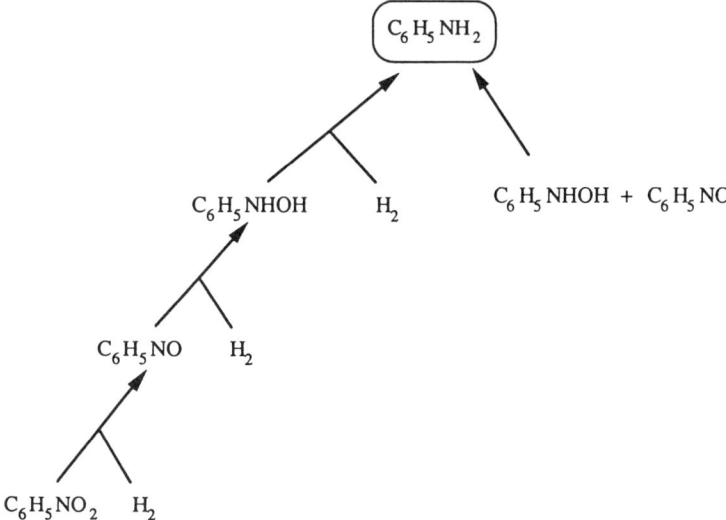

FIG. 13. Reaction path to aniline.

These enabling conditions can be related, in turn, to the operating conditions of the particular processing unit(s), using the *parent-equipment* attribute of *system-description* (an attribute of **node**). Through this mechanism the procedure FIND-ENABLING-CRITERIA traces out the sequences of events that lead to the reaction of interest. Figure 13 shows the path that leads to the formation of aniline. Because of the information contained in the LCR modeling objects that constitute this path, we know precisely the conditions that led to the creation of each element contained in the path. For example, by accessing the *heat-of-formation* attribute associated with the modeling element **reaction**, we identify that heat liberation is nonuniform during hydrogenation. This implies that an accumulation of the intermediate, φ-NHOH, could lead to an uncontrollable energy release.

More importantly, the tree spanning the generation pathways for aniline production (see Fig. 13) makes explicit the fact that hydrogen injection is an insufficient means for controlling reaction temperature. This results from the disproportion reaction identified by the procedure, FIND-ALL-PATHWAYS. As a consequence, a reduction in the hydrogen injection rate and regulation of the cooling fluid rate are necessary to control the heat released by the reaction. The realization that the heat released by the reaction cannot be controlled through hydrogen addition, alone, brings out an important parameter for the safe operation of the reactor, because the experimental detection of the N-phenyl hydroxylamine disproportionation with nitrosobenzene is difficult without prior knowledge of its occurrence (Stoessel, 1989).

Since LCR specifies the preconditions associated with each reaction generated by the procedure, FIND-ALL-PATHWAYS, the conditions that enabled a specific chemical transformation can be easily identified. This knowledge allows us to make explicit such information as the decomposition temperature of φ-NO_2, the disproportionation temperature of φ-NHOH, and the preconditions for the exothermic formation of such products as azobenzene and diazobenzene. By expanding the node description to include the atmosphere around the reactor (e.g., air), the initial CSS can be expanded to include *oxygen*. Then, the procedure, FIND-ALL-PATHWAYS, can identify additional reactions involving oxygen, e.g., combustion reactions with hydrogen, phenol, and Raney nickel as well.

The analysis of the inductively synthesized chemical pathways, described above, indicates that, if one started with only an understanding of the overall reaction, i.e., the hydrogenation of nitrobenzene to form aniline and water, the analyst may not have recognized the need to control reaction temperature, while moderating both cooling rates and hydrogen injection rates. Lacking this information, the loss of recirculation or the need to provide emergency cooling to satisfy additional duty requirements resulting from *N*-phenyl hydroxylamine accumulation, or potentially the need for redundant recirculation pumps may not have been understood and appreciated.

IV. Deductive Determination of the Causes of Hazards

The fundamental premise in all of the approaches that attempt to mitigate or control hazards lies on the assumption that they have the ability to both (1) identify accurately and (2) pinpoint precisely the location of a potential future hazard. We have shown that, although understanding the set of enabling conditions is essential for safe plant operation, the identification of the entire set of enabling conditions is an intractable task. It is impossible to completely identify the set of enabling conditions, leading to a hazardous state for the following reasons. The physicochemical conditions (e.g., presence of certain chemicals within predefined ranges of compositions, temperatures, pressures, flowrates within certain ranges of values) enabling the activation of particular chemical transformations form a set whose members possess variable values. When these variable-valued conditions are substituted into the equations representing the behavior of a chemical plant, the variables defining the conditions for the occurrence of a top-level event (TLE) (e.g., mixing of two chemicals forming an explosive mixture) achieve values necessary for the occurrence of a hazard. Since the variable-valued preconditions for a hazard are drawn from a range of, theoretically, infinite real-valued alternatives, we conclude that it is intractable to deduce all possibilities leading to a hazard. [*Note:* Expressed in different words, the analytic intractability in determining all conditions leading to hazards comes from the following weakness: There are no analytic (numerical) procedures that can take a set of equalities (e.g., the equations modeling the behavior of a plant) and a set of inequalities (e.g., preconditions for the activation of chemical pathways), solve them together, and produce a consistent set of new inequalities.]

Shifting to an interval representation, where we bracket the variable-valued preconditions into arithmetic intervals, over which the behavior is safe or unsafe, is only a partial solution for two reasons:

- First, the boundaries of the intervals, defining the conditions for the occurrence of a TLE are essentially functions of the values that the preconditions for the accomplishment of a reactive pathway can take on.
- Second, if the equations modeling the dynamic behavior of a plant display chaotic solutions (inherent in many non-linear systems), then we cannot be certain that the process behaves the same way when the preconditions defining a reactive pathway take on any values within given arithmetic intervals of values.

As a result, we have focused on the interpretation of the pathway *leading to a hazardous state* and its topography, and how these relate to the inherent safety of the process design technology rather than on the elucidation of pathways *leading to top-level events*. In the absence of a methodology for the complete identification of all conditions enabling the occurrence of a TLE, such an approach is essential if the safety of a chemical operation is to be enhanced.

A. METHODOLOGICAL FRAMEWORK

In the previous section we discussed how an inductive approach can be used to generate all the chemical reaction pathways and the associated thermodynamic states, which lead to top-level hazardous events. A potential hazard is said to exist when the thermodynamic state or sequence of thermodynamic states leading to the hazard cannot be prevented, or the impact of these states cannot be dissipated by the design or the operating capabilities of the process. Using the precursor thermodynamic state(s) as a goal state(s), we can then deductively identify the sequences of changes in the structure of the processing system and/or its operation that led to the creation of the hazardous state. Consequently, root causes and their temporal preconditions, whether equipment and/or operational failures, can be identified and associated with the known hazard. Evaluation of the paths, leading to the hazard, permits the quantitative assessment of the top-level event and affords the risk assessment of the potential hazard in the context of the process design technology. In this section, we will show how the structure of the path itself provides a metric for assessing the inherent safety of the process design technology being evaluated.

The deductive identification of paths leading to a top-level event is accomplished through a *recursive tracing of the variable-influence links*, which describe the transition from state to state on the way to the hazardous event. These *variable-influence* links describe how the various physiochemical variables affect each other, and are generated from the network of modeling relationships describing the behavior of a plant. By propagating the values of the enabling/promoting conditions through the network of modeling relationships, one can assess the ability of the process to dissipate potentially hazardous effects. Obviously, the completeness and correctness of the results obtained through such an approach depend on the completeness and correctness of the modeling relationships. The modeling languages, LCR and MODEL.LA., are used to capture the requisite modeling relationships, which in addition to the first-principles-based equations, may include heuristics, empirical knowledge, experimental correlations, and design decisions. For example, LCR is used to capture the modeling of chemical kinetics, description of chemical structures for the computation of physical properties, whereas MODEL.LA. is used to capture the topology of the unit operations in a plant, the material and energy balances around the unit operations, and the description of operating variables (e.g., temperature, flow, pressure, composition).

Although the knowledge required to assess the potential for the prevention or dissipation of hazards is often held by different abstractions of the overall representation, the semantic relationships of the modeling languages afford efficient access to this information. The effect of protective processes, equipment restraints, sensors and control systems, emergency procedures, etc., are captured as constraints. Constraints may be embedded in the underlying representation as equations, or be associated directly to it via a constraint list, i.e., a collection of explicit process restraints. These restraints may be *passive*, such as materials of construction, or *active*, such as protection processes, sensors and control systems, and operating procedures. *Constraints limit the variable value of an enabling condition. They may achieve this task in one of two ways: demand mitigation or demand prevention.*

Mitigation requires no corrective action to limit a variable value. It is accomplished through the appropriate design of an equipment that (1) leads to the restraint of the top-level event, given an enabling condition, or (2) limits the value that an extensive (e.g., flow, total inventory of a material) or intensive variable (e.g., composition, temperature) can achieve. For example, the value of an intensive variable can be limited by the physical phenomena that are allowed to occur within a process. Orchestration of these phenomena, or where and how they occur, can place a restriction on the type of species that may be generated, the rates at which reactions occur, the separation of constituents into phases, etc. Similarly, extensive variable values such as equipment volume, maximum accumulation, or total mass, can be limited by process equipment design. Since the equations describing a process are a manifestation of these phenomena, select modification of the *variable-influence links* (i.e., of the cause-and-effect pathways) can maximize the inherent safety of the design technology. More importantly, *modification of the topology of the variable-influence links or limitation of the achievable variable values by the enabling preconditions are the only means by which the inherent safety of a process design technology can be altered.*

Preventing the occurrence of an enabling condition requires the undertaking of a corrective action. These actions are designed to limit the variable value of an enabling condition; however, they do so through active participation of protective equipment (e.g., release vents), control loops, or operating procedures. Like the mitigation, the prevention of enabling conditions necessitates an understanding of the *variable-influence links* (i.e., cause-and-effect links) and consequently, of the underlying physicochemical phenomena. We begin the elucidation of the *variable-influence links*, leading to a potentially hazardous states, by first identifying the $(n + 2)$ independent variables that describe the potentially hazardous state. Since process design technologies are describable by the network of process modeling relationships (i.e., topology of the process flowsheet, material and energy balances, chemical and phase equilibrium relationships, kinetic and transport rate expressions, constitutive equations) that define the interactions of various variable quantities, we can trace out the influence of each variable that is associated with the state preceding the TLE. A trace of the variable-influence links defines the pathway of cause-and-effect interactions, which lead to the TLE.

B. Variables as "Causes" or "Effects"

The variable-influence pathway leading to the top-level event is constructed from the structure (Boolean) form of the incidence matrix, representing the network of process modeling relationships. But, the variable-influence pathways encompass certain information on directional causality. Consequently, it is important to identify the role of any variable in the set of modeling relationships: is a variable an input (cause), an output (effect), or of indeterminate directionality? In order to assign a role to each variable, we have developed a specific methodology that will be described in the following paragraphs. Consider the Boolean form of the incidence matrix that represents the modeling relationships, determining the behavior of a specific plant. These relationships capture all available knowledge about the plant; i.e., they are not limited to first-principles modeling equations, but include qualitative, order-of-magnitude, empirical correlations, etc. The fact that the matrix is in a Boolean form allows the simultaneous presence of relationships with inhomogeneous variables (i.e., real-valued, interval-valued, qualitative, or logical variables).

An *input-variable*, i.e., a variable that indicates the influence of the surrounding world on the process is clearly a potential *cause* and never an *effect* of the process' behavior. Typical examples of such input variables are design specifications, setpoints of control systems, characteristics of process feeds. The values of the input variables are established by factors external to the process. Input variables are generally extensive variables. An exception occurs when invariant intensive variables are associated with a feed. For example, the oxygen concentration in air may be considered invariant, when it is being used as a feedstock (e.g., formaldehyde production from air and methanol). Similarly feedstocks delivered to the process with concentration specifications can become inputs (e.g., 100% methanol). Manual settings by operators are input variables. These include manual valve manipulations, setpoints of controllers, structure of control loops, and starting or shutting of pumps. Furthermore, *all potential failures must be characterized as input variables*. This result occurs because potential failures are sources in the set of process equations that drive the causality in a particular direction. Therefore, we can use the following rules (see also Nagel, 1991) for the unambiguous characterization of certain variables as input variables (causes):

Rule 1. All influences of the surrounding world on a particular process are characterized as input variables, and are considered as causes of subsequent evolutions in the thermodynamic state of the process.

Rule 2. Process design specifications, and manual setting of operating variables and controller parameters are considered to be input variables, i.e., causes of subsequent events.

Once the set of input variables, associated with the influence of the surrounding world, has been identified, a systematic procedure examines the remaining variables in an effort to determine their unambiguous role as inputs or outputs. It should be clear at the outset of the subsequent discussion that *the unique and unambiguous characterization of all the process variables as inputs (causes) or outputs (effects) is in general impossible, and corresponds to an undecideable proposition*.

The assignment of a variable as *output* variable is always associated with a particular modeling relationship and implies that the variable takes on its value from the solution of the corresponding equation. Thus, it is identical to the concept of an output variable within the scope of the *input/output set assignment* for the solution of a set of nonlinear algebraic equations. But, unlike the solution of algebraic equations, an *output* variable within the scope of the deductive identification of hazards indicates a *physical consequence*, i.e., an *effect*, resulting from a specific set of causes (i.e., input variables). Consequently, we cannot use the variety of algorithms which have been developed for the identification of input/output set assignments, but we can employ some of the same ideas. Here are some of the rules which guide the selection of output variables (for detailed discussion, see Nagel 1991):

Rule 3. A variable occurring in a relationship, whose remaining variables have been characterized by Rules 1 and (or) 2 as input variables, is an output variable. It represents an effect of the process' behavior and can never be a cause.

Rule 4. A variable can be an output from only one relationship. Therefore, a variable already assigned as the effect (i.e., output variable) of a particular relationship, will be treated as cause (i.e., input variable) in subsequent relationships.

Rule 5. When a variable is the only variable in a particular relationship, then it must be characterized as output.

Rule 6. When a variable occurs in only one relationship and has not been characterized as input variable by Rules 1 and (or) 2, then it must be an output variable.

It is clear from the Rules 1–6 that the assignment of a variable as a *cause* or *effect* is guided by strict and unambiguous physical causality arguments. For example, if all the variables except one in a physical relationship have

INDUCTIVE AND DEDUCTIVE REASONING: IDENTIFYING HAZARDS 227

been characterized unambiguously as causes, the remaining variable *must* be the effect of the particular relationship. Also, it is obvious that if a particular variable has been characterized as the effect of a particular relationship, it can only be a cause in subsequent relationships. Finally, the variables that specify the state preceding a top-level event must be output variables.

It is also clear that Rules 1–6 do not resolve the character of all variables appearing in a set of modeling relationships. Thus, after a repeated application of Rules 1–6, two or more relationships with the corresponding unassigned variables, remain to be characterized. Such subsets of relationships should be "solved" simultaneously and can produce a number of alternative sets of variables, which could have been assigned as output. Clearly, any of these alternative assignments is arbitrary and does not reflect any unambiguous physical causality. What is more important is the fact that *the determination of unambiguous cause-and-effect links among the variables of the remaining relationships corresponds to an undecideable proposition, which can only be resolved through additional independent knowledge.*

Under conditions of incomplete assignment and ambiguity on the role of various variables in the cause-and-effect links, we have adopted a conservative attitude, exploring all potential causalities that may be produced from the modeling relationships.

C. CONSTRUCTION OF VARIABLE-INFLUENCE DIAGRAMS

Let us now see how the ideas of the previous section can be used to construct the variable-influence diagram, that defines the paths leading to top-level event. Consider a set of four modeling relationships, represented by the structural matrix shown below.

	x_1	x_2	x_3	x_4	x_5	x_6	x_7
1		x		x		x	
2		x			x	x	x
3	x	x	x	x		x	
4		x		x		x	

The columns represent variables in the defining process relationship and rows represent the relationships themselves. In accordance with the definitions given above, construction of the variable-influence path proceeds as follows:

1. Assignment of inputs:
   Step 1—assign constants as inputs (Rule 2)
   Step 2—assign invariant intrinsic properties of feeds as inputs (Rule 1)
   Step 3—assign setpoints as inputs (Rule 2)
2. Assignment of outputs
   Step 1—assign variables occurring in only one equation as outputs (Rule 6)
   Step 2—assign single variables in equations as outputs (Rules 3, 5)
   Step 3—eliminate assigned variables and corresponding equations (Rule 4)
   Step 4—assign dependent variables of scientific equations and definitions as outputs
   Step 5—repeat

In the structural incidence matrix given above, there are seven variables (*columns:* terms on abscissa, i.e., on horizontal axis; viz., x_1–x_7) and four relationships (*numbered rows:* terms on ordinate, i.e., on horizontal axis; viz., 1–4). Therefore, three variables must be assigned as inputs to specify the system. They are x_2, x_3, and x_5. Rewriting the structural matrix by moving the input variables to the right-hand side and delineating them with a vertical line, we have

	x_1	x_3	x_6	x_7	x_2	x_4	x_5
1		x		x			x
2			x	x	x		x
3	x	x	x		x	x	
4			x		x	x	

where the left-hand side of the structural matrix represents outputs and the right-hand side represents inputs. Using the definitions shown above, the following variables are characterized as effects (i.e., outputs): x_1 appears only in relationship 3 and therefore receives its value from that equation; similarly, equation 4 contains a single variable on the left hand side of the structural matrix, therefore x_6 takes its value from relationship 4. After the elimination of rows and columns associated with each assigned output variable, x_3 is identified as taking its value from relationship 1, and x_7 taking its value from relationship 2. Thus, x_3 and x_7 are characterized as effects (i.e., outputs) from relationships 1 and 2, respec-

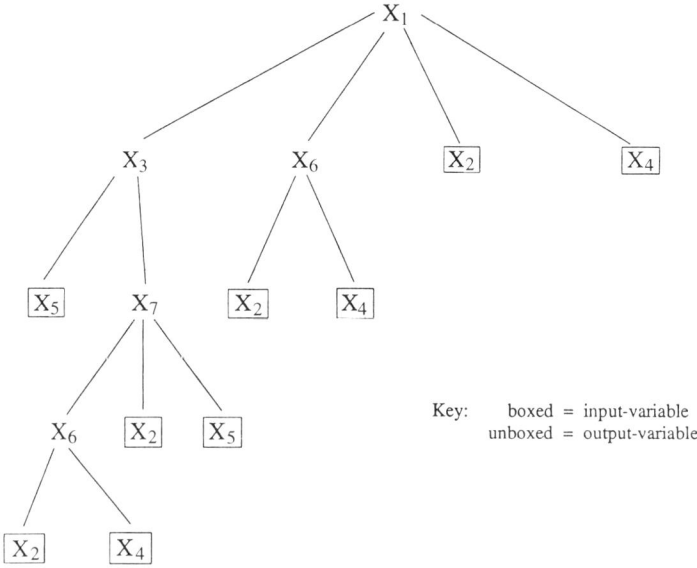

FIG. 14. Variable-influence pathway.

tively. The following structural incidence matrix indicates all assignments:

	x_1	x_3	x_6	x_7	x_2	x_4	x_5
1		__x__		x			x
2			x	__x__	x		x .
3	__x__	x	x		x	x	
4				__x__	x	x	

The variable influence path leading to x_1 is shown in Fig. 14. As a consequence, x_1 can only be influenced by variables that appear on the path; causality is established by the input variables. This occurs because the output set assignment given to the set of relationships in the structural matrix is unique, as it is made evident by rearranging the tabulation order of relationships and variables comprising the structural matrix:

	x_6	x_7	x_3	x_1	x_2	x_4	x_5
4	__x__				x	x	
2	x	__x__			x		x .
1		x	__x__				x
3	x		x	__x__	x	x	

Unique output set assignment occurs when the structural incidence matrix is triangular, such as the above. As a consequence, the variable-influence path leading to the top-level output variable (e.g., x_1) is also unique and can be affected only by inputs that occur along the path.

Alternatively, the network of relationships may not be represented by a triangle structural matrix. In this instance, the structural matrix may not have a unique output set assignment, and there may be alternative paths leading to the variable that describes the state preceding the top-level event. Alternative paths are created when loops exist in the set of process modeling relationships necessitating simultaneous solution. For example, in the structural matrix shown below there are three distinct output set assignments, leading to three distinct potential variable-influence paths. (*Note:* Underlined entries in bold indicate the assigned outputs to each relationship.)

	x_1	x_2	x_3	x_4	x_5	x_6	x_7	x_8	x_9	x_{10}	x_{11}	x_{12}
1		**x**	x		x				x			x
2					**x**		x		x	x		
3				**x**			x	x	x		x	
4			**x**	x		x		x	x		x	
5					x	**x**				x		x
6					x	x	**x**					x
7								**x**				
8									**x**	x		x
9	**x**	x		x	x		x				x	

	x_1	x_2	x_3	x_4	x_5	x_6	x_7	x_8	x_9	x_{10}	x_{11}	x_{12}
1		**x**	x		x				x			x
2					x		**x**		x	x		
3				**x**			x	x	x		x	
4			**x**	x		x		x	x		x	
5					**x**	x				x		x
6					x	**x**	x					x
7								**x**				
8									**x**	x		x
9	**x**	x		x	x		x				x	

	x_1	x_2	x_3	x_4	x_5	x_6	x_7	x_8	x_9	x_{10}	x_{11}	x_{12}
1		**x**	x		x				x			x
2					x		**x**		x	x		
3				**x**			x	x	x		x	
4			**x**	x		x		x	x		x	
5					x	**x**				x		x
6					**x**	x	x					x
7								**x**				
8									**x**	x		x
9	**x**	x		x	x		x				x	

Alternatively output set assignments, and consequently variable-influence paths, result from the block structure established by variables x_5, x_6, and x_7. Thus, variable x_5 can take its value from any one of the three relationships, since there is no way that x_5 can be uniquely and unambiguously assigned as an effect to any one of these three relationships.

Construction of the variable-influence diagrams, indicating the paths of the cause-and-effect links, which lead from the external causes to the variables describing the potentially hazardous state, is achieved by using the procedure CONSTRUCT-VARIABLE-INFLUENCE-PATHWAYS, a procedure called from the procedure, IDENTIFY-POTENTIAL-HAZARD. This method traces the influence of various variables defining the state preceding the top-level event. It constructs variable-influence paths from the set of process modeling relationships contained in the structural incidence matrix. The algorithm used for the generation of variable influence pathways is given below:

Algorithm 3: ⟨CONSTRUCT-VARIABLE-INFLUENCE-PATHWAYS⟩.

```
input:              process-flowsheet
     enabling-criteria   ;output from IDENTIFY-POTENTIAL-HAZARD
initialize
process-flow-equations ← apply IDENTIFY-PROCESS-FLOW-EQUATIONS
                    to process-flowsheet
VIM ← apply IDENTIFY-INFLUENCE-MATRIX to process-flow-equations
TLE-vars ← apply IDENTIFY-TLE-VARS to enabling criteria   ;identify top-
investigation-list ← TLE-vars                              level event
for each tree in investigation-list                        variables
if next-available-node ← apply IDENTIFY-NEXT-EXPANDABLE-NODE to tree
    then
    expansion-list
        ← apply EXPAND-NODE to VIM, tree, next-available-node
    endif
return
for each expansion in expansion-list
    apply CLASSIFY-BRANCH to next-available-node    ;classify
                                                     according to
                                                     technology-type
    if (apply CONTINUE-EXPAND-BRANCH-P to expansion,
       next available-node)
    then            ;predicate test to evaluate
                     branch expansion
       (append tree to investigation-list)
    endif
return
```

Whenever the procedure CONSTRUCT-VARIABLE-INFLUENCE-PATHWAYS has to generate a number of alternative variable-influence paths, it employs a breadth-first search strategy. The procedure, EXPAND-NODE, is used to establish the expansion list through the following algorithm:

Algorithm 4: ⟨EXPAND-NODE⟩.

    input: VIM, tree, var
    initialize
        expansion-list ← nil
    apply CLEAR-ALL-ASSIGNMENTS to VIM
    for each node in (**apply** PATHWAY-NODES to tree, var)
        apply ASSIGN-OUTPUT-VARIABLE to VIM, node
        return
    for each equation in (**apply** EQNS-CONTAINING-VAR to VIM, var)
        if (**apply** VALID-ASSIGNMENT-P to VIM, eqn, var)
            then
                new-vars ← **remove** var from (**apply** VARS-IN-EQN to eqn)
                apply CLASSIFY-VARS to new vars
                new-tree ← **apply** APPEND-TREE to tree, node, new-vars
                append new-tree to expansion-list
            endif
        return
    return

D. Characterizing of Variable-Influence Pathways

The procedure CONSTRUCT-VARIABLE-INFLUENCE-PATHWAYS applies to each node of the variable-influence pathway the procedure, CLASSIFY-BRANCH, which in turn classifies each branch of the pathway according to the type of the technology, which can be used to control the variable-value specifying the node. The algorithm of this procedure is given below:

Algorithm 5 ⟨CLASSIFY-BRANCH⟩.

    input: node
    for each child-node in (**apply** CHILDREN to node)
    select case for child-node
        terminal-variable
    is apply TAG-NO-EXPANSION to child-node   ;no node expansion
        endcase
        already-in-pathway

```
            is apply TAG-NO-EXPANSION to child-node
         endcase
      endselect

      select case for node
         inherent-controllability
            is apply TAG-TYPE-1-TECHNOLOGY to node    ;see definition below
         endcase
         Type-2-controllability and process-modification-applied
            is apply TAG-TYPE-2-TECHNOLOGY to node
         endcase
         Type-3-controllability and control-loop-applied
            is apply TAG-TYPE-3-TECHNOLOGY to node
         endcase
      endselect
      return node
```

Nodes are classified in this manner in order to make explicit the protective system responsible for mitigating a disturbance and its relationship to the top-level event. The procedure CONSTRUCT-VARIABLE-INFLUENCE expands each node of the tree and classifies each branch according to the technology type that could be used to control the variable-value, using the following definitions:

1. A *Type-1 technology* is capable of reducing the number of TLEs associated with the specific design, or is capable of modifying the topology of the path leading to the TLE through a change in process chemistry or the structure of the designed process flowsheet, provided design changes do not involve the introduction of *Type-2* or *Type-3 technologies*. Examples include reactant substitution, catalyst introduction or substitution, byproduct reduction, chemical character modification, solvent substitution, and intermediate chemicals elimination. Industrial processes with *Type-1 technology* changes include the following: (a) substitution of the reactant raw materials, hydrogen cyanide and acetylene, with propylene, ammonia, and air; (b) production of butyl lithium in dilute solutions, which prevents spontaneous combustion in the presence of air; and (c) direct hydroxylation of ethylene forming ethylene glycol, rather than oxidation of ethylene forming ethylene oxide, which in turn is hydrated to form ethylene glycol, thus eliminating ethylene oxide as an intermediate. Changes in the pathway topology, leading to the TLE, require design modification of the unit operations, changes in the piping, substitution of unit operations, or layout modification.

2. A *Type-2 technology* limits the value that a variable, which is involved in the pathway leading to a TLE, can achieve without requiring an action. Examples involving *Type-2 technologies* include the following:

(a) reduction of inventories, (b) minimization of operating (process condition) extrema, (c) reduction of intermediate accumulation. Industrial processes incorporating *Type-2 technologies* include the following: utilization of in situ reactions to limit intermediate accumulation (e.g., methyl isocyanate consumption via in situ reaction reduces the need for storage), utilization of continuous processes to minimize process hold up (e.g., continuous nitration processes eliminate the need for batch manufacturing of nitroglycerine), and reduction in the intensity of processing conditions can lead to smaller material and energy releases (e.g., catalyst improvements have enabled the Oxo process to produce aldehydes from syngas and olefins at lower pressure).

3. A *Type -3 technology* limits the value that a variable, involved in the path leading to a TLE can achieve *and require an action to do so*. *Type-3 technologies* involve active control of process variables in a direct or indirect manner, in order to limit the value that they can achieve. Examples include the following: (a) manipulation of feed rates, (b) use of cooling water, (c) temperature control and pressure control, (d) catalyst activity assessment through the monitoring of conversion and selectivity.

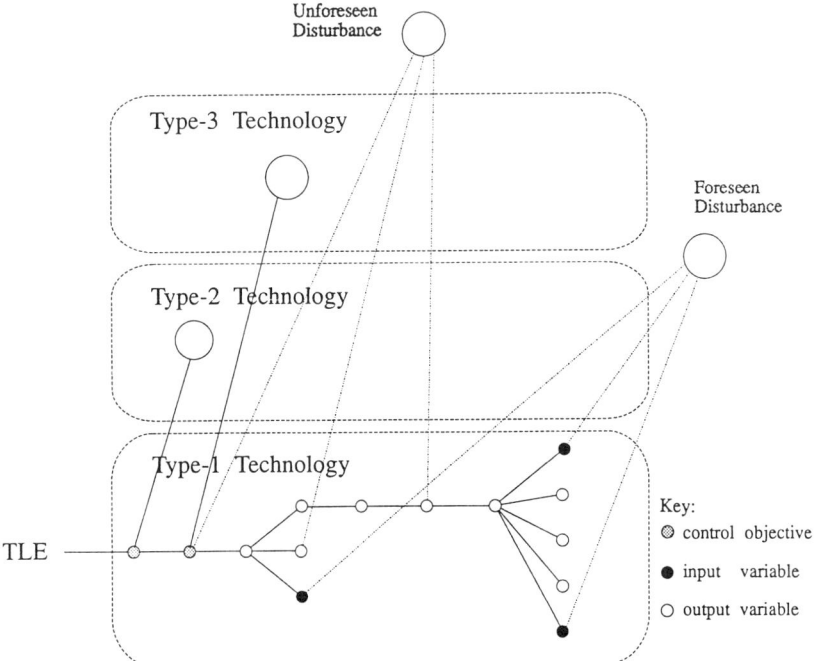

FIG. 15. Mitigation of disturbances through the specification of the control points.

The result of tagging the paths with the type of hazards-preventive technology is shown in Fig. 15. Having such a diagram one can then associate the different technology types with various parts of the variable-influence diagram, for the purpose of directly or indirectly controlling the variable values, composing the variable-influence pathway to the top-level event. Such an approach not only provides an understanding on how to control the values of each variable along the variable-influence path, but, also permit the designer to evaluate the feasibility of the process and its safety devices with respect to various operational criteria.

E. ASSESSMENT OF HAZARDS-PREVENTIVE MECHANISMS

The strategy we use for the specification of the most attractive hazards-preventive control objectives is based on *closeness*, i.e., hazards-preventing control objectives, which affect a variable that is at a minimum distance from the top-level event is preferred over those that affect more distant variables. However, the point on the variable-influence path, where the actual manipulation (e.g., design modification, controller or safety device) takes place, remains unspecified; it depends on whether the origin, type and intensity of disturbances are known ahead of time or not.

For disturbances that can be *foreseen*, it is preferred that the hazards-controlling objectives be placed near their origin (i.e., entry point of an input), in order to mitigate the effect of the disturbance before it has the opportunity to be amplified (see example in the diagram of Fig 15). *Foreseen disturbances* enter the process through the set of external variables or inputs. These include pump failures, valve failures, controller malfunctions, etc.

Notice, though, that the greater the distance between the location of the hazards-controlling objective and the top-level event, the greater the opportunity (i.e., the higher the probability) for the appearance of an unknown (*unforeseen*) disturbance along the pathway that connects the control objective to the TLE. *Unforeseen disturbances* can enter the process anywhere along the pathway and often change the pathway leading to the top-level event. Since the thermodynamic state preceding a top-level event is specified by the procedure, IDENTIFY-POTENTIAL-HAZARD, any disturbance, foreseeable or unforeseeable, must act as an input to these variables, if the TLE is to be enabled. Consequently, the only effective mechanism for the mitigation of unforeseen disturbances is the establishment of control objectives at *level 1*; the last defensive line before enabling the top-level event. This specification is translated into a series of conditional statements on the values of the variables at *level 1* of the variable-influence diagram (see Fig. 15). The methodology used by the procedure, GLOBAL-HAZARD-IDENTIFICATION, enables the identification of TLEs independently of the pathways that lead to them. Additional TLEs are identified by the procedure GLOBAL-HAZARD-IDENTIFICATION as it expands the scope of the process description in its search for potential hazards. Expansion of the process description includes the reformulation of the boundary defining the process, thus allowing the incorporation of disturbances (e.g., as new inputs) that were *"unforeseen"* by the earlier formulation of the process description.

Type-2 technologies can be particularly effective in mitigating unforeseeable disturbances, because they do not require active control of the disturbance. Nevertheless, whenever possible, *Type-1 technologies* offer the best mechanisms for the mitigation of unforeseeable disturbances. This is accomplished through the addition or subtraction of relationships from the structural incidence matrix; operations that describe the recommended changes in chemistry, materials, type of unit operations, or structure of the process flowsheet. On the other hand, control of foreseeable disturbances is better handled by *Type-3 technologies*. Each variable along the variable-influence pathway is subsequently characterized by the type of hazards-controlling technologies, i.e., *Type-1, Type-2,* or *Type-3 technologies*, which could be employed for controlling the value of the specific variable. Such characterizations allow fast and effective screening of the potential disturbance mitigation alternatives.

The procedure IDENTIFY-NONDISSIPATIVE-PATHWAYS constructs the set of enabling conditions that lead to a top-level event. It takes as its input the variable-influence pathways, which it obtains from the procedure, CONSTRUCT-VARIABLE-INFLUENCE-PATHWAYS. TLE variables are then identified, and each variable contained in the set is traced to identify potentially feasible roots for causing the TLE. When an achievable root is identified (i.e., an input disturbance, controlled or uncontrolled, which can enable the top-level event), it is collected into root causes and returned. The algorithm used by this procedure is shown below:

Algorithm 6: ⟨IDENTIFY-NONDISSIPATIVE-PATHWAYS⟩.

   input: classified-influence-paths
   initialize: TLE-variables ← (collect-TLE-variables)
   for each variable in TLE-variables
      when (unique-path-p classified-influence-paths)   ;predicate test for identifying unique paths

         (**apply** CONSTRUCT-FEASIBLE-ROOTS   ;constructs pathways or trees
           to TLE-variables classified-
           influence-paths)   which support

                    and collect into root-causes
                    return
            else
                    for each path in classified-influence-paths
                            to (**apply** CONSTRUCT-FEASIBLE-ROOTS   ;constructs a
                                    to TLE-variables path)   feasible root to
                            and collect into root-cause   the goal state
                            return   (TLE variable)
                                                                                    path
            return goal state (TLE)
end

                                                                            enabling criteria

Figure 16 shows how the procedure IDENTIFY-NONDISSIPATIVE-PATHWAYS has identified a potential path, starting from a given TLE and deducing a specific (*foreseen*) disturbance as a potential cause for the given TLE. Furthermore, in Fig. 16 we can also see the type of the preventive technologies (as identified by the procedure, CLASSIFY-BRANCH) that are appropriate for each variable on the pathway, which leads from the foreseen disturbance to the specific TLE.

The procedure IDENTIFY-NONDISSIPATIVE-PATHWAYS calls the procedure CONSTRUCT-FEASIBLE-ROOTS, which assesses the feasibility of a particular variable trace to enable the TLE, given a disturbance as its input. The

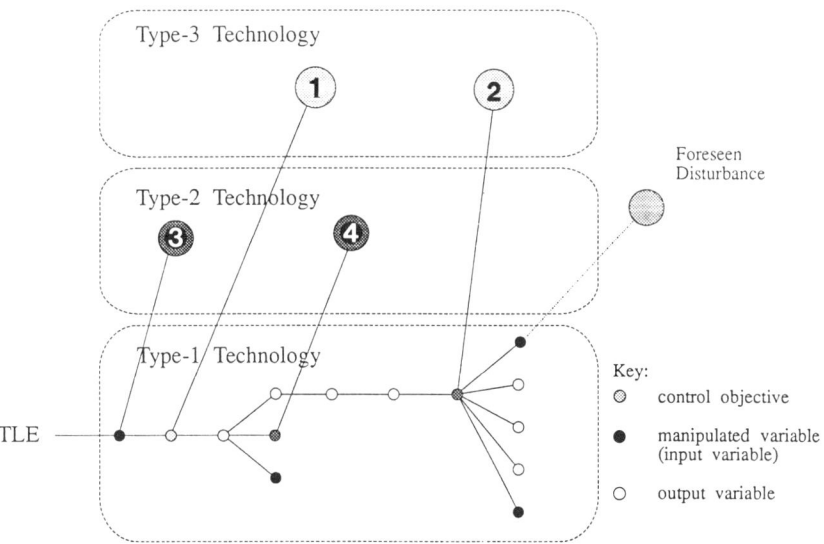

FIG. 16. Placement of technologies for the prevention of a specific TLE.

procedure CONSTRUCT-FEASIBLE-ROOTS also identifies the control mechanism responsible for the mitigation of the input disturbance. When a controlling mechanism is not associated with the input, the variable and its pathway are collected directly into roots. The algorithm defining the logic of the procedure, CONSTRUCT-FEASIBLE-ROOTS is shown below:

Algorithm 7: ⟨CONSTRUCT-FEASIBLE-ROOTS⟩.

> **input**: variable-set
>   pathway
> **when** (enabling-feasibility-p variable-set pathway)
>   **for** each variable in variable-set
>   **when** (controlled-variable-p variable pathway)
>   controller ← (get-controller variable pathway)
>   collect variable, pathway, and controller into roots
>   **return**
>   **else**
>   collect variable into roots
>   **return**
>   **return**
> **return**
> **end**

F. FAULT-TREE CONSTRUCTION

The construction of the variable-influence pathways and the classification of its nodes in terms of the type of technology that can be used to mitigate the propagation of a hazard, constitute the essential basis for the generation of topological fault trees, a tool that can guide the evaluation of hazards preventive mechanisms. The procedure CONSTRUCT-TOPOLOGICAL-FAULT-TREE is at the core of this construction. It takes as its input the top-level event, the associated hazardous state preceding the TLE, and the root causes for each variable defining the hazardous state. Using these inputs, it determines the reactions responsible for the hazardous state and the enabling criteria of the reaction. With this information, it constructs a *level-1-gate:* a logical gate immediately preceding the top-level event. This gate establishes the demands (i.e., inputs) of the top-level event. By looping through the set of demands associated with the *level-1-gate*, we can construct qualitative logical gates from a given demand input and its root causes. Since each qualitative gate has associated with it an input and an output, these can be linked together to form a tree. By linking the tree to the *level-1-gate* and in turn appending the *level-1-gate* to the TLE, a topological fault-tree can be constructed. The algorithm used is as follows:

Algorithm 8: ⟨CONSTRUCT-TOPOLOGICAL-FAULT-TREE⟩.

   input: TLE
           hazardous-state
           root causes
   initialize: hazard-variables ← (get-hazardous-variables hazardous-state)
               reaction ← (get-reaction hazardous-state)
               enabling-criteria ← (get-enabling-criteria reaction)
               level-one-gate ← (construct-level-one-gate
                                       hazard-variables enabling-criteria)
   for each input in level-one-gate
      gates ← (CONSTRUCT-TOPOLOGICAL-GATES input   ;gate construction
   (get-root-causes ;input))
         tree ← (LINK-GATES input gates)   ;connects gates
         append tree to level-one-gate   via their logical
         return   outputs
   append TLE to level-one-gate
   return
   end

Basic gates are constructed using the procedure, CONSTRUCT-TOPOLOGI-CAL-GATES. This procedure builds *or-gates*, *and-gates*, and *special-gates*. Special-gates are based on first-order predicate logic and require a certain amount of quantitative analysis prior to Boolean assignment (*i.e., and-gates, or-gates, and special-gates*). The procedure receives an input variable and the root causes associated with that variable. Then, it uses the variable trace associated with the root causes to establish the pathway leading to the external input and the protective devices associated with that input. By collecting the controllers associated with the various controlled variables, the procedure identifies the necessary *and-gates, or-gates*, and *special-gates*. The algorithm used for the implementation of the procedure, CONSTRUCT-TOPOLOGICAL-GATES, is shown below:

Algorithm 9: ⟨CONSTRUCT-TOPOLOGICAL-GATES⟩

       input: variable
               root-causes
       initialize: and-gates ← nil
                   or-gates ← nil
                   special-gate ← nil

```
        initial-internal-input
            ← (get-internal-input variable      ;network
                root-causes)                    starting point
    for each variable in (get-output-assignment root-causes)
        up from initial-internal-input
        when (unbranched-node-p variable)       ;predicate test for
            controller                          determining if node
                ← (get-controller variable      is branched
                    root-causes)
            when controller                     ;constraint
                collect (construct-and-gate     ;identification
                    variable controller         of variable
                    (get-input variable))
                    into and-gates
                and
                collect (construct-or-gate variable controller (get-input
                    variable))
                    into or-gates
                return
            collect (construct-special-gate variable (get-inputs variable))
                into special-gates
            return
        gates ← append and-gates or-gates special-gates
        return
    return
end
```

Complex topological fault trees are constructed using the gates identified by CONSTRUCT-TOPOLOGICAL-GATES. Since each logical gate has a set of inputs and an output, all the gates can be linked together by matching gate outputs to gate inputs.

Quantitative analysis of the pathway, leading to the TLE, is required to establish the logical basis for converting *special-gates* into structures (i.e., trees) containing *or-gates* and *and-gates*. By associating averaged probability and failure rate data with the resulting topological fault tree (i.e., with the variables and the technology type of preventive mechanisms), the latter can be transformed into a conventional fault-tree. Construction of the fault tree in this manner has particular utility because it is complete; all pathways leading to the TLE will be identified within the scope of the relationships describing the particular process design technology. This prevents incompleteness in the pathways leading to the TLE and minimizes errors in the frequence of that event, often by more than three orders of magnitude (ICI, 1988). (*Note*: ICI has shown that incompleteness can result in estimates of the TLE that are off by as much as three to five orders of magnitude; whereas errors in probability and failure rate data lead to estimates that are often within a factor of 2–5.)

G. AN EXAMPLE OF REACTION-BASED HAZARD IDENTIFICATION: REACTION QUENCH

Consider a process that consists of a reactor used for the processing of a highly unstable chemical that is sensitive to small increases in temperature. The reactor is equipped with a quench tank to protect the system against a runaway reaction and is monitored by two temperature sensors (see Fig. 17): T_1 and T_2. Sensor T_1 automatically activates the quench tank outlet valve when it detects a temperature rise above the specified upper limit. Sensor T_2 sounds an alarm in the control room to alert the operator to the process upset. When the alarm sounds, the operator closes the reactor inlet valve. The operator also pushes a quench tank valve button in the control room in case the quench valve fails to open. Note that A is the reactant; B, the product; and C, the quench.

The analysis for the deductive determination of root causes begins by constructing the set of relationships that describe the process flowsheet.

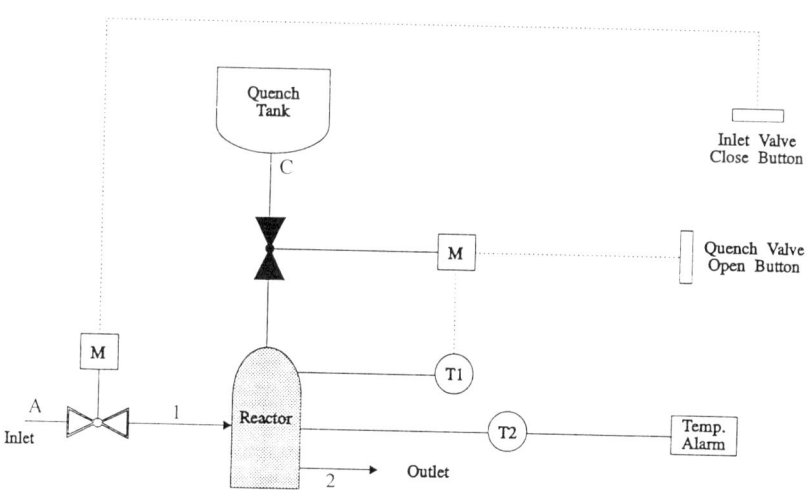

FIG. 17. The elements of the reactor-quench example.

These are shown below:

$$F_2 = f_2^A + f_2^B + f_2^C, \quad (1)$$

$$f_2^C = Q + F_1^C, \quad (2)$$

$$f_1^A = f_2^A + r_A, \quad (3)$$

$$f_1^B = f_2^B - r_A, \quad (4)$$

$$F_1 = f_1^A, \quad (5)$$

$$f_1^B = 0, \quad (6)$$

$$f_1^C = 0, \quad (7)$$

$$r_A = k[C_A]V_{RE}, \quad (8)$$

$$k = A * \exp(-E^*/RT_R) = f(T_R), \quad (9)$$

$$[C_A] = f(f_1^A, Q, r_A), \quad (10)$$

$$Q_r = f(\Delta h_r, r_A), \quad (11)$$

$$0 = f(Q, T_0, F_1, T_F, Q_r, F_2, T_R), \quad (12)$$

$$Q = f(SP_1), \quad (13)$$

$$SP_1 = f(T_R, T_1), \quad (14)$$

$$F_1 = f(SP_2), \quad (15)$$

$$SP_2 = f(Op_2), \quad (16)$$

$$OP_2 = f(T_R, T_2), \quad (17)$$

where F denotes a feed, f denotes a molar flowrate, Q denotes the quench feed, r_A is a reaction rate expression, V_{RE} is the volume of the reactor, SP denotes valve stem position, T denotes temperature, and Op denotes an operator. Complex relationships involving these variables show only the corresponding functional dependence [e.g., relationships (9) through (17)]. The *structural matrix* constructed from this set of equations and relationships is shown in Fig. 18. Each numbered row describes the relationship between the variables that are contained in it. The variables are represented by the columns of the matrix. The occurrence of a variable in a particular relationship is signified by a nonzero entry (e.g., x in Fig. 18) at the intersection of the column representing the variable with the row representing the relationship.

INDUCTIVE AND DEDUCTIVE REASONING: IDENTIFYING HAZARDS

	F_1	F_2	Q	f_1^A	f_1^B	f_1^C	f_2^A	f_2^B	f_2^C	r_A	k	C_A	T_R	Q_R	V_{RE}	T_2	T_1	O_{p2}	SP_1	SP_2	Δh_r	T_Q	T_F
1		x					x	x	x														
2				x			x			x													
3					x		x			x													
4						x		x		x													
5	x				x																		
6						x																	
7							x																
8										x	x	x			x								
9											x	x											
10				x	x					x		x											
11										x				x							x		
12	x	x	x							x	x											x	x
13				x												x							
14													x				x	x					
15	x																		x				
16																		x	x				
17										x							x	x					

Fig. 18. The structural incidence matrix for the reactor-quench example.

The process of constructing the variable-influence pathway, which leads to the state preceding the top-level event, begins with the identification of the potential input variables. Following the procedures outlined in Section IV.B, the assignment of the input variables is initiated. But, the identification of the *input specifications* is intricately related to the scope of the assumed *system's boundary*. Thus, by restricting our attention to the reactor only (see shaded unit in Fig. 17), examination of the process modeling relationships identifies two system constants: Δh_r (the heat of reaction) and V_{RE} (the reactor volume); these become input variables in the structural matrix. Similarly, if the scope of the process' boundary is expanded to include external feeds, then the boundary of the system encloses all the shaded area of Fig. 19. This expansion identifies the invariant intensive properties describing the inlet feed (T_F) and the quench feed (T_Q) as external or input variables. Step 3 of the input assignment procedure (see Section IV.B) assigns the setpoints T_2 and T_1 as input variables, thus further expanding the scope of the process boundary (see Fig. 20).

Once inputs are assigned, output assignment begins using the method described in Section IV.B. Using this procedure, f_1^B is identified as taking its value from Eq. (6) and f_1^C is identified as taking its value from Eq. (7).

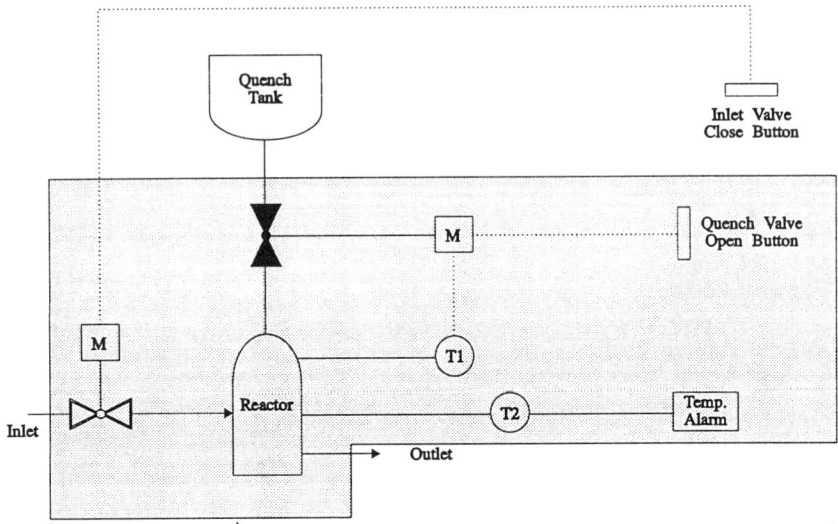

FIG. 19. The system boundary defining the expanded input specifications for the reactor-quench example.

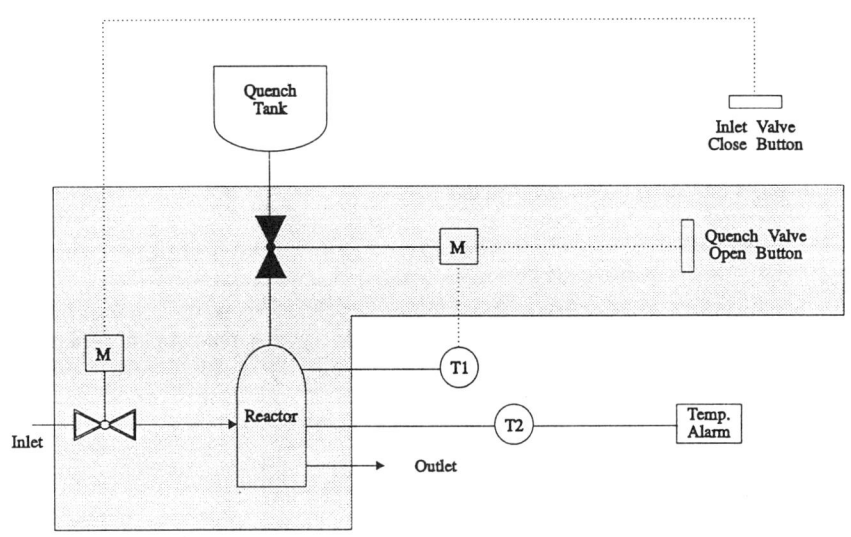

FIG. 20. The final system boundary indicating further expansion of the input specifications.

The respective rows and columns of the structural matrix are then eliminated. The value of the reaction rate constant can be given only by the definitional relationship (9), and thus it is assigned as output from Eq. (9). On elimination of the row and column corresponding to Eq. (9) and variable k, no other output assignments can be made. The block structure that results allows the remaining variables to take their output from any one of several equations.

To break the loop of cause-and-effect relationships, we make the unqualified assumption that Op_2, the variable denoting the human operator, obtains its value from relationship (17); i.e., the operator responds to the temperature in the reactor (T_R) and not vice versa. This assumption is then catalogued, as *Assumption-1*, so that it may be retracted at a later stage, as required. Such retraction allows the causality that emanates from the set of process equations to be modified. Once we have eliminated Eq. (17) and the column corresponding to variable Op_2, we can repeat the steps of the output set assignment (see Section IV.B), and find that F_1 obtains its value from Eq. (15). The set of input and output assignments, made so far, establish the causality between

- Reaction temperature and human operator
- Human operator and feed valve stem position
- Feed valve stem position and feed rate.

Elimination of the respective rows and columns identifies further that f_1^A takes its value from Eq. (5).

At the end of the preceding assignments, a block structure of "*simultaneous relationships*" still remains. Again, though, an unqualified assumption can be made regarding the causality of variables in Eq. (14). The assumption is that SP_1, *the variable describing the feed valve stem position, is driven by reactor temperature*. This assumption is catalogued and denoted as *Assumption-2*.

The variable-influence diagram resulting from the set of the two assumptions, i.e., *Assumption*-1 and *Assumption*-2, is shown in Fig. 21. This diagram makes explicit the pathways leading to T_R and illustrates how disturbances can propagate through the network of relationships and enable the occurrence of the TLE. If we now retract *Assumption*-1, and replace it by its opposite (which will be labeled *Assumption*-3), i.e., the operator does not react to the temperature of the reactor but *the value of the reactor temperature is defined by the actions of the operator*, then the variable-influence diagram changes and is now represented by that of Fig. 22. By collecting the entire set of graphs that describe the resulting set of cause-and-effect relationships, we can guarantee complete identification of pathways, leading to the TLE, within the scope of the modeling effort,

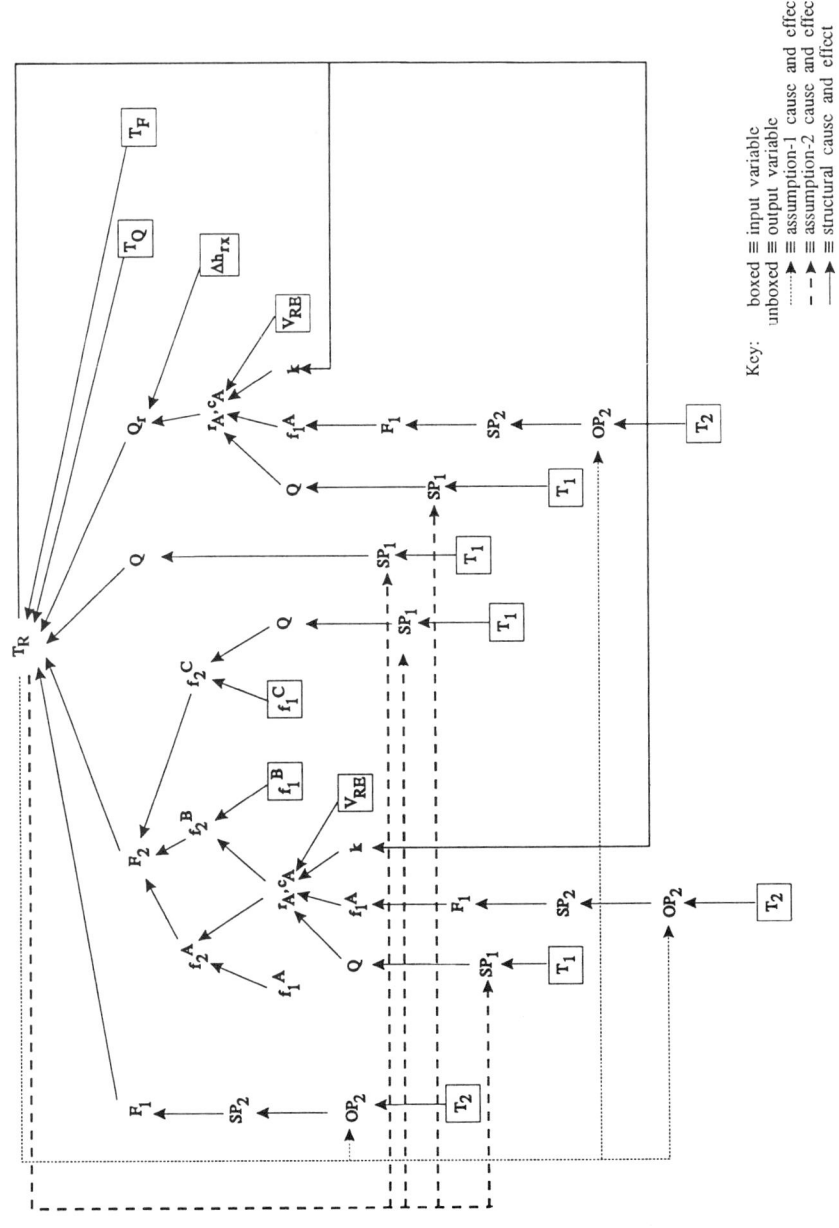

FIG. 21. The variable-influence diagram resulting from *Assumption-1* and *Assumption-2*.

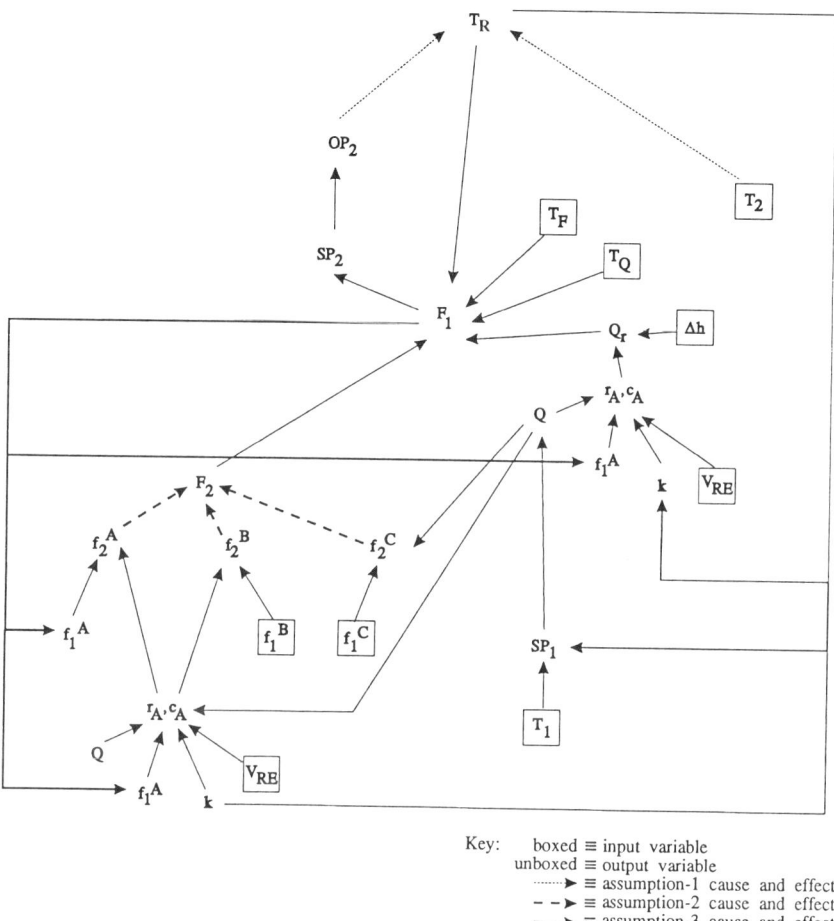

Key: boxed ≡ input variable
unboxed ≡ output variable
······▶ ≡ assumption-1 cause and effect
– –▶ ≡ assumption-2 cause and effect
——▶ ≡ assumption-3 cause and effect
——▶ ≡ structural cause and effect

FIG. 22. The variable-influence diagram resulting from *Assumption-2* and *Assumption-3*.

in conjunction with the set of the assumptions made. By associating the variables, contained in the variable-influence diagram, with the appropriate types of hazards-preventing technologies (i.e., *Type-1*, *Type-2*, *Type-3* technologies), the pathway of root causes is constructed. Figure 23 illustrates the association of the variables with specific *Type-1*, *Type-2*, and *Type-3 technologies*, for the case where the variable-influence diagram has been constructed assuming *Assumption-1* and *Assumption-2* to be true. Note that the heat of reaction Δh_r is the only Type-1 technology associated with the TLE. *This implies that a potential pathway for mitigating the*

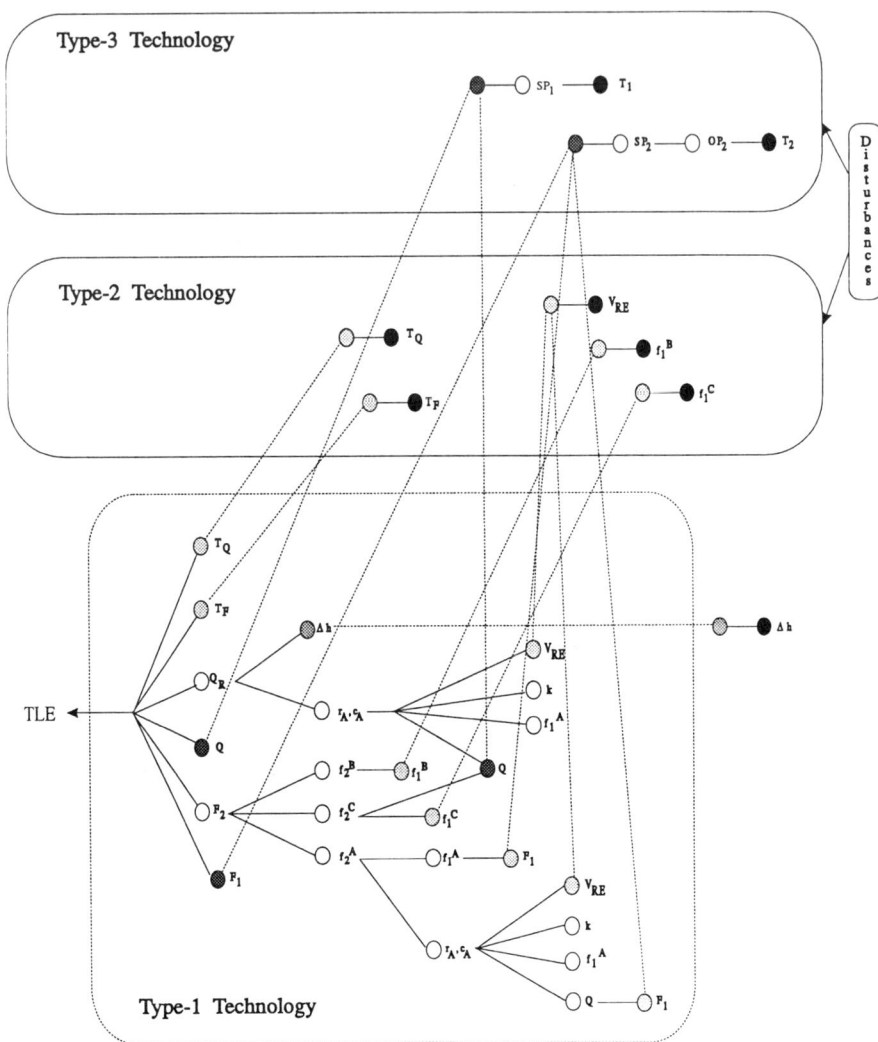

Fig. 23. Construction of root cause diagram.

TLE is to change the heat of reaction and obviously implies a change in chemistry. Furthermore, the variable-influence diagram makes explicit that top-level event prevention necessitates the placement of control objectives on every variable that immediately precedes the top-level, i.e., variables $T_Q, T_F, Q_R, Q, F_2, F_1$, should come under control.

Notice, also, that *the value of the variables described by Type-2 technologies are generally design specifications*; they affect the magnitude of the top-level event. For example, V_{RE} specifies the volume of the reactor. Similarly, feed and quench temperatures vary the rate at which a TLE is achieved; likewise, so may f_1^B and f_1^C as the values that describe the molar flowrates of the product and the quench, respectively, contained in the feedstream, vary. This is particularly so if the reaction described is autocatalytic. Trace amounts of product in the feed can catalyze the reaction leading to a thermal disturbance that ultimately could trigger a reaction runaway.

Type-3 technologies require active control to mitigate a disturbance and consequently are protective systems associated with the process flowsheet. Notice that certain control actions cannot be modeled easily without overspecifying the system such as the manual override of the quench valve by the operator. For this reason, the constraint list, as described earlier, is associated with each piece of process equipment. This allows us to associate in an *a posteriori* manner additional control structures that are available to the process.

How the different technology types are used to identify potential root causes that can enable the top-level event is illustrated in Fig. 23. This illustration makes explicit the fact that disturbances, which enter into the system, can enter only through the inputs identified by Type-2 and Type-3 technologies. These disturbances can either be known a priori (*foreseen*) or not (*unforeseen*). Modeled disturbances can affect the process only through its input variables and are therefore identified by the pathways leading to the top-level event. Unmodeled disturbances can act in one of two ways: they can affect the value of a variable through an unmodeled relationship, or they can change the casuality described by the pathway, negating the assumption set from which it was built. However, since various assumption sets are used to construct the alternative pathways leading to the top level event, the impact of unmodeled disturbances that enabled some pathways can be contained. Furthermore, because of the manner in which the top-level event has been identified, it can be enabled only through the set of variables that describe the state immediately preceding it. Since any disturbance must eventually pass through that state, if it is to enable the top-level event, control objectives designed around these variables can mitigate foreseen and unforeseen disturbances that lead to a top-level event. Using the knowledge associated with different technology types, and the pathways that lead to a top-level event, we can construct a topological fault tree. This is illustrated in Fig. 24. *Note the need for a special gate.* This type of gate is required because without quantitative assessment, it cannot be determined whether the gate is a single *and-gate*, a single *or-gate*, or a structure containing *and-gates* and *or-gates*.

The topological fault tree shown in Fig. 24 suggests that the top-level event can be enabled through a change in the input of T_Q (e.g., this could imply no quench). Using the algorithm presented in Section IV.C, the

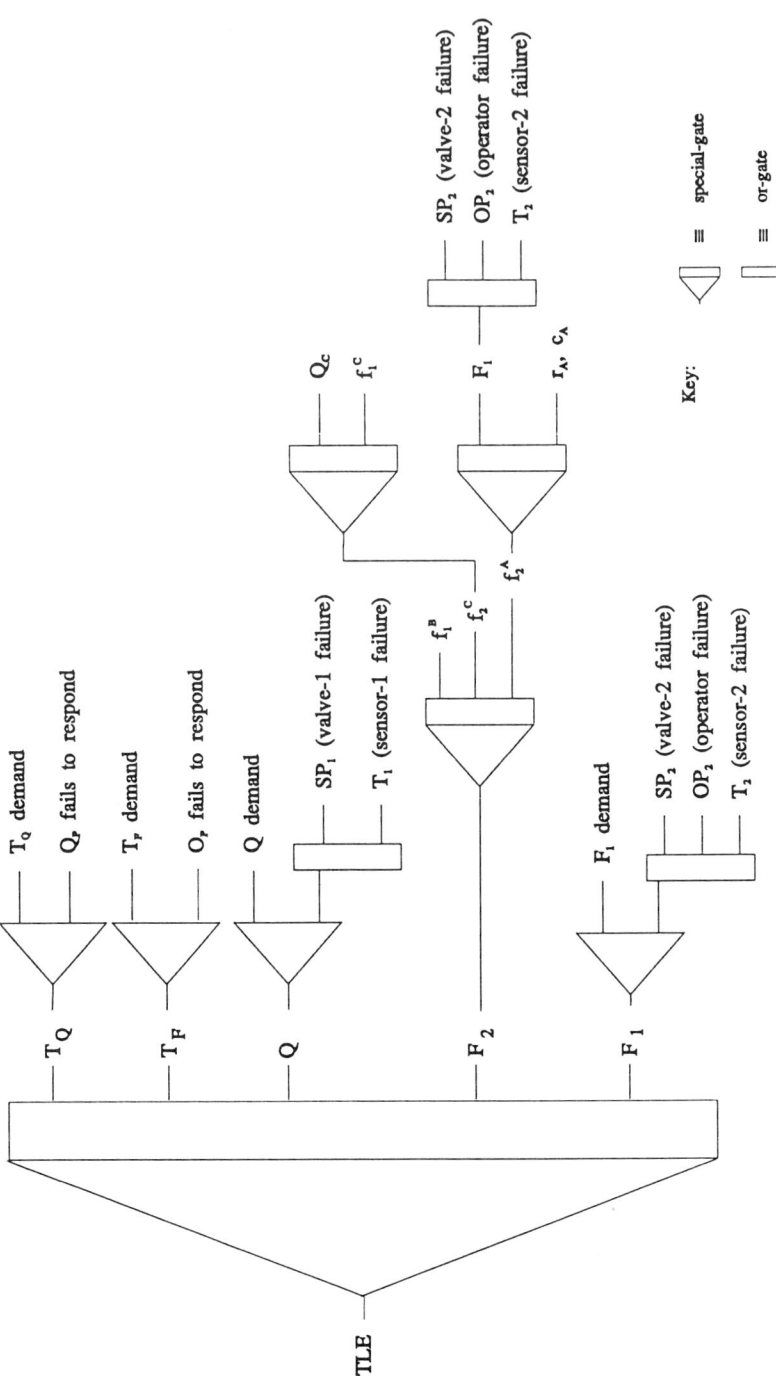

Fig. 24. The topological fault tree resulting from *Assumption-1* and *Assumption-2*.

logical gate constructed from this input variable is an *and-gate*, the result of a controlled variable involving an unbranched node. T_F, feed temperature, is the next variable with a pathway to a top-level event. Similarly, an *and-gate* is constructed for feed temperature, T_F, representing a thermal deviation demand in the feed and the failure of the operator to notice the upset, or take the required action. Quench flow rate, Q, also has a pathway to the top-level event (see Fig. 23). Following the algorithm proposed, construction of logical gates associated with this pathway illustrates that a demand by Q can be enabled in one of two ways: (1) the stem position of valve 1 fails to obtain the necessary position (i.e., the valve fails to close), or (2) the sensor setpoint T_1 is in error.

Unlike T_2, T_F, and Q, the outlet feed, F_2, requires a special gate to model the logical consequences of root causes passing through it. This gate takes as its input f_2^B, f_2^C, and f_2^A. Figure 23 illustrates that f_2^B obtains its input directly from f_1^B, a Type-2 technology. Consequently, any disturbance in f_1^B could potentially enable F_2. Since f_2^C and f_2^A are branched nodes, they in turn require construction of special-gates. The special-gate constructed from f_2^C takes as its input Q, and f_1^C. Similar to f_1^B, f_1^C is a Type-2 technology that takes its input directly from the

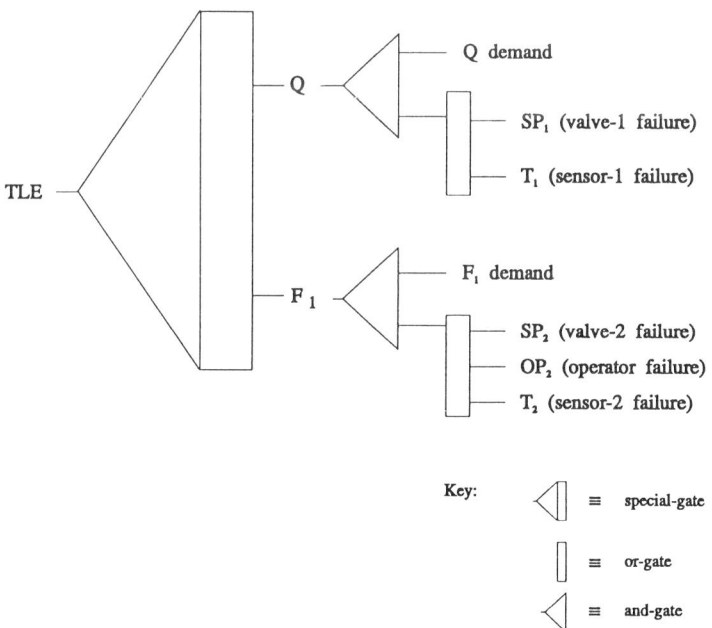

FIG. 25. The topological fault tree resulting from *Assumption-2* and *Assumption-3*.

external world. Therefore, any disturbance in the external surroundings that can lead to a change in the value of f_1^C could potentially also enable the top-level event.

Variable f_2^A, which is a branched node, also requires the construction of a special-gate. This gate requires as its input F_1 and $\{r_A, C_A\}$. Although the expansion for this node is not shown, what is important to recognize is the fact that with the exception of V_{RE}, each of the inputs leading to this set have been covered previously. The implications of changing V_{RE}, a Type-2 technology, is that a change in reactor volume could enable a top-level event. Although the mechanism for this process has not been modeled, it could be the result of an improper design or the accumulation of material internal to the reactor. *What is important is that the effect of V_{RE} has been made explicit.*

The final variable with a pathway leading to the top-level event is F_1, an unbranched node. Since F_1 has associated with it a Type-3 technology as its protective system, a demand on F_1 can be mitigated by the proper positioning of the stem controlling valve 2, proper action by the operator, and a proper setpoint (i.e., T_2) on the temperature controller. A con-

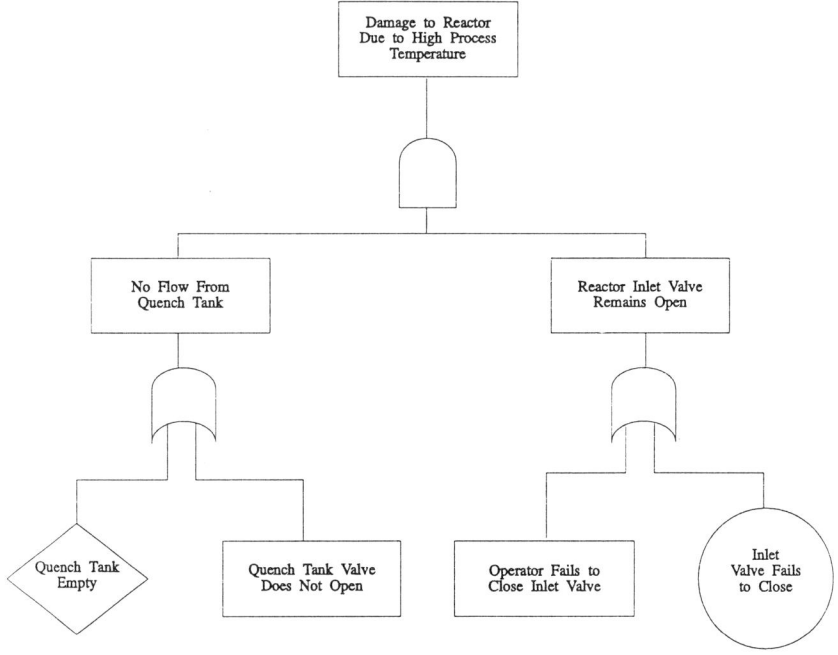

FIG. 26. The industrial implementation of a fault tree for the reactor-quench example.

densed version of the topological fault tree shown in Fig. 24, involving only Type-3 technologies, is shown in Fig. 25. It identifies two principal pathways to the top-level event: Q and F_1. Notice that a special-gate is still needed to connect these inputs to the output, T_R. The necessity of the special-gate will remain until *quantitative* analysis can determine whether the top-level event can be enabled by either Q or F_1, or whether both are required. This depends on the dynamics of the system: reaction rates, time requirements for runaway, heat release rates, heat removal rates, reactor volume, quench rates, etc.

The fault tree cited in literature for this process is shown in Fig. 26 (Battelle, 1985). Notice the similarity between Figs. 26 and 25, particularly in the structure of the two trees, and recognize that as a result of quantitative analysis, Fig. 26 has an *and-gate* as its top-level gate. More importantly, recognize that without complete quantification of the root causes, the fault tree given in Figure 26 may be incomplete.

V. Conclusion

By using the domain-specific modeling languages, LCR and MODEL.LA., we have developed a process-based methodology to identify potential hazards in a chemical process and generate mechanisms for the prevention or mitigation of their effects, through the identification and correction of inherent design weaknesses, or the adaptation of the operating procedures. The methodology is based on an interplay of inductive and deductive reasoning. *Inductive* reasoning has been used in order to identify (1) all potential chemical reactions, which could lead to a hazard, and (2) the requisite conditions that would enable the occurrence of these reactions. *Deductive* reasoning has been used to convert the enabling conditions needed for a reaction into process design or operational "faults" that would constitute the causes for the occurrence of a hazard. The methodology is more efficient, complete, and cost-effective than are current hazard analysis approaches. It contributes three important achievements: (1) formalization of the hazards identification problem, (2) systemization of hazard analysis through all phases of the design process, and (3) construction of a methodology that completely identifies all potential hazards within the scope of the modeling efforts. Furthermore, it establishes a formal strategy for the integration of safety into a design technology at any point in the design process and provides a means for discriminating among design alternatives with respect to disturbance mitigation. Finally, the methodology also provides the basis for the optimization of a design technology with respect to the parameters that describe its inherent safety.

References

Abelson, H., Sussman, G. J., and Sussman, J., "Structure and Interpretation of Computer Programs." MIT Press, Cambridge, MA, 1985.

Atallah, S., Assessing and managing industrial risk. *Chem. Eng.*, September, p. 8 (1980).

Batstone, R., *in* "Proceedings of the International Symposium on Preventing Major Chemical Accidents," February, p. 5.126. AIChE, Washington, DC, 1987.

Battelle, "Guidelines for Hazard Evaluation Procedures." AIChE Press, Washington, DC, 1985.

Boykin, R. F., and Kazarians, M., Quantitative risk assessment for chemical operations. *In* "Proceedings of the International Symposium on Preventing Major Chemical Accidents," February, p. 1.87. AIChE, Washington, DC, 1987.

Brachman, R. J., and Levesque, H. J., "Readings in Knowledge Representation." Morgan Kaufmann Publishers, Los Altos, CA, 1985.

Brannegan, D. P., "Hazards Evaluation in Process Development," Chemical Process Hazards Review, p. 18. American Chemical Society, Washington, DC, 1985.

Bretherick, L., "Bretherick's Handbook of Reactive Chemical Hazards." Butterworth, London, 1990.

Carson and Mumford, Analysis of incidents involving major hazards in the chemical industry. *J. Hazard. Mat.* **3**, 149 (1979).

Cormen, T. H., Leiserson, C. E., and Rivest, R. L., "Introduction to Algorithms." MIT Press, Cambridge, MA, 1990.

Cox, R. A., An overview of hazard analysis. *In* "Proceedings of the International Symposium on Preventing Major Chemical Accidents," February, p. 1.37. AIChE, Washington, DC, 1987.

Culbertson, T. L., and Searson, A. H., "Exxon Facility Design Assessment and Control of Hazards," Exxon internal publication. Exxon, 1983.

Dale, S. E., Cost effective design considerations for safer chemical plants. *In* "Proceedings of the International Symposium on Preventing Major Chemical Accidents," p. 3.79. AIChE, Washington, DC, 1987.

de Groot, J. J., and Van der Elst, F. H., Thermal properties of peroxides. *Inst. Chem. Eng. Symp. Ser.* **68**, 3/V:1 (1981).

Haastrup, P., Design errors in the chemical industry. *Inst. Chem. Eng. Symp. Ser.* **80**, J15 (1983).

Hendrikson, J. B., *J. Am. Chem. Soc.* **108**, 6748 (1986).

Hoffmann, J. M., Chemical process hazard review. *Am. Chem. Soc.*, p. 1 (1985).

ICI, "Process Safety," Course Notes. ICI, 1988.

Kletz, T. A., Make plants inherently safe. *Hydrocarbon Process.*, September, p. 72 (1985).

Kritikos, T., A model for process design automation. Ph.D. Thesis, Massachusetts Institute of Technology, Cambridge, MA (1991).

Lakshmanan, R., and Stephanopoulos, G., Synthesis of operating procedures for complete chemical plants. *Comput. Chem. Eng.* **12**, 985 (1988).

Lees, F. P., "Loss Prevention in the Process Industries." Butterworth, London, 1980.
Lees, F. P., Hazards warning structure: Some implication and applications. *Inst. Chem. Eng. Symp. Ser.* **80**, J1 (1983).
Lowe, D. R., and Solomon, C. H., Hazards identification procedures. *Inst. Chem. Eng. Symp. Ser.* **80**, G8 (1983).
Maher, M. L., *in* "Expert Systems in Engineering" (D. T. Phan, ed.). IFS Publications/Springer-Verlag, Berlin, 1988.
Mosleh, A., Bier, V. M., and Apostolakis, G., A critique of current practice for the use of expert opinion in probabilistic risk assessment. *Reliab. Eng. Syst. Saf.* **20**, 63 (1988).
Nagel, C. J., Identification of hazards in chemical process systems. Ph.D. Thesis, Massachusetts Institute of Technology, Cambridge, MA (1991).
Ozog, H., Hazard identification analysis and control. *Chem. Eng. (N.Y.).* February 18, p. 161 (1987).
Ozog, H., and Bendixen, L. M., Hazard identification and quantification. *Chem. Eng. Prog.*, April, p. 55 (1987).
Perkins, J. D., and Barton, G. W., Modelling and simulation in process operation. *In* "Foundations of Computer-Aided Process Operations" (G. V. Reklaitis and H. D. Spriggs, eds.). CACHE Corp., Austin, TX, and Elsevier, New York, 1987.
Sheil, B., Power tools for programming. *Datamation*, February, p. 131 (1983).
Sheil, B., The artificial intelligence tool box. *In* "Artificial Intelligence Applications in Business" (W. Reitman, ed.), p. 113. 1984.
Slater, C., and Pitblado, "Major Industrial Hazards Project Report." The Warren Centre for Advanced Engineering, University of Sidney, Sidney, Australia, 1987.
Sriram, D., and Maher, M. L., *in* "Applications of Artificial Intelligence in Engineering Problems" (D. Sriram and R. Adey, eds.), Vol. 1. Southhampton University, UK 1986.
Stephanopoulos, G., The future of expert systems. *Chem. Eng. Prog.*, September, p. 44 (1987).
Stephanopoulos, G., Johnston, J., Kritikos, T., Lakshmanan, R., Mavrovouniotis, M., and Siletti, C., Design-kit: An object-oriented environment for process engineering. *Comput. Chem. Eng.* **11**, 655 (1987).
Stephanopoulos, G., Johnston, J., and Lakshmanan, R., An intelligent system for planning plant-wide process control strategies. *Journal A* **29**(3), 81 (1988).
Stephanopoulos, G., Henning, G., and Leone, H., MODEL.LA. A modeling language for process engineering. Part I. *Comput. Chem. Eng.* **14**, 813 (1990a).
Stephanopoulos, G., Henning, G., and Leone, H., MODEL.LA. A modeling language for process engineering. Part II. *Comput. Chem. Eng.* **14**, 847 (1990b).
Stoessel, F., Experimental study of thermal hazards during the hydrogenation of aromatic nitro compounds. *Proc. Int. Symp. Loss Prev. Saf. Promot. Process Ind.*, 6th, p. 77-1 (1989).
Walling, C., "Free Radicals in Solution." Wiley, New York, 1957.

SEARCHING SPACES OF DISCRETE SOLUTIONS: THE DESIGN OF MOLECULES POSSESSING DESIRED PHYSICAL PROPERTIES

Kevin G. Joback[1] and George Stephanopoulos

Laboratory for Intelligent Systems in Process Engineering
Department of Chemical Engineering
Massachusetts Institute of Technology
Cambridge, MA 02139

I. Introduction	258
A. Brief Review of Previous Work	260
B. General Framework for the Design of Molecules	264
II. Automatic Synthesis of New Molecules	267
A. The Generate-and-Test Paradigm	267
B. The Search Algorithm	271
C. Case Study: Automatic Design of Refrigerants	283
D. Case Study: Automatic Design of Polymers as Packaging Materials	284
III. Interactive Synthesis of New Molecules	290
A. Illustration of Interactive Design	291
B. Case Study: Interactive Design of Refrigerants	296
C. Case Study: Interactive Design of an Extraction Solvent	299
D. Case Study: Interactive Design of a Pharmaceutical	301
IV. The *Molecule-Designer* Software System	304
A. General Description	304
B. Interactive-Design-Relevant Sections	305
V. Concluding Remarks	307
References	309

Strings of letters form words. From words to verses and stanzas, a poet composes a work with its own dynamic behavior, such as emotional impact on the reader, that transgresses the character of its components. In an analogous manner, atoms form functional groups, and these, in turn, yield molecules with distinct behavior, e.g., physical properties. It takes a

---

[1] Present address: Molecular Knowledge Systems, Inc., Nashua, New Hampshire, USA.

Homeric or Shakespearean genius to convert letters to an epic with a predefined desired impact. It suffices to efficiently search a space of combinatorial alternatives, in order to identify the molecules that satisfy the desired constraints on a set of physical properties. Often, the requisite scientific knowledge is fragmented, dispersed, and nonformalized, making the deductive search for the desired molecules inefficient or impossible. The inductive "genius" of a scientist or engineer is needed to break the impasse in such cases. By evolution or revolution one needs to respond to tighter and shifting product specifications and identify new solvents, pharmaceuticals, imaging chemicals, herbicides and pesticides, refrigerants, polymeric materials, and many others. In this chapter we will sketch the characteristics of an intelligent, computer-aided tool to support the synthetic search for the desired molecules. With functional groups as the "letters" of an alphabet, automatic and interactive procedures compose and screen classes of potential molecules. The automatic synthesis algorithm defines and searches the space of discrete solutions (molecules) through a hierarchical sequence of the space's representations. At each level of detail, a set of explicit constraints is used to depict restrictions on the structure of molecules that can be generated from various combinations of functional groups. In addition, interval arithmetic is employed to test the satisfaction of design specifications, leading to the elimination of large classes of infeasible molecules. One, though, should never overestimate the effectiveness of search algorithms in locating the desired solution(s). Quite frequently we need to resort to human-driven, abductive jumps. In this chapter we will also describe how the automatic search can become interwoven with effective human–machine interaction. Thus, the resulting computer-aided tool, the *Molecule-Designer*, constitutes a paradigm of an intelligent system with two distinct but integrated and complementary capabilities. Examples on the synthesis of refrigerants, solvents, polymers, and pharmaceuticals will illustrate the logic and features of the design procedures in the *Molecule-Designer*.

I. Introduction

Physical properties have a major impact on the economics of many processes and the viability of many products. The refrigerant in a refrigeration cycle, the working fluid in a power cycle, and the solvent used in an azeotropic distillation all determine the physical and economic feasibility of the corresponding processes. Chemical products such as artificial sweeteners, lubricants, and textiles all must exhibit specific physical properties if they are to be accepted.

Until recently, the identification of compounds possessing desired physical property values required extensive experimental search through vast numbers of candidate molecules. Estimates indicate that 3000–5000 compounds need to be tested before a new, useful pharmaceutical is identified, and 5000–8000 to find a new pesticide (Verloop 1972). Computational estimation of the physical property values for candidate molecules can reduce significantly the need for experimentation. In this spirit, Horvath (1992) published a tremendous thesaurus of techniques and approaches for molecular design, by focusing on the estimation of physical properties from molecular structures. Numerous techniques are available for the computational estimation of thermodynamic properties (Reid et al., 1987), environmental properties (Lyman et al., 1982), polymer properties (van Krevelen 1976), biological activity (Martin, 1978; Hansch and Leo, 1979; Franke, 1984), phase equilibria (Fredenslund et al., 1977), and others. However, these analytic methods can not be directly used in a synthetic manner for the design of molecules, as the following example illustrates:

Consider the design of a molecule whose physical properties, PP_1, PP_2, and PP_3, should have the values α, β, and γ, respectively. If functions f_1, f_2, and f_3 relate the three physical properties to the molecular structure, then

$$PP_1 = f_1(molecular\ structure) = \alpha,$$

$$PP_2 = f_2(molecular\ structure) = \beta, \qquad (1)$$

$$PP_3 = f_3(molecular\ structure) = \gamma.$$

The molecular structure of the unknown chemical could be found by inverting these three relationships. However, an explicit inversion is not analytic (the molecular structure is described by integer variables denoting the presence or absence of specific atoms and bonds), and it accepts multiple solutions (there may be several molecules satisfying the constraints). Implicit inversion of Eqs. (1) is possible through the formulation of appropriate optimization problems. However, in such cases the complexity and nonlinear character of the functional relationships used to estimate the values of physical properties in conjunction with the integer variables description of molecular structures, yield very complex mixed-integer optimization formulations.

Thus it is not surprising that the first efforts in systematically designing molecules possessing desired physical properties were heuristic in character, focusing on specific classes of chemical products. Godfrey (1972) used an empirical miscibility scale to determine whether two liquids were miscible. Francis (1944) used critical solution temperatures to choose solvents for the selective extraction of hydrocarbons. Berg (1969) developed a hydrogen bond classification scheme used to identify azeotropic distillation solvents. In recent years, the sophistication of the methods has improved (Venkatasubramanian *et al.*, 1994; Constantinou *et al.*, 1994; Gani *et al.*, 1991; Gani and Fredenslund, 1993; Nielsen, *et al.*, 1995), but the basic character of the various approaches has remained the same, specifically, all design-oriented techniques, (1) are *problem-specific*, i.e., for solvents, polymers, or pharmaceuticals; (2) are based on the *generate-and-test* paradigm with experiential heuristics employed to reduce the search space of potential alternatives; and (3) cannot utilize efficiently all available knowledge in a given area of molecular design.

A. Brief Review of Previous Work

Research findings from the areas of physical property estimation and chemical products selection are applicable to the design of molecules. A brief review of research in these areas is presented along with previous work in molecular design.

1. Estimation of Physical Properties

Every approach developed for the design of molecules with desired properties requires the estimation of physical properties. Some property estimation techniques lend themselves easily to the identification of the requisite molecular structures, starting from the desired design specs, and others do not. Depending on how various approaches attempt to relate molecular structure to physical properties, they can be grouped into five categories: pattern recognition, topological, group contribution, equation-oriented, and molecular-modeling-based techniques.

a. Pattern Recognition. Discriminant analysis and classification are the two statistical techniques used most often in pattern recognition. Both are multivariate techniques concerned with *separating* distinct sets of objects into classes and *allocating* new objects to previously defined classes. A discriminant function is developed from a set of experimental data called the "training set." This function is then used to classify new compounds. In many applications the number of classes equals 2; e.g., carcinogenic or noncarcinogenic, toxic or nontoxic.

Pattern recognition techniques lend themselves to synthetic designs of molecules for those problems for which the design specs require that a molecule is a member of a certain class.

b. Topological Techniques. These techniques ignore the actual three-dimensional shape of a molecule, the nature and lengths of the chemical bonds connecting its atoms, the angles between the bonds, and sometimes even atom types (Rouvray, 1986). Typically only the number of atoms and their interconnections are considered. This information is reduced to an *index* such as the Wiener Path Number (Wiener, 1947), Alternburg Polynomial Index (Alternburg, 1966), Gordon–Scantlebury Index (Gordon and Scantlebury, 1964), Hosoya's Z Index (Hosoya and Murakami, 1975), or Randić's Branching Index (Randić, 1975). These indices are then used to correlate the values of physical properties. The applicability of these techniques for property estimation has been extensively discussed in Kier and Hall (1986). By their nature, topological techniques require detailed information about the molecular connectivity of a compound and are difficult to incorporate into synthetic design procedures.

c. Group Contribution Techniques. These assume that each fragment of a molecule contributes a certain amount to the value of its physical properties. Contributions for each group are statistically regressed from large sets of experimental data. Techniques can become very complex, including nonlinear effects and interactions among groups. They are very appropriate for the design of molecules with desired physical properties and they constitute the basis for both *interactive* and *automatic* design procedures described in this chapter. Group contribution techniques have been used by other researchers for molecular design, as we will see in the next section.

d. Equation-Oriented Techniques. These techniques correlate estimated physical properties to properties more easily available or measured using empirical or theoretical models. Not relating a compound's molecular structure to its properties, these techniques cannot be used directly for molecular design. However, used in conjunction with group contribution techniques, rendering the values of the correlated physical properties, they broaden the list of physical properties which yield the specifications of the desired molecule.

e. Molecular-Modeling-Based Techniques. These techniques start with an atomic model of a molecule and use quantum and statistical mechanics to estimate its physical properties. Many of these estimates are more accurate than those obtained by any other estimation technique. Molecular-modeling-based techniques offer fairly complex and implicit relationships between molecular structure and physical properties. A straightforward generate and test is the only way such techniques are employed.

2. Selection of Desired Chemicals

Selecting a chemical product from a set of candidates is a two-step procedure. The first step is the most critical and involves the identification of those physical properties that are important to the performance of the chemical product and their values that give optimal performance. The second step involves a search through a database for existing compounds that possess these physical properties values. Unknown property values must be estimated. Such an approach has the advantage of being fast. Additionally, compounds in the database are, typically, commercially available or can be readily synthesized. The drawback of such an approach is that new compounds cannot be found.

3. Design of a Desired Chemical

This is also a two-step procedure similar to that of selecting a desired chemical from a list of candidate molecules. Unlike the selection from an existing set of compounds, the design of compounds implies the synthetic stipulation of molecules for which there are no experimental data of their physical properties. Therefore, using some of the available estimation techniques, different approaches have been proposed in the past and will be discussed in the following paragraphs.

a. Design of Solvents. Gani and Brignole (1983) and Brignole *et al.* (1986) used the UNIFAC (Fredenslund *et al.*, 1977) group contribution method to synthesize molecular structures with specific solvent properties for separation processes. Their synthesis procedure is divided into three steps:

1. Select the groups considered to be suitable building blocks for the molecular structures.
2. Combine the groups into candidate molecules according to specified combination rules.
3. Screen the candidate molecules using UNIFAC to evaluate their usefulness for a particular separation task.

To reduce the number of potential group combinations to a tractable number, several additional constraints are placed on the candidate solvents. For example, a high boiling point is required in order to facilitate simple separation of the solvent by distillation. In similar spirit are the works of Gani et al. (1991) and Gani and Fredenslund (1993), but the efficiency of search has improved with heuristic knowledge, while techniques for discrete optimization have been used to design optimal solvent mixtures. The works of Macchietto et al. (1990) and Odele and Macchietto (1993) have focused on the selection (rather than design) of optimal solvents for extractive separation processes.

b. Design of Polymers. Derringer and Markham (1985) proposed a generate and test methodology for designing polymers possessing desired physical properties, using the van Krevelen (1976) group contribution estimation techniques. Recognizing that the number of candidate polymers may be large, Derringer and Markham devised a ranking procedure that includes a desirability measure for each predicted property.

c. Design of Polymer Coatings. Tortorello and Kinsella (1983a, b) used the solubility parameter concept to design high-performance aircraft coatings resistant to water, fuels, hydraulic fluids, and lubricating oils. Additional design specs included resistance to high temperature and flexibility at low temperature.

d. Design of Drugs. Drug design has been the most active area for the development of systematic procedures to identify new chemical products. Beginning with a small set of experimental data on the efficacy of candidate compounds, one derives a statistical relationship between the drug's potency and a set of physicochemical properties such as Hammett's constant (1935), Taft's (1956) steric parameter, and the octanol–water partition coefficient, whose use was made popular by Hansch and coworkers (1963). These physicochemical properties are then related to structural characteristics. Such statistical relationships are called *quantitative structure activity relationships* (QSARs).

QSARs provide great insight into the drug design problem. Examination of the derived relationships often indicates how a drug's potency is being affected by reactive, transport, and steric considerations. To improve the potency, a drug designer can search for substituents, which when added to the candidate molecule will reduce steric hindrances or increase the rate of transport. Extensive tabulations exist (Hansch and Leo, 1979) listing the effect that specific substituents have on the various physical-chemical properties, typically used in drug designs.

B. General Framework for the Design of Molecules

The previous works on selecting and designing molecules with desired properties share certain common characteristics. Beginning with these characteristics, a general methodology was developed for designing molecules (Joback and Stephanopoulos, 1990). The overall philosophy of the design methodology is described in the following six paragraphs.

1. Problem Formulation

The first step in any design is to identify the target (Stephanopoulos and Townsend, 1986). A molecular design target consists of physical property, chemical, and structural constraints. Physical property constraints are typically concerned with the performance of the chemical product, such as its ability to perform as an aircraft coating. Chemical constraints are often related directly to molecular structure, restricting or requiring the occurrence of functional groups, such as the desire to design a diol with desired properties. Structural constraints are required when a molecule is constructed by assembling functional groups. The groups must have the correct type and occurrence of bonds so that they can be assembled into feasible molecules. Brignole *et al.* (1986) developed an extensive set of rules to constrain the choice of groups and ensure the structural feasibility of the resulting molecule. The effective formulation of the molecular design problem is crucial to the success of the design.

2. Target Transformation

For the computer to evaluate the performance of a candidate molecule, it must be able to estimate the values for those physical properties identified during problem formulation. The target transformation step develops *estimation procedures*, which enable the evaluation of the target constraints' physical properties in terms of the values of new physical properties (i.e., the transformed target). For example, if we want to design a molecule with vapor pressure P_{vp}, in a given range of values, we can use the Riedel–Plank–Miller correlation [Eq. (2)] and transform the target into these new properties, as shown by Equations (3a), (3b), and (3c):

$$P_{vp} = P_{vp}(T_b, T_c, P_c) \quad \text{(correlation by Riedel–Plank–Miller)}, \quad (2)$$

where

$$T_b = T_b(\text{molecular structure})$$

(group-contribution technique by Joback), (3a)

$$P_c = P_c(\text{molecular structure})$$

(group-contribution technique by Lydersen), (3b)

$$T_c = T_c(\text{molecular structure})$$

(group-contribution technique by Fedors). (3c)

Starting with a compound's molecular structure, the T_b, P_c, and T_c are estimated first, using the three group contribution techniques indicated above. These values are then used in an equation-oriented technique to yield the final estimate for P_{vp}.

3. Design Procedure

The previous design approaches are based on the generate-and-test paradigm. This paradigm consists of two parts: the *generator* and the *tester*. The generator enumerates candidate solutions, whereas the tester evaluates each candidate and either accepts or rejects it. When the number of candidates becomes very large, exhaustive enumeration becomes impractical. Although not all previous molecular design approaches have explored them, several strategies are available to manage the search space (Hayes-Roth *et al.*, 1983), such as (1) move the tester into the generator, (2) prune partial solutions, and (3) abstract the search space. The last strategy is extremely powerful in managing the combinatorics of the design problem and constitutes the basis of the automatic design methodology to be discussed in Section II of this chapter.

4. Representation and Enumeration of Molecules

Designing molecules through the use of group-contribution estimation techniques results in candidate molecules which are represented as a collection of functional groups. To form complete molecules, it is necessary to connect these groups together. At times more than one way of connecting the groups is possible. For example, the following collection of groups

$$-CH_3 \quad >C< \quad -F \quad -F \quad -Cl \quad >C= \quad =CH_2$$

TABLE I
FOUR ENUMERATED MOLECULES

$$CH_3-\underset{\underset{Cl}{|}}{\overset{\overset{F}{|}}{C}}-\overset{\overset{F}{|}}{C}=CH_2 \qquad CH_3-\underset{\underset{F}{|}}{\overset{\overset{Cl}{|}}{C}}-\overset{\overset{F}{|}}{C}=CH_2$$

$$CH_3-\underset{\underset{F}{|}}{\overset{\overset{F}{|}}{C}}-\overset{\overset{Cl}{|}}{C}=CH_2 \qquad F-\underset{\underset{Cl}{|}}{\overset{\overset{F}{|}}{C}}-\overset{\overset{CH_3}{|}}{C}=CH_2$$

can be combined, ignoring stereoisomers, to form the four different molecules shown in Table I. Molecule enumeration has been extensively investigated by researchers doing work in structure elucidation (Gray, 1986). Before more rigorous estimation techniques, such as molecular modeling, or chemical constraints can be used, it is necessary to provide techniques for the representation and enumeration of all potential molecular structures.

5. Screening of Molecules

Chemical constraints are applied once the satisfactory candidate molecules have been enumerated. Typically, chemical constraints prevent the generation of unstable substructures within the structure of generated molecules. For example, if the substructure —O—O— occurs in a compound desired to be stable, that compound is pruned. For design procedures based on group-contribution techniques, the application of chemical constraints must occur after enumeration, since the relative locations of the various groups within a molecule remain unspecified at earlier stages.

6. Final Evaluation

It is sometimes necessary to modify the physical properties estimation techniques employed by generate-and-test design procedures. Often this modification is introduced in order to remove computational steps in the estimation techniques that require knowledge of the global molecular structure. Using groups as the design basis, only the partial and local structure of a molecule is known during the design. However, once the candidate molecules are enumerated and screened, global molecular structure is known and more accurate estimation techniques can be used to further prune the candidates.

II. Automatic Synthesis of New Molecules

This section presents an algorithmic strategy for the automatic generation of molecules and their screening against a set of constraints, which represent the physical properties' values that the desired molecules should satisfy. *Functional groups* are the essential building blocks for the construction of molecules, allowing the use of *group-contribution estimation techniques* for the testing of physical property constraints. These techniques are simple, fast, and yield estimates of sufficient accuracy for preliminary screening purposes. The use of more complicated estimation techniques, such as molecular modeling, is unwise at this stage of design. The probability of an ab initio automatic identification of satisfactory molecules at this stage is low, implying that the effort expended for the examination of each candidate molecule should be kept to a minimum. Thus, a hierarchical approach has been adopted for the synthesis of desired molecules:

Phase 1. Simple estimation techniques are used to rapidly screen large number of candidate molecules, generated by a hierarchical search algorithm.

Phase 2. Candidate molecules satisfying the design constraints are reported to the human designer, who orders them using subjective preferences.

Phase 3. The retained molecules are evaluated through the use of more detailed estimation techniques, e.g. molecular modeling, complex equations of state.

In this chapter we will deal only with phase 1.

A. The Generate-and-Test Paradigm

The generate-and-test search paradigm, used by the automatic design algorithm for the synthesis of molecules, is composed of two modules. The first module, the *generator*, enumerates candidate molecules. The second, the *tester*, evaluates each molecule, estimating its physical properties and checking for structural feasibility, and either accepts or rejects it. We represent molecules as collections of groups, e.g., chloropropane is represented as (—Cl —CH$_3$ —CH$_2$— —CH$_2$—). The generator simply constructs candidate molecules by selecting a collection of groups from an initial set. The representation of groups allows the tester to use group-contribution techniques to estimate physical property constraints and check the design constraints.

The combinations of groups that can be selected is infinite. However, from practical considerations, molecules for a typical application fall within some size range, which can be translated into an upper limit on the number of groups chosen. For example, refrigerants are generally of a small molecular weight. Placing a limit of 15 on the number of groups that can be used to form a molecule is a reasonable bound. A lower limit is established from the fact that at least two groups must be used to form a structurally feasible molecule. With limits on the minimum and maximum number of groups that can be chosen, the generator selects collections of groups beginning with all combinations of two groups, then all combinations of three groups, and so on until the upper limit is reached.

1. Design Constraints

The tester module checks each candidate molecule for satisfaction of the design constraints. Three types of constraints are used: (a) physical property constraints, (b) structural constraints, and (c) chemical constraints.

a. Physical Property Constraints. Estimation procedures are established for each physical property used in the design constraints. An estimation procedure is a collection of estimation techniques that determine physical property values when only the molecular structure is known. It is composed of group-contribution and equation-oriented estimation techniques. For example, using the constraint on vapor pressure, we obtain

$$P_{vp}(273 \text{ K}) > 1.01 \text{ bar}.$$

We need to employ an estimation procedure for P_{vp} and the associated independent physical properties, such as those described in Section I.B [see Eqs. (2) and (3a–c)], in order to determine each candidate molecule's vapor pressure at 273 K. Starting with a compound's molecular structure, T_b, T_{br}, and P_c are first estimated using the three group-contribution techniques [Eqs. (3a), (3b), and (3c), respectively]. These values are then used in an equation-oriented technique to yield the final estimate for P_{vp} [Eq. (2)].

b. Structural Constraints. Structural constraints determine whether a collection of groups can be connected in some manner to form a feasible molecule. Three requirements define the conditions for structural feasibility:

1. All groups in the collection should be able to be joined into a single connected component. The collection of groups (—F —F —F —F) is not feasible because it does not form a single molecule.
2. The single connected molecule formed from a set of groups cannot have any unconnected bonds. Connecting the groups (—CH_2— —F) gives us a single structured entity with one single bond unconnected.
3. The connections made in the single connected entity, formed from a set of groups, must all be between bonds of the same type. Single bonds may only connect with single bonds, double with double, etc.

Given a set of groups, it is possible to enumerate all ways in which they could be connected verifying that at least one candidate satisfies the three requirements. However, a graph theoretic examination of molecular structures provides a set of structural constraints that are much easier to apply. Such structural constraints have been developed and are shown in Table II.

c. Chemical Constraints. Chemical constraints are heavily dependent on the global connection of atoms within a molecule. Representing molecules as collections of groups does not provide knowledge about global connectivity. Chemical constraints are thus better used at later stages of the search methodology, where the complete structure of a molecule is known.

2. Combinatorial Explosion

Given a reasonable number of groups from which molecules can be constructed, the number of candidates that can be generated is extremely large. Allowing repetition of groups and ignoring the order of selection, the number of candidate molecules that can be generated by selecting n groups from a set of k groups is given by Eq. (4):

$$C^R(k,n) = \frac{(k+n-1)!}{n!(k-1)!}. \qquad (4)$$

The total number of candidate molecules that can be selected from a set of k groups in which each candidate molecule has between 2 and n_{max}

TABLE II
Structural Constraints on Forming Feasible Molecules

1. If G is a collection of n groups, then $n \geq 2$.
2. If G is a collection of n groups with n_c cyclic groups, n_m mixed groups, and n_a acyclic groups, and $n_a > 0$ and $n_c > 0$, then $n_m > 0$.
3. If G is a collection of n groups with n_c cyclic groups, n_m mixed groups, and n_a acyclic groups, then $n_m > 0$ implies that $n_a > 0$ or $n_c > 0$.
4. If G is a collection of n groups with n_c cyclic groups and n_m mixed groups, then either $n_c + n_m \geq 3$ or $n_c + n_m \geq 2$.
5. If G is a collection of groups, then the number of groups having an odd number of free bonds must be even.
6. If G is a collection of n groups with b free bonds, then $\frac{b}{2} \geq n - 1$.
7. If G is a collection of n groups with b free bonds, then $\frac{b}{2} \leq \frac{1}{2} n(n-1)$.
8. If a collection of groups contains more than one bond type, then there must be a transition group containing each bond type. A transition group is one that contains more than one bond type.
9. If G is a collection of groups with $n_{a,i}$ denoting the number of acyclic groups with a valence i and v_{mj} denoting the valence of some jth mixed group, then $n_1 \leq \Sigma_{\text{mixed}}(v_{m,j} - 2) + n_{a,3} + 2n_{a,4} + \cdots + (i-2)n_{a,i} + \cdots$.
10. If G is a collection of n groups with n_i denoting the number of groups with a global valence i and all n groups are acyclic, then $n_1 = 2 + n_3 + 2n_4 + \cdots + (i-2)n_i + \cdots$.
11. The number of occurrences of each bond type in a collection of groups must be even.

groups is given by Eq. (5):

$$\text{Total candidates} = \sum_{n=2}^{n_{\max}} C^R(k,n) = \sum_{n=2}^{n_{\max}} \frac{(k+n-1)!}{n!(k-1)!}. \qquad (5)$$

Table III shows how this total number of candidates can quickly grow to very large values. Managing this combinatorial explosion is the major focus of the automatic design algorithm.

TABLE III
Combinatorics of Group Selection ($k = 40$ Groups)

$n_{\max}$	# Molecules
4	135,710
5	1,221,718
6	9,366,778
7	62,891,458
8	377,348,953
9	2,054,455,593

TABLE IV
INITIAL SET OF GROUPS

$>$CH$_3$	—CH$_2$—	$>$CH—	$>$C$<$
=CH$_2$	=CH—	=C$<$	=C=
≡CH	≡C—	—F	—Cl
—Br	—I	—OH	—O—
$>$CO	—CHO	—COOH	—COO—
=O	—NH$_2$	$>$NH	$>$N—
—CN	—NO$_2$	—SH	—S—

B. THE SEARCH ALGORITHM

The magnitude of the combinatorial problem, resulting from a large number of functional groups, can be reduced only by reducing the number of groups. This reduction is done by abstracting the groups into families of groups, called *metagroups*. Table V shows the groups of Table IV clustered into four metagroups.

Instead of generating molecules by choosing from an initial set of groups, we choose from an initial set of metagroups. The candidate molecules formed from a collection of metagroups are called *metamolecules*, which are sets of molecules. Using the metagroups from Table V, the metamolecule (2 1 0 0) is the set of all molecules that can be formed by taking any two groups from *metagroup 1* and any one group

TABLE V
EXAMPLE METAGROUPS

Metagroup 1 $\{$ —CH$_3$, =CH$_2$, ≡CH, —F, —Cl, —Br, —I, —OH, —CHO, —COOH, =O, —NH$_2$, —NO$_2$, —CN, —SH $\}$

Metagroup 2 $\{$ $>$CH$_2$, =CH—, =C=, ≡C—, $>$CO, —COO—, —O—, $>$NH, —S— $\}$

Metagroup 3 $\{$ =C$<$, $>$CH—, $>$N— $\}$

Metagroup 4 $\{$ $>$C$<$ $\}$

from *metagroup* 2. The number of molecules contained in metamolecule (2 1 0 0) is

$$C^R(15,2) \times C^R(9,1) = \frac{(15+2-1)!}{2!(15-1)!} \times \frac{(9+1-1)!}{1!(9-1)!},$$

or 1080 molecules.

1. Evaluation of Metamolecules

Abstracting groups into metagroups reduces the combinatorics of candidate molecule generation. However, we must be able to efficiently evaluate whether a metamolecule satisfies the design constraints.

a. Structural Constraints. As long as all the groups within each metagroup have a consistent molecular characteristic such as global valence, the structural constraints are still applicable. Metagroups 1 and 2 are consistent in ring class and global valence. Structural constraint 10 of Table II is thus applicable. Applying the constraint to metamolecule (2 1 0 0) yields

$$2 = 2 + 0 + 2(0) = 2,$$

i.e. the constraint is satisfied. This implies that each of the 1080 molecules contained in (2 1 0 0) satisfies the constraint.

b. Physical Property Constraints. Associated with each of the groups in a metagroup is a contribution toward a particular physical property. The contribution of a metagroup is called a *metacontribution* and is defined by a set of values. Table VI shows the contributions toward the value of the boiling point, T_b, for each of the groups in metagroup 2 (see Table V) toward T_b. The metacontribution of metagroup 2 toward T_b is thus the following set of values:

(22.42 22.88 24.96 26.15 27.38 50.17 68.78 76.75 81.10).

To use metacontributions in the calculation of physical properties, it is necessary to find a representation that can capture the set value of the metacontributions and can be manipulated by mathematical operators. Interval numbers were chosen as the representation.

TABLE VI
T_b GROUP CONTRIBUTIONS FOR METAGROUP 2 (ACYCLIC GROUPS)

Groups	Contribution
—CH$_2$—	22.88
=CH—	24.96
=C=	26.15
≡C—	27.38
—O—	22.42
\>CO	76.75
—COO—	81.10
\>NH	50.17
—S—	68.78

2. Interval Arithmetic and Meta-Contributions

The generalization of ordinary arithmetic to closed intervals is known as *interval arithmetic*. An interval is defined as a closed bounded set of real numbers (Moore, 1979):

$$X = [\underline{X} \ \overline{X}] = \{x | \underline{X} \leq x \leq \overline{X}\}. \tag{6}$$

Thus, intervals have a *dual* nature as both a number and a set. The basic interval arithmetic operations are

$$[\underline{X} \ \overline{X}] + [\underline{Y} \ \overline{Y}] \equiv [\underline{X} + \underline{Y} \ \ \overline{X} + \overline{Y}],$$

$$[\underline{X} \ \overline{X}] + [\underline{Y} \ \overline{Y}] \equiv [\underline{X} - \overline{Y} \ \ \overline{X} - \underline{Y}],$$

$$[\underline{X} \ \overline{X}] * [\underline{Y} \ \overline{Y}] \equiv [\min(\underline{X} * \underline{Y}, \ \underline{X} * \overline{Y}, \ \overline{X} * \underline{Y}, \ \overline{Y} * \overline{Y}),$$

$$\max(\underline{X} * \underline{Y}, \ \underline{X} * \overline{Y}, \ \overline{X} * \underline{Y}, \ \overline{X} * \overline{Y})],$$

$$[\underline{X} \ \overline{X}] \div [\underline{Y} \ \overline{Y}] \equiv [\underline{X} \ \overline{X}] * [1/\overline{Y} \ \ 1/\underline{Y}] \quad \text{iff } 0 \notin [\underline{Y} \ \overline{Y}].$$

The metacontribution of the metagroup 2 (see Table VI) in interval representation is [22.42   81.10]. Thus, we can construct Table VII, which shows the metacontributions for each metagroups, displayed in Table V for boiling point T_b, reduced boiling point T_{br}, and heat of vaporization ΔH_{vb} (Joback and Reid, 1987). Using the group-contribution estimation

TABLE VII
METACONTRIBUTIONS

Metagroup	T_b	T_{br}	ΔH_{vb}
1	[−10.50 169.09]	[0.0027 0.0791]	[−0.670 19.537]
2	[22.42 81.10]	[0.0020 0.0481]	[2.205 9.633]
3	[11.74 24.14]	[0.0117 0.0169]	[1.691 2.138]
4	[18.25 18.25]	[0.0067 0.0067]	[0.636 0.636]

models

$$T_b = 198.18 + \sum_{\text{all groups}} n_i \Delta_{i, T_b}, \tag{7}$$

$$T_{br} = 0.584 + 0.965 \sum_{\text{all groups}} n_i \Delta_{i, T_{br}} - \left(\sum_{\text{all groups}} n_i \Delta_{i T_{br}} \right), \tag{8}$$

$$\Delta H_{vb} = 15.30 + \sum_{\text{all groups}} n_i \Delta_{i, \Delta H_{vb}}, \tag{9}$$

we can estimate the values of T_b, T_{br}, and ΔH_{vb} for metamolecule (2 1 0 0) as follows:

$$T_b = 198.18 + 2[-10.50 \quad 169.09] + [22.42 \quad 81.10]$$
$$= [199.6 \quad 617.46] \text{K}, \tag{10}$$

$$T_{br} = 0.584 + 0.965(2[0.0027 \quad 0.0791] + [0.0020 \quad 0.0481])$$
$$- (2[0.0027 \quad 0.0791] + [0.0020 \quad 0.0481])^2$$
$$= [0.549 \quad 0.783], \tag{11}$$

$$\Delta H_{vb} = 15.30 + 2[-0.670 \quad 19.537] + [2.205 \quad 9.633]$$
$$= [16.165 \quad 64.007] \text{ kJ/mol.} \tag{12}$$

The intervals given by Eqs. (10)–(12) span the range of physical property values possessed by each of the 1080 molecules in metamolecule (2 1 0 0).

Interval values for these fundamental properties can be used in equation-oriented estimation techniques. The Watson relation (Watson, 1943)

$$\Delta H_v = \Delta H_{vb} \left(\frac{1 - T/T_c}{1 - T_{br}} \right)^{0.38} \tag{13}$$

is used to estimate the enthalpy of vaporization at 250 K for metamolecule

(2 1 0 0), where T_c is obtained from

$$T_c = \frac{T_b}{T_{br}} = \frac{[199.6 \quad 617.46]}{[0.549 \quad 0.783]} = [254.9 \quad 1124.7].$$

Inserting the value of T_c into Eq. (13), we obtain the interval value of the enthalpy of vaporization:

$$\Delta H_v = [16.165 \quad 64.007]\left(\frac{1 - 250/[254.9 \quad 1124.7]}{1 - [0.549 \quad 0.783]}\right)^{0.38}$$

$$= [0.688 \quad 229.40] \text{ kJ/mol}.$$

3. Searching through Successive Molecular Abstractions

The generate-and-test search paradigm described earlier must now be modified to deal with the abstractions introduced by the metamolecules. Instead of generating and testing individual molecules, we generate and test metamolecules. Those metamolecules satisfying the test are reduced in abstraction, by dividing a metagroup into more meta-groups. This refinement produces a new generation of metamolecules that are retested.

Let us demonstrate the logic of the procedure using the metagroups of Table V and the meta-contributions of Table VII. Consider the following constraint on boiling point: $T_b > 500$ K.

Limiting the number of groups contained in a molecule between 2 and 4, the following 65 metamolecules are generated:

(2 0 0 0) (0 2 0 0) (0 0 2 0) (0 0 0 2) (1 1 0 0)
(1 0 1 0) (1 0 0 1) (0 1 1 0) (0 1 0 1) (0 0 1 1)
(3 0 0 0) (0 3 0 0) (0 0 3 0) (0 0 0 3) (2 1 0 0)
(2 0 1 0) (2 0 0 1) (0 2 1 0) (0 2 0 1) (0 0 2 1)
(1 2 0 0) (1 0 2 0) (1 0 0 2) (0 1 2 0) (0 1 0 2)
(0 0 1 2) (1 1 1 0) (1 1 0 1) (1 0 1 1) (0 1 1 1)
(4 0 0 0) (0 4 0 0) (0 0 4 0) (0 0 0 4) (3 1 0 0)
(3 0 1 0) (3 0 0 1) (0 3 1 0) (0 3 0 1) (0 0 3 1)
(1 3 0 0) (1 0 3 0) (1 0 0 3) (0 1 3 0) (0 1 0 3)
(0 0 1 3) (2 2 0 0) (2 0 2 0) (2 0 0 2) (0 2 2 0)
(0 2 0 2) (0 0 2 2) (2 1 1 0) (2 1 0 1) (2 0 1 1)
(0 2 1 1) (1 2 1 0) (1 2 0 1) (1 0 2 1) (0 1 2 1)
(1 1 2 0) (1 1 0 2) (1 0 1 2) (0 1 1 2) (1 1 1 1)

TABLE VIII
T_b Values for Four Metamolecules

Metamolecule	T_b
(2 0 0 0)	[177.18 536.36]
(2 1 0 0)	[199.60 617.46]
(3 0 1 0)	[178.42 729.59]
(2 2 0 0)	[222.02 698.56]

Recall that the metamolecule (1 0 1 2) represents the set of all molecules that can be formed by taking any one group from metagroup 1, no group from metagroup 2, any one group from metagroup 3, and any two groups from metagroup 4.

Structural constraint 10 of Table II is used to prune the candidate metamolecules. The maximum valence any group has is 4. Therefore, each metamolecule is checked to ensure that it satisfies the constraint $n_1 = 2 + n_3 + 2n_4$; 61 metamolecules are pruned using this constraint. The remaining four metamolecules are

$$(2\ 0\ 0\ 0)\quad (2\ 1\ 0\ 0)\quad (3\ 0\ 1\ 0)\quad (2\ 2\ 0\ 0).$$

Physical constraints are applied next. The boiling point value T_b was estimated using the metacontributions of Table VII and Eq. (7). Table VIII shows these estimates for each of the remaining four metamolecules. These values show that all four metamolecules satisfy the constraint on boiling point T_b.

The next step of the search algorithm is to reduce the level of abstraction. Groups were abstracted into metagroups to reduce the combinatorics of the problem. However, this same abstraction reduced the effectiveness of property constraints. As the abstraction is reduced, this effectiveness is regained. Metagroup 1 is divided into two new metagroups:

$$\begin{matrix} -CH_3 & =CH_2 & =CH & -F & -Cl & -Br \\ -I & -OH & -CHO & =O & -NH_2 & -SH \end{matrix}, \quad \text{(metagroup 1,1)}$$

$$\{-COOH \quad -NO_2 \quad -CN\}. \quad \text{(metagroup 1,2)}$$

This division of metagroup 1 is propagated to the metamolecules. Metamolecule (2 0 0 0) was the set of all molecules that could be formed by taking any two groups from metagroup 1. With metagroup 1 divided into

TABLE IX
T_b Values for 13 Metamolecules

[(2 0) 0 0 0]	[177.18 385.86]
[(0 2) 0 0 0]	[449.50 536.36]
[(1 1) 0 0 0]	[313.34 461.11]
[(2 0) 1 0 0]	[199.60 466.96]
[(0 2) 1 0 0]	[471.92 617.46]
[(1 1) 1 0 0]	[335.76 542.21]
[(3 0) 0 1 0]	[178.42 503.84]
[(0 3) 0 1 0]	[586.90 729.59]
[(2 1) 0 1 0]	[314.58 579.09]
[(1 2) 0 1 0]	[450.74 654.34]
[(2 0) 2 0 0]	[222.02 548.06]
[(0 2) 2 0 0]	[494.34 698.56]
[(1 1) 2 0 0]	[358.18 623.31]

two new metagroups there are three possibilities:

1. Take any two groups from metagroup 1,1.
2. Take any two groups from metagroup 1,2.
3. Take any one group from metagroup 1,1 and any one group from metagroup 1,2.

These possibilities correspond to an expansion of the metamolecule (2 0 0 0) into three new metamolecules:

$$[(2\,0)\,0\,0\,0] \quad [(0\,2)\,0\,0\,0] \quad [(1\,1)\,0\,0\,0].$$

Table IX displays the 13 new meta-molecules resulting from the expansion of all four metamolecules.

The metacontributions toward T_b are also divided; e.g.

$$T_b \text{ of metagroup } 1,1 = [-10.50 \quad 93.84],$$

$$T_b \text{ of metagroup } 1,2 = [125.66 \quad 169.09].$$

T_b is estimated for each meta-molecule and the property constraint is applied. Table IX shows estimated T_b values for the 13 metamolecules.

Applying the property constraint prunes metamolecules [(2 0) 0 0 0] [(1 1) 0 0 0] and [(2 0) 1 0 0]. Additionally, the estimate of T_b for metamolecule (0 3 0 1 0) shows that all the molecules it contains have T_b values that satisfy the property constraint. None of the metamolecules resulting from further expansion of metamolecule [(0 3) 0 1 0] need to be checked.

The search continues with the expansion of metagroups until all metagroups contain only one group. At that point the abstraction has been removed, and the metamolecules generated represent individual molecules.

4. Strategies for the Formation of Molecular Abstractions

Metagroup division can be accomplished in many ways. Given a set of k objects, the number of ways these can be portioned into p sets is given by

$$S(k,p),$$

which is the Stirling number of the second kind. Starting with a set of hypothetical groups, $(a\ b\ c\ d)$, and dividing them into two metagroups, yields $S(4,2) = 7$ possibilities. These are [$(a)(bcd)$] [$(b)(acd)$] [$(c)(abd)$] [$(d)(abc)$] [$(ab)(cd)$] [$(ac)(bd)$] [$(ad)(bc)$]. For a reasonable number of groups, the possible choices of metagroups is very large.

Two approaches can be used to add back detail: *expansion* and *division*. Expansion adds back knowledge about the metagroups, which is used by the structural constraints. Division focuses on reducing the width of metacontributions, thus improving the screening power of physical property constraints.

a. Division In Half. Dividing a metagroup in half is the simplest strategy. However, division without regard to the metacontributions could prove inefficient. Given the following set of groups with the corresponding contributions

Groups   [g_1 g_2 g_3 g_4],

Contributions   [10 60 11 61],

our initial meta-group [g_1 g_2 g_3 g_4] would have a metacontribution of [10 61]. Dividing the meta-group in half would result in the two metagroups [g_1 g_2] and [g_3 g_4]. The metacontributions for these new metagroups would be [10 60] and [11 61]. These metacontributions are almost identical to the original. The division thus added to the combinatorics without improving the possibility for pruning.

Meta-contributions should be considered when dividing metagroups in half. The midpoint of the initial metacontribution is $(61 - 10)/2 = 25.5$. All groups whose contributions are less than $25.5 + 10 = 35.5$ are collected into the first new metagroup, and all those with contributions greater than 35.5 are collected into the second new metagroup. The resulting metagroups have tighter interval representation of their contributions, leading to more efficient screening.

b. Division by Largest Gap. The interval representation of metacontributions ignores their discrete nature. Dividing a metagroup at the largest gap in its contributions attempts to take advantage of the distribution of contributions and produce two new metagroups whose metacontributions are distributed over a much more narrow range than the original metacontribution. Using the same example set of groups and contributions given above, the initial metacontribution, [10 61], has a width of 51. Dividing the metagroup at the largest gap in the contributions produces two new metagroups with metacontributions [10 11] and [60 61]. The total width of these two intervals is 2, a considerable reduction from 51.

c. Division to Isolate Groups. Extreme values of the contributions by some groups can greatly affect the interval value of the calculated properties. The contribution toward the boiling point T_b, from Joback's method (Joback and Reid, 1987) for the group $=O$, is -10.5. If we are searching for low values of T_b, then it is desirable to have many $=O$ groups in our molecules. However, from chemical considerations it is unlikely that a molecule with a large number of $=O$ groups would be stable. Isolating $=O$ onto its own metagroup enables the designer to pose constraints on the maximum number of occurrences of the metagroup in any metamolecule.

5. *Evaluation of the Search Algorithm: Taming the Combinatorial Explosion*

Tables X and XI summarize the results of the application of the search algorithm in two case studies. Let us look at the highlights of each one of them:

Case 1. In this case study we want to synthesize molecules that have a vapor pressure, at 273 K, larger than 1.0 bar. The molecules are to be composed from a set of 44 functional groups, and they can contain up to 3 functional groups, i.e., $k = 44$ and $n = 3$. There exist 15,180 molecules that can be created from various combinations of the functional groups (Table X). The search algorithm generates 460 metamolecules and rejects 104 of them. The refinement of the metagroups and the pruning of infeasible molecules is guided by a series of constraints, as indicated in the footnotes of Table X.

TABLE X

Pruning Results for $k = 44$, $n = 3$ Automatic Design
[Constraint $= P_{vp}(273\ K) > 1.0$ bar]

# Metagroups	# Metamolecules	Kept	Pruned
1	1	1	0
3^a	10	4	6
4^b	7	4	3
10^c	51	1	50
11	3	1	2
12^d	3	2	1
13	5	3	2
14	6	3	3
15	6	4	2
16	8	7	1
17	11	11	0
18	12	11	1
19	21	18	3
20	28	26	2
21	33	29	4
22	44	44	0
27	109	85	24
28^e	102	102	0
Total^f	460	356	104

^{a} Expanded by ring class.
^{b} Isolated — COOH, — NO$_2$, and — CN.
^{c} Expanded by global valence.
^{d} Isolated = O. Restricted = O occurrences to 1.
^{e} 12 metagroups never occurred in any metamolecules.
^{f} There are 15,180 molecules contained in the search.

Case 2: The premises are the same as in case 1, but here we allow the formation of molecules with up to five functional groups. The total number of potential molecules is 1,712,304. The search algorithm has generated 4131 metamolecules and rejected 2,094 of them (Table XI). See footnotes of Table XI for constraints guiding the pruning of infeasible metamolecules.

From both these examples is clear the advantage of using a search with successive molecular abstractions; *the number of metamolecules needed to be evaluated is far smaller than the number of individual molecules.*

The algorithm was also analyzed in order to identify some of its "bounding" properties. Assume that an initial metagroup is divided into two children metagroups. The metamolecules formed from these metagroups are tested, and all those that contain occurrences of the second

TABLE XI
PRUNING RESULTS FOR $k = 44$, $n = 5$ AUTOMATIC DESIGN
[CONSTRAINT = $P_{vp}(273 \text{ K}) > 1.0$ BAR]

# Metagroups	# Metamolecules	Kept	Pruned
1	1	1	0
3[a]	21	8	16
4[b]	23	8	15
5[c]	23	12	11
6[d]	30	27	3
7	58	27	31
8	58	32	26
9	37	35	2
10	53	35	18
11	71	42	29
12	87	86	1
13	110	86	24
14	165	107	58
15	206	179	27
24[e]	1675	185	1490
30[f]	625	479	146
37[g]	888	688	200
Total[h]:	4131	2037	2094

[a] Expanded by ring class.
[b] Isolated — COOH, — NO$_2$, and — CN.
[c] Isolated = O. Restricted = O occurrences to a maximum of 1.
[d] Isolated — F.
[e] Expanded by global valence.
[f] Expanded several metagroups containing two or three groups in half.
[g] Expanded all nonzero occurring metagroups to individual groups. Ten metagroups never occurred in any metamolecule.
[h] There are 1,712,304 molecules contained in the search.

group are pruned away. This scenario is repeated with the surviving metagroups, and so on.

Dividing a metagroup (MG) containing k groups into two metagroups, MG$_1$ and MG$_2$, containing k_1 and k_2 groups, respectively, allocates the

$$\frac{(k+n-1)!}{n!(k-1)!}$$

possible molecules into

$$\frac{(2+n-1)!}{n!(2-1)!} = n+1$$

metamolecules. One metamolecule contains only occurrences of MG_1, one metamolecule contains only occurrences of MG_2, and the remaining $n - 1$ metamolecules contain occurrences of both metagroups. If no metamolecules containing MG_2 survive the testing, then the percentage of molecules pruned is given by

$$\left[1 - \frac{(k-1)!}{(k_1-1)!} \frac{(k_1+n-1)!}{(k+n-1)!} \right] * 100\%.$$

Repeating this expansion and pruning process r times, until the final metagroup contains only one group, requires the generation and testing of

$$r(n+1) + 1$$

metamolecules. The advantage of abstraction, as measured by the total number of molecules contained in the search divided by the number of metamolecules needed to be generated and tested, is given by

$$\text{Advantage of abstraction} = \frac{1}{r(n+1)+1} \frac{(k+n-1)!}{n!(k-1)!}.$$

Considering the worst-case scenario, in which MG_2 and all subsequent second metagroups contain only a single group, we have $r = k - 1$ leading to

$$\text{Advantage of abstraction} = \frac{1}{(k-1)(n+1)+1} \frac{(k+n-1)!}{n!(k-1)!}.$$

Table XII shows this advantage of abstraction for several values of k and

TABLE XII

ADVANTAGE OF ABSTRACTION

$k \backslash n$	2	3	4	5	6
15	5.6	11.9	43.1	136.8	391.5
20	7.2	20.0	92.2	369.6	1,321.6
25	8.9	30.2	169.2	819.0	3,513.5
30	10.6	42.4	280.3	1,590.0	7,956.7
35	12.2	56.7	431.7	2,808.6	16,060.2
40	13.9	73.1	629.6	4,621.3	29,726.5
45	15.6	91.6	880.5	7,195.8	51,426.2
50	17.2	112.2	1,190.3	10,720.4	84,272.3

n. In particular, Table XII shows that if we have an automatic design involving 40 groups with an occurrence value of 5, then the number of metamolecules needed for exhaustive searching will be 4621.3 times smaller than the number of potential molecules.

C. CASE STUDY: AUTOMATIC DESIGN OF REFRIGERANTS

Automotive air conditioners are a major source of refrigerant emissions, which contribute to the depletion of the Earth's protective ozone layer. This case study generates replacement refrigerants for automotive air conditioners.

Identifying the target set of constraints is the first step of the methodology. Constraints are derived from performance considerations and in an evolutionary manner attempting to find a compound with properties better than refrigerant 12. The design temperatures between which the refrigerant must operate are 110°F (43.3°C) maximum and 30°F (-1.1°C) minimum (Langley, 1986). The following constraints form the design target:

- $P_{vp}(T = -1.1°C) > 1.4$ bar
  The lowest pressure in the cycle should be greater than atmospheric (Dossat, 1981). This reduces the possibility of air and moisture leaking into the system. Douglas (1988) recommends a safety factor of 5 psig (pounds per square inch gauge).
- $P_{vp}(T = 43.3°C) < 14$ bar
  A high system pressure increases the size, weight, and cost of equipment (Dossat, 1981). A pressure ratio of 10 is considered to be the maximum for a refrigeration cycle (Perry and Chilton, 1973).
- $\Delta H_v(T = -1.1°C) > 18.4$ kJ/g-mol
  The value for refrigerant 12's enthalpy of vaporization at -1.1°C is 18.4 kJ/g-mol [American Society of Heating, Refrigerating and Air-Conditioning Engineers (ASHRAE), 1972]. A higher value reduces the amount of refrigerant required.
- $C_{p_L}(T = 21.1°C) < 32.2$ cal/g-mol·K
  It is desirable to have a low liquid heat capacity to reduce the amount of refrigerant that flashes on passage through the expansion valve (Dossat, 1981). Refrigerant 12's liquid heat capacity at 21.1°C is 32.2 cal/g-mol·K (ASHRAE, 1972).

Estimation procedures were developed for each physical property used in the target constraints: P_{vp}, H_v, C_{p_L}. These estimation procedures were based on correlations that require the evaluation of seven physical properties, given by group-contribution techniques:

(a) Reduced boiling point, T_{b_R}
(b) Normal boiling point, T_b
(c) Critical pressure, P_c
(d) Coefficients for cubic feet of ideal gas heat capacity with temperature, $C^°_{p,a}, C^°_{p,b}, C^°_{p,c}, C^°_{p,d}$

In the automatic design, 44 functional groups were used; molecules containing 2–7 groups were allowed; the number of group occurrences was limited to 7 to account for the fact that most refrigerants are of small molecular size, and 47 molecules were designed that satisfy the four physical property constraints and the structural constraints listed in Table II. Table XIII shows some of these 47 molecules along with their estimated values for P_{vp}, ΔH_v, and C_{p_L}.

Molecules 19 and 20 are of particular interest. These are two ringed compounds that possess physical properties satisfying the design constraints. Although the chemical stability of these compounds still needs to be verified, it is considered a success that the automatic design was able to design such not obvious compounds.

D. Case Study: Automatic Design of Polymers As Packaging Materials

In this second example we demonstrate the applicability of the automatic design methodology to the design of polymers with desired properties. The problem is to design polymers for use as integrated-circuit (IC) encapsulants.

Electronic packages are sealed to prevent gross contamination, handling damage, and the entry of corrosive gases (Mih, 1984). To package microelectronic circuitry so that it is useful and functions properly under various environmental conditions, it is essential to select the correct packaging materials (Fogiel, 1972). Polymeric coatings are widely used in the electronics industry because of their excellent properties and low cost (Goosey, 1985). Polymers used for semiconductor encapsulation must protect against moisture, chemical agents, wide temperature variations, and mechanical shock. The polymeric material must be able to satisfy these requirements with a minimal effect on device parameters over an

TABLE XIII
Refrigerant Design—Automatic Results

Molecule	P_{vp}(272.05)	P_{vp}(316.45)	H_v(272.05)	C_{p_L}(294.35)
1. 1(—CH$_3$)1(—Cl)	1.59	6.20	21.58	20.00
2. 2(—F)1(>NH)	2.69	10.96	19.06	22.26
3. 1(—Cl)1(—F)1(—CH$_2$—)	1.65	6.61	20.71	22.89
4. 1($\equiv$CH)1(—CH$_3$)1($\equiv$C—)	1.67	6.30	21.44	24.06
5. 2(—F)2($\equiv$C—)	2.09	7.87	19.61	21.97
6. 1(=CH)1(—F)1(—CH$_2$—)1($\equiv$C—)	1.74	6.72	20.56	26.95
7. 1(=O)2(—CH$_3$)1(=C<)	1.71	7.29	27.14	30.06
8. 1(—Cl)2(—F)1(>N—)	2.65	10.41	19.04	24.83
9. 1(—Cl)2(—F)1(>CH—)	1.74	6.96	19.45	25.28
10. 2(—CH$_3$)1(—F)1(>N—)	1.81	7.29	20.42	30.04
11. 3(—F)1(>NH)1(>N—)	1.70	8.12	20.88	29.98
12. 1(—CH$_3$)2(—F)1(—O—)1(>N—)	1.96	8.35	19.70	31.60
13. 1(=CH$_2$)2(—F)1(=CH—)1(>N—)	2.15	8.61	18.73	30.15
14. 1(=CH$_2$)2(—F)1(—CH$_2$—)1(=C<)	1.41	5.78	19.59	30.78
15. 1(=CH$_2$)2(—F)1(=CH—)1(>CH—)	1.42	5.82	19.12	30.62
16. 1($\equiv$CH)2(—F)1($\equiv$C—)1(>N—)	2.77	10.54	18.90	28.88
17. 1($\equiv$CH)2(—F)1($\equiv$C—)1(>CH—)	1.83	7.07	19.31	29.34
18. 3(—F)2(=CH—)1(>N—)	1.66	7.13	18.92	31.79
19. 2(^{r}>CH—)1(=C^r<^{r})1(=O)2(—F)	1.53	7.11	26.08	31.93
20. 3(^{r}>CH—)3(—F)	1.40	5.80	18.67	31.11

extended period of time, and be relatively inexpensive and easy to process. Some of the important physical properties of packaging material are (Dillinger, 1988)

1. Permeability to water vapor at high temperatures.
2. Thermal conductivity.
3. Outgassing in plastics at elevated temperatures and the resulting impact on water vapor permeability.
4. Thermal expansion coefficient and mismatch between expansion coefficients of package and chip interconnect.

The following constraints are the design specifications of a good encapsulant:

- $T_g > 400°C$
  The glass transition temperature must be $> 400°C$. The encapsulant must keep its structural integrity during use. The high temperatures at which microelectronic circuits operate place a restriction on T_g.
- $R > 10^{16}$ Ω-cm
  The volume resistivity of the solid polymer must be $> 10^{16}$ Ω-cm. Since the packaging material will make contact with the metal leads of the microelectronics device, it is essential that the compound have a high-volume resistivity.
- $\lambda > 0.16$ w/m·K
  The thermal conductivity of the solid polymer is desired to be greater than the thermal conductivity of the currently used polyimide. A high thermal conductivity is desirable allowing the microelectronic circuitry to be cooled more effectively.
- $P(O_2) < 1.0$ cm^3-mil 100 in.$^{-2}$ day^{-1} atm^{-1}
  The permeability of the polymer to oxygen should be < 1.0 cm^3-mil 100 in.$^{-2}$ day^{-1} atm^{-1}. Diffusion of oxygen and water through the polymer to the microelectronic circuitry could cause corrosion and is thus undesirable. To establish a value for this physical property constraint, we examined polymers used as barriers. Polymers with a permeability to oxygen of ≤ 1.0 cm^3-mil 100 in.$^{-2}$ day^{-1} atm^{-1} are considered high-barrier materials.

Estimation techniques from van Krevelen (1976) and Salame (1986) were combined into estimation procedures for the properties of each target constraint: $T_g, R, \lambda, P(O_2)$. These procedures are detailed in the following paragraphs:

1. *Thermal conductivity.* This is estimated by the following correlation (van Krevelen, 1976)
$$\lambda(298 \text{ K}) = \lambda(C_p^s, V, U),$$
where

$C_p^s = C_p^s$ (groups)—specific heat by the van Krevelen group-contribution technique

$V = V$(groups)—specific volume by the modified van Krevelen group-contribution technique

$U = U$(groups)—Rao function by the van Krevelen group-contribution technique

2. *Electrical resistivity.* This is estimated by the following correlation (van Krevelen, 1976)
$$R = R(P_{LL}, V),$$
where

$P_{LL} = P_{LL}$(groups)—dielectric polarization by the van Krevelen group-contribution technique,

$V = V$(groups)—specific volume by the modified van Krevelen group-contribution technique.

This estimation procedure requires two properties to be estimated by group-contribution techniques: P_{LL} and V.

3. *Glass transition temperature.* This is estimated by the following correlation (van Krevelen, 1976)
$$T_g = T_g(Y_g, M),$$
where

$Y_g = Y_g$(groups)—glass transition function by the van Krevelen group-contribution technique,

$M = M$(groups)— repeat unit weight by the van Krevelen group-contribution technique.

This estimation procedure requires two properties to be estimated by group-contribution techniques: Y_g and M.

4. *Permeability to oxygen.* This is estimated by the following correlation

TABLE XIV
Polymer Design—Automatic Results

Molecule	T_g	R	L	P_i
1. 1(—⬡—CH₂—⬡—)	479.5	20.1	0.164	2.40e−03
2. 1(—OCONH—)	423.5	122.6	0.214	3.25e−15
3. 1(—CONH—)2(—C(CH₃)(C₆H₅)—)	445.7	19.6	0.172	6.42e−02
4. 1(—CONH—)1(—C(CH₃)(C₆H₅)—)1(—CH(C₆H₅)—)	408.8	19.4	0.181	1.74e−02
5. 1(—O—)1(—CHF—)2(—⬡—CH₂—⬡—)1(—OCONH—)	453.7	19.6	0.163	1.33e−04
6. 1(—CH₂—)1(—CHF—)2(—C(CH₃)(C₆H₅)—)1(—OCONH—)	442.6	19.8	0.161	1.62e−01
7. 3(—CH₂—)1(—CONH—)2(—⬡—CH₂—⬡—)	429.9	19.8	0.166	7.73e−02

#	Structure				
8.	2(—O—)1(—CH$_2$—)1(—CONH—)2(—C$_6$H$_4$—CH$_2$—C$_6$H$_4$—)	432.0	19.6	0.165	9.92e—03
9.	1(—O—)1(—CH(CH$_3$)—)4(—C$_6$H$_4$—CH$_2$—C$_6$H$_4$—)	466.6	20.2	0.160	7.59e—02
10.	2(—C(CH$_3$)(C$_6$H$_5$)—)3(—C$_6$H$_4$—)1(—OCONH—)	445.9	19.8	0.168	2.24e—01
11.	1(—C(CH$_3$)(C$_6$H$_5$)—)2(—CH(C$_6$H$_5$)—)2(—C$_6$H$_4$—)1(—OCONH—)	421.7	19.8	0.174	1.73e—01
12.	1(—CH$_2$—C$_6$H$_4$—CH$_2$—)3(—C$_6$H$_4$—CH$_2$—C$_6$H$_4$—)2(—OCONH—)	453.0	19.5	0.168	6.15e—07
13.	3(—CH(C$_6$H$_5$)—)1(—C$_6$H$_4$—CH$_2$—C$_6$H$_4$—)2(—OCONH—)	423.2	19.3	0.180	5.74e—05
14.	2(—C(CH$_3$)(C$_6$H$_5$)—)2(—CH(C$_6$H$_5$)—)2(—OCONH—)	434.3	19.4	0.176	1.27e—03

(Salame, 1986)

$$P = P(\pi),$$

where

$$\pi = \pi(\pi_i, N_b)$$—is given by the Salame correlation,

with

$\pi_i = \pi_i(\text{groups})$—permachor by the Salame group-contribution technique,

$N_b = N_b(\text{groups})$—number of backbone groups by the Salame group-contribution technique.

This estimation procedure requires two properties to be estimated by group-contribution techniques: π_i and N_b. These procedures result in seven physical properties that are estimated by group-contribution techniques: C_p^s, V, U, P_{LL}, Y_g, M, and π.

In the automatic design of new polymers, 21 groups were used. Polymers having between one and six group occurrences were allowed. The design procedure produced over 18,000 feasible polymers. Table XIV shows several polymers randomly selected from the set of feasible candidates along with the estimated values of their important physical properties. The large number of retained molecules indicates that the design specs are fairly "loose" and could be tightened by either tightening the bounding values of the physical property constraints or introducing additional physical properties constraints.

III. Interactive Synthesis of New Molecules

The automatic synthesis of molecules, described in the previous section, attempts to generate feasible candidates without resorting to the detailed, fragmented, often informal, but nevertheless sound and valuable knowledge possessed by the human designer. The search for new molecules is carried out efficiently, but it is entirely built on a limited amount of knowledge, which is represented by the sets of constraints on (1) physical properties, (2) the feasibility of molecular structures, and (3) chemical stability and other chemical properties of the resulting molecules.

The interactive design attempts to support the articulation of and incorporate the designer's informal knowledge and abductive capabilities in postulating "promising" molecular structures. In this manner, the resulting computer-aided tool can capture and utilize the best of two worlds: (a) the computer's ability to locate feasible solutions after extensive search of the solution space and (b) the human designer's "intelligence" in expanding and guiding the search in an efficient manner. Therefore, although this section will deal only with the features of the interactive design procedures, the *Molecule-Designer*, the computer-aided tool that implements both automatic and interactive procedures, has integrated both in a seamless manner (see Section IV).

The human-driven character of interactive design allows the use of (1) problem-specific subjective preferences, (2) informal, qualitatively stated scientific knowledge, (3) rapid evaluation of alternatives, and (4) evolutionary design of new molecules, starting from known and existing alternatives. It provides the designer with the following facilities:

1. Visualization of the target constraints, helping the designer in the selection of the most "promising" functional groups to be incorporated in a molecule.
2. Extensive databases for the extraction of patterns in the structural evolution of known molecules.
3. Interactive definition of new property estimation techniques and qualitative scientific rules, which are to be incorporated in the automatic design algorithm.

A. ILLUSTRATION OF INTERACTIVE DESIGN

The framework of interactive design is entirely built on the premise of *additivity of group contributions* for the estimation of physical properties. Let us look at some typical illustrations. Table XV shows the estimation of normal boiling points T_b, and normal melting points T_m, using Joback's group contribution techniques (Joback and Reid, 1987), along with the

TABLE XV

EXAMPLE OF LINEAR GROUP CONTRIBUTION ESTIMATION TECHNIQUES

Groups	Contributions[a]	
	Δ_{i,T_b}	Δ_{i,T_m}
—CH$_3$	23.58	−5.10
—CH$_2$—	22.88	11.27
>CH—	21.74	12.64
>C<	18.25	46.43
—F	−0.03	−15.78
—Cl	38.13	13.55

[a] $T_b = 198.18 + \Sigma n_i \Delta_{i,T_b}$; $T_m = 122.5 + \Sigma n_i \Delta_{i,T_m}$.

contributions of a few select groups. The additivity of group contributions allows the molecules to be assembled group by group, while offering a "partial" estimate of the desired physical properties. Given a constraint, such as, $T_b > 300$ K, a group is added to the molecule and then the constraint is evaluated. Choosing one $—CH_3$ group results in $T_b = 221.76$ K. The constraint is not satisfied. Choosing a second $—CH_3$ group results in $T_b = 245.34$. The constraint is still not satisfied. Adding three $—CH_2—$ groups results in $T_b = 313.98$, which satisfies the constraint. Adding a second constraint, $T_m < 250$ K, makes the monitoring of the numerical values difficult. Graphical representations can simplify this problem.

Figure 1 shows a two-dimensional design space formed from T_b and T_m. The two imposed constraints form a feasible region denoted by the shaded area. The contributions of each group toward T_b and T_m form a two-dimensional (2D) vector. These are called *group vectors*. The intercepts of the group contribution models shown in Table XV establish the starting point for the first group vector. Beginning at the intercept point an appropriate set of group vectors are selected that terminate in the feasible region and produce a structurally feasible molecule. Figure 1 shows the group vectors for chloropropane.

Complex constraints on T_b and T_m are easily handled by the interactive design procedure. All that is required is to identify a feasible region or regions. The constraints need not be linear or convex.

The following heuristics on group selection are often helpful in selecting groups that satisfy the physical property constraints and are structurally feasible.

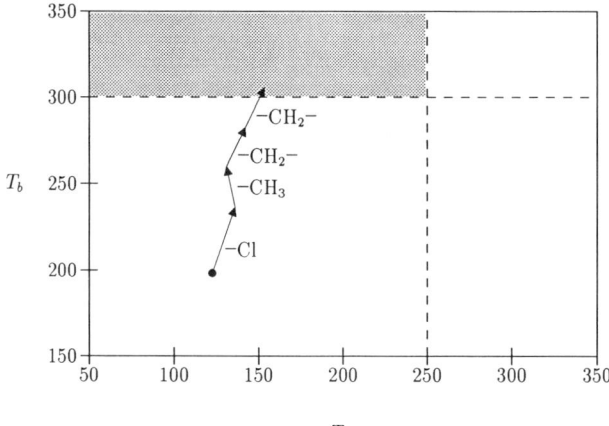

FIG. 1. An interactively designed molecule: chloropropane.

Rule 1. Separate the groups into three sets:

1. *Terminators*: all groups having one free bond.
2. *Extenders*: all groups having two free bonds.
3. Branchers: all groups having more than two free bonds.

Rule 2. Select the following initial group sets:

1. If a pure acyclic molecule is to be designed, choose two terminators.
2. If a pure cyclic molecule is to be designed, choose two cyclic extenders.

Rule 3. Continue by first considering extenders. If a brancher is to be added, follow it immediately with terminators.

These heuristics ensure that when a molecule reaches the feasible region, it is either structurally feasible or requires the addition of only one or two groups for feasibility.

1. Reduction in the Dimensionality of the Search Space

The dimension of the design space is equal to the number of fundamental physical properties needed to evaluate the property constraints. Using the estimation procedure for the vapor pressure P_{vp}, given by the Riedel–Plank–Miller correlation

$$P_{vp} = P_{vp}(T_b, T_{br}, P_c),$$

where

$T_b = T_b(\text{groups})$ is given by Joback's group contribution,

$T_{br} = T_{br}(\text{groups})$ is given by Lydersen's group contribution,

$P_c = P_c(\text{groups})$ is given by Ambrose's group contribution,

we obtain three fundamental properties: T_b, T_{br}, and P_c. Designing for constraints on P_{vp} thus requires a three-dimensional (3D) design space. This is unfortunate because representing and manipulating 3D objects graphically is more complex than 2D manipulation.

The dimensionality of the design space can become more than a mere complication. One estimation procedure for the liquid heat capacity requires seven fundamental physical properties. To interactively design for constraints on C_{p_L} would require a seven-dimensional physical property space to display each of the seven fundamental properties.

Studies on *factor analysis* (Cramer, 1980a, b; Klincewicz, 1982; Joback, 1984) show that a number of physical properties are highly intercorrelated.

TABLE XVI

Physical Property–Factor Relationships

$1/\sqrt{P_c} =$	0.157	$- 0.019 F_1$	
$V_c =$	296.1	$- 89.66 F_1$	$- 36.68 F_3$
$n_A =$	14.50	$- 5.35 F_1$	$- 1.18 F_3$
$C^o_{p,298} =$	25.70	$- 8.72 F_1$	$- 2.44 F_3$
$T_c =$	545.9	$- 24.65 F_1$	$- 87.92 F_3$
$T_b =$	358.4	$- 25.26 F_1$	$- 64.94 F_3$
$\Delta H_{vb} =$	7686.6	$- 432.3 F_1$	$- 1614.3 F_3$

High correlations between two properties indicate the possibility of replacing one with a function of the other. This would enable us to reduce the dimensionality of a design space. One study (Joback, 1984) found that nine physical properties were well approximated by three new properties called *factors*. Table XVI shows several of the derived physical property estimation techniques, using essentially two of these factors, F_1 and F_3. Group contribution estimation techniques were developed for both factors, F_1 and F_3.

Incorporating these equation-oriented estimation techniques into a new estimation procedure for P_{vp} yields the following correlation, proposed by Riedel, Plank, and Miller

$$P_{vp} = P_{vp}(T_b, T_c, P_c),$$

where the following properties can be estimated from their correlations to factors F_1 and F_3 (Table XVI):

$$T_b = T_b(F_1, F_3),$$

$$T_c = T_c(F_1, F_3),$$

$$P_c = P_c(F_1, F_3),$$

with F_1 and F_2 computed by group-contribution correlations established by Joback:

$$F_1 = F_1(\text{groups}),$$

$$F_3 = F_3(\text{groups}).$$

The fundamental properties are F_1 and F_3. Two fundamental properties enable design in a 2D physical property space.

2. Utilization of Interactive Design

The graphical representation of physical property constraints allows the designer to quickly gain insight into the feasibility of the problem and the relative importance of each constraint. Excessively large or small feasible regions, redundant constraints, or open feasible regions may indicate a need to respecify design constraints.

Once satisfied with the feasible region, design may proceed in one of the following three ways:

a. Evolutionary Design. Starting from an existing molecule, interactive design allows the evolution of the molecule to new altenatives through guided structural modifications. For example, Fig. 2a shows the feasible region defined by constraints on three physical properties and how two functional groups of the initial (infeasible) molecule were replaced by two

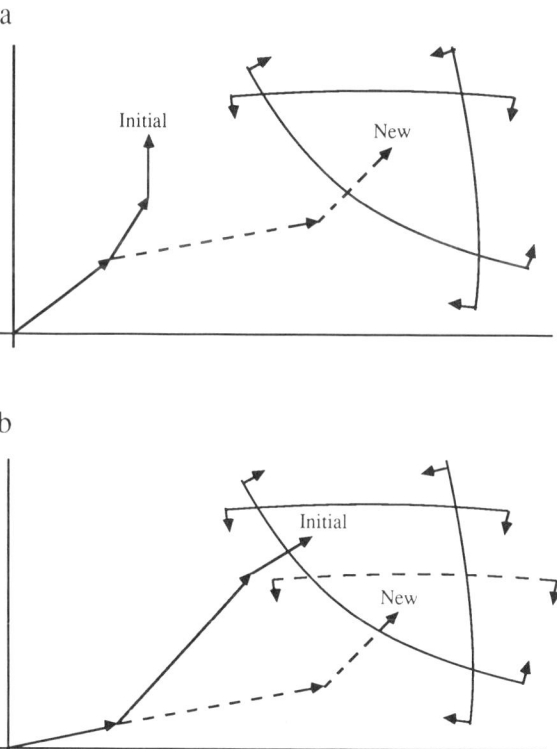

FIG. 2. Interactive design approaches: (a) evolutionary design of improved molecules; (b) evolutionary tightening of design constraints.

other groups to yield a feasible molecule. Figure 2b, on the other hand, shows how interactive design can be employed to tighten the specifications of a physical property constraint (e.g., by moving the location of a constraint; see dashed-line constraint), requiring the evolution of the initial molecule to a new one (satisfying the new set of constraints).

b. Grass-Roots Design. Building the molecule from scratch, group by group, using pure interactive or in conjunction with automatic search.

c. Combination of Evolutionary and Grass-Roots Designs. A molecule is designed from scratch so that it meets the initial specifications. Subsequently, improvements are searched for by tightening the design specifications and carrying out an evolutionary design. Figure 2b depicts such a situation.

B. CASE STUDY: INTERACTIVE DESIGN OF REFRIGERANTS

The specifications of this case study are identical to those described in Section II.C and will not be reproduced here.

To design in a 2D space, the factor relationships shown in Table XVI are used to reduce the fundamental properties to the factors F_1 and F_3. Each physical property constraint, which is a function of F_1 and F_3, is plotted in a 2D $\{F_1-F_3\}$ design space shown in Fig. 3. The region in which all constraints are satisfied is shaded. The displayed symbols correspond to the constraints as follows:

Symbol	Constraints
Circle	$P_{vp}(T = -1.1°C) > 1.4$ bar
Triangle	$P_{vp}(T = 43.3°C) < 14$ bar
Square	$\Delta H_v(T = -1.1°C) > 18.kJ/\text{g-mol}$
Circle	$C_{p_L}(T = 21.1°C) < 32.2$ cal/g-mol·K

The graphical representation used in the interactive design procedure immediately provides insights into the design problem. The first insight is that the chosen constraints yield a feasible solution space. Although each constraint was justified on its own, there was no guarantee that the set of four constraints would produce a feasible space. If the feasible region were too large or small, this would indicate the need to respecify the design target. Additionally, it is seen that for a major portion of the design space the $P_{vp}(T = 43.3°C) < 14$ bar constraint is redundant. It is superseded by the $\Delta H_v > 18.4$ constraint.

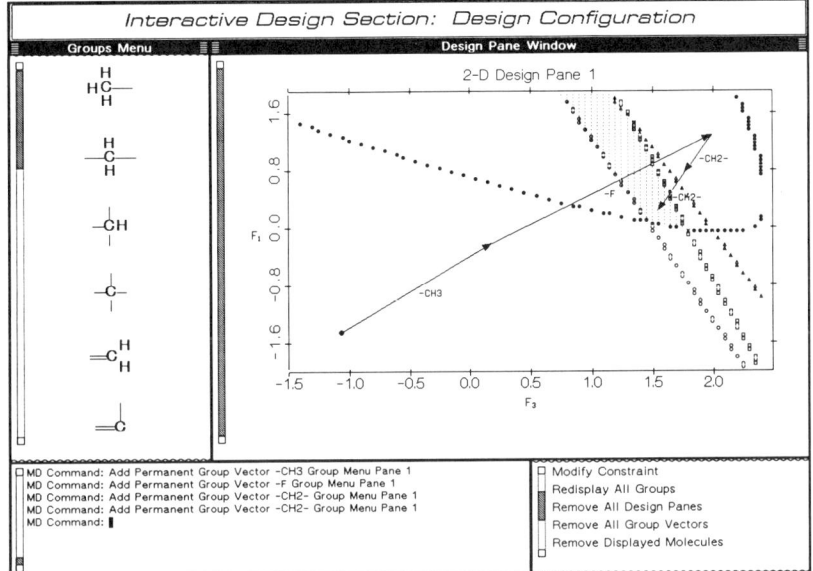

Fig. 3. Refrigerant design: a solution.

TABLE XVII

Estimated Property Values for Designed Refrigerants

Compound	P_{vp} 272.05 K	P_{vp} 316.4 K	ΔH_v 272.05 K	C_{p_L} 294.3 K
1. CH_3-CH_3	2.710	9.486	18.67	22.49
2. CH_3-Cl	1.590	6.193	21.58	19.99
3. CH_3-NH_2	0.375	2.152	29.78	23.50
4. $CH_3-CH_2-CH_2-F$	1.195	5.013	21.21	31.10
5. $CH_2=CH-CH_3$	1.310	5.195	21.24	25.33
6. $CH_2=CH-Cl$	0.750	3.340	24.14	23.01
7. $CH_2=CH-O-CH_3$	0.543	2.648	24.83	29.83
8. $F-CH_2-CH_2-CH_2-F$	1.236	5.330	20.34	34.03
9. $F-CH=CH-CH_2-F$	1.045	4.573	20.54	29.89
10. CCl_2F_2	0.440	2.213	24.70	27.87
11. $CBrClF_2$	0.122	0.837	28.10	29.02
12. $CHBrF_2$	0.538	2.855	22.96	26.04
13. $CH(COOH)F_2$	0.002	0.041	42.56	37.58
14. $CH(HCO)F_2$	0.417	2.421	25.91	30.42
15. NH_2-NH_2	0.027	0.311	41.41	26.03
16. $CH\equiv C-Cl$	0.968	4.084	24.33	21.72
17. $CH\equiv C-CH_3$	1.670	6.297	21.44	24.06
18. $CH\equiv C-NH_2$	0.214	1.353	32.56	25.42

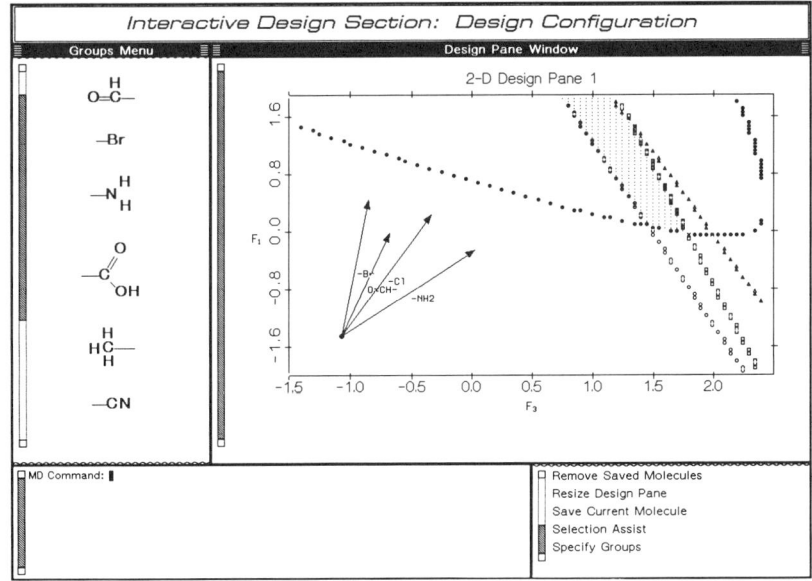

FIG. 4. Candidate groups for chlorine replacement.

Figure 3 also shows the group vectors for a designed molecule. The location of the vectors' end point can be directly interpreted into physical property information. For example, the molecule satisfies all constraints but has a liquid heat capacity near the constraint. Table XVII shows several interactively designed molecules with estimated values for their physical properties. Some of the molecules may be chemically unstable. At this stage of the design methodology only physical property and structural constraints are being explicitly considered. When choosing the groups, the designer implicitly considers chemical constraints. Once molecules are designed, the next steps of the methodology, molecule enumeration and screening, would identify and remove chemically unstable compounds.

1. Replacing Chlorine

Removing chlorine from current refrigerants would seem to reduce the hazard of ozone depletion. The interactive design procedure is well suited for searching for group replacements. The group vector for the chlorine group is shown in Fig. 4. The target is to find one or more groups making nearly equivalent contributions to that of chlorine.

The search for single group replacements is begun by restricting the possible groups to those with one single free bond. These groups are then sorted with respect to distance. The three closest groups, —CHO, —Br, and —NH$_2$, are displayed in the design space shown in Fig. 4.

Again the graphical representation used by the interactive design enables us to evaluate the effect of any substitution. Replacing, —Cl by O=CH—, would result in a compound having reduced vapor pressure, an increased enthalpy of vaporization, and an increased liquid heat capacity. These alterations of physical properties are derived from the relative positions of the group vectors and the locations of the constraints.

C. Case Study: Interactive Design of an Extraction Solvent

The success of a liquid–liquid extraction process depends on the selection of the most appropriate solvent (Lo et al., 1983). This case study examines the design of a solvent for the extraction of acetic acid from water. Lo et al.'s (1983) procedure for solvent selection was adapted for interactive design use.

The physical properties of the solvent used to facilitate separation in liquid–liquid extraction have major impact on process performance. Two important physical properties are (1) solvent selectivity for the solute and (2) the solute's partition coefficient between the solvent and the parent liquor.

Lo used the three-term solubility parameter (Barton, 1983) and a graphical procedure to identify solvents for liquid–liquid extraction. In a 2D space constructed from the polar component of the solubility parameter δ_p and the hydrogen bonding component of the solubility parameter δ_H, the distribution coefficient of the solute B, m_B, is given by

$$m_B \propto r_{B,S}^{-2},$$

and the selectivity β_B is given by

$$\beta_B \propto \left(\frac{r_{A,S}}{r_{B,S}}\right)^2,$$

where $r_{i,j}$ is the distance between points i and j. The optimal solvent for extraction thus lies on a line passing through the solute and parent liquor close to the solute for large distribution but far from the parent liquor for high selectivity.

Figure 5 shows the design space for the problem of designing a solvent to extract acetic acid from water. The target area is the lower left-hand portion of the acetic acid–water line. Figure 5 shows the group vectors for acetone near the target region. Table XVIII shows several solvents that

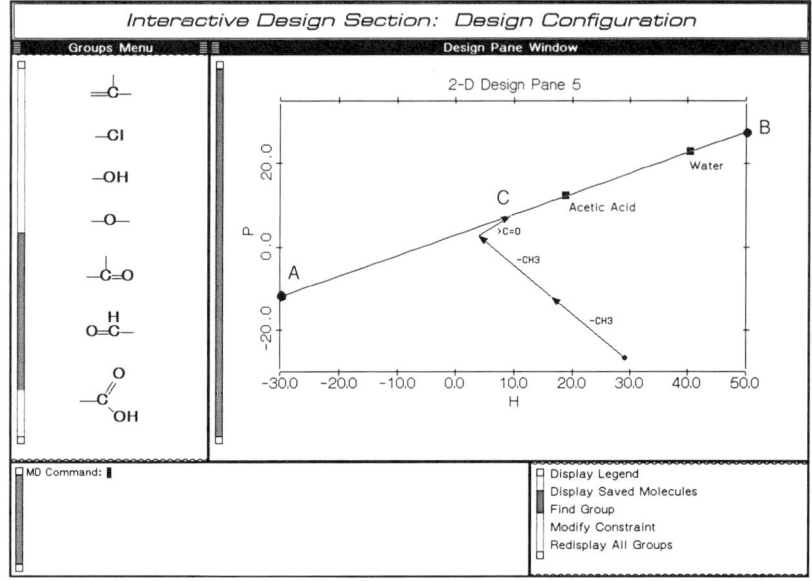

FIG. 5. Solvent design: a solution.

were interactively designed along with estimated values for their solubility parameters.

The $-CH_2-$ group has a δ_p contribution of $-.328$, and δ_H contribution of $-.512$. The slope of the $-CH_2-$ group vector in the interactive design space is .64. This is very close to the acetic acid–water target line slope of .49. Adding several $-CH_2-$ groups to any of the solvents shown in Table XVIII thus results in an acceptable new solvent.

TABLE XVIII

LIQUID–LIQUID EXTRACTION SOLVENTS

Solvent	δ_p	δ_H
$CH_3 - Cl$	7.2	6.2
$CH_2 = CH - CH_3$	4.7	4.0
$CH_2 = CH - Cl$	9.2	6.3
$CH_3 - O - CH_3$	3.8	5.7
$CH_3 - CO - CH_3$	7.5	9.5
$CH_2 = C(CH_3)_2$	4.6	4.1
$C(CH_3)_4$	0.5	1.2
$CCl_2(CH_3)_2$	9.5	5.8

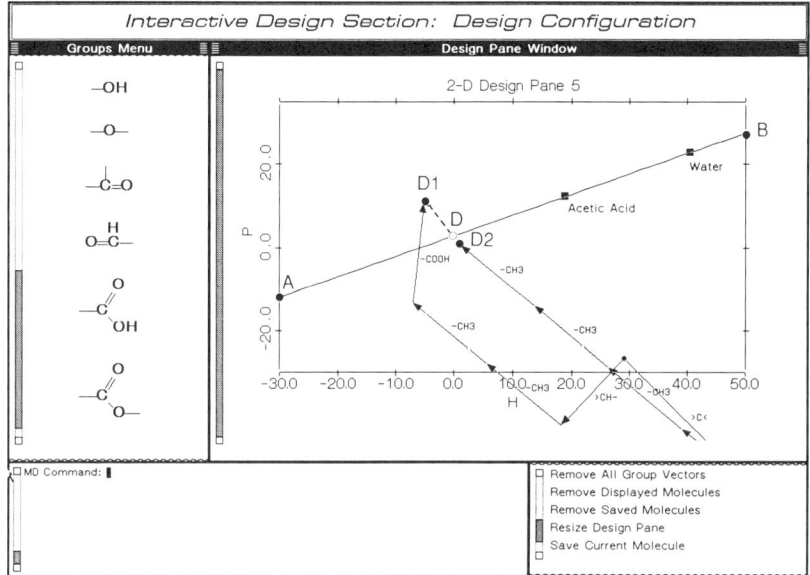

FIG. 6. Interactive design of solvent mixtures.

The direction of the $-CH_2-$ group vector is toward lower δ_p and δ_H values. This reduces the distribution coefficient with increasing number of $-CH_2-$ groups. This result is in accordance with experimental observation (Lo et al., 1983).

Linear mixing rules for solubility parameters enable the design of solvent mixtures. In a δ_p-δ_H design space a binary mixture's solubility parameters lie on a straight line joining the components' solubility parameters. Figure 6 shows the group vectors for the $[CH_3-CH(COOH)-CH_3]-[C(CH_3)_4]$ solvent pair (point D resulting from the mixing of the two solvents represented by points D_1 and D_2). Such an approach can be used for any mixture property approximated by a linear mixing rule. Research is continuing on the use of nonlinear mixing rules for the interactive design of mixtures.

D. CASE STUDY: INTERACTIVE DESIGN OF A PHARMACEUTICAL

This case study presents an example taken from work done by Cramer (1980c; Cramer et al., 1979). It demonstrates how the interactive design procedure assists in the second step of the quantitative structure activity relationship (QSAR) approach to drug design.

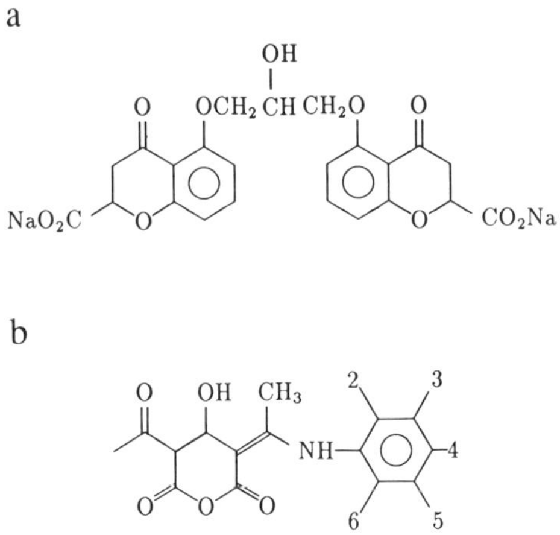

FIG. 7. Current and lead compounds for the drug design case study: (a) existing antiallergic, sodium chromoglycate; (b) a lead compound.

The QSAR approach to drug design takes a two-step approach to the correlation of biological activity with structure. The first step correlates a measure of biological activity, usually the reciprocal concentration having a 50% effect, with a number of physical properties. This correlation yields a model that is used to determine the optimal value of the physical properties giving the maximum potency. The second step of the approach is to identify group substitutions that lead to structures possessing these optimal physical property values.

Sodium chromoglycate (Fig. 7a) is effective at warding off asthmatic attacks. However, it must be administered by inhalation. Cramer et al. (1980c) performed a QSAR study to find a more potent pharmaceutical that could be administered orally. Searching through a database of approximately 1000 compounds they selected the pyranenamies (Fig. 7b) as their lead compound. The pyranenamines have biological properties sufficiently promising to merit a synthetic search for structurally related compounds having improved properties. The task was to develop highly active compounds by varying phenyl ring substituents.

Potency of the developed candidates was measured by the log of the concentration causing a 50% inhibition in asthmatic activity. This measure is represented as pI_{50}. To relate pI_{50} to molecular structure Cramer et al. (1980c) followed the two step QSAR approach. Nineteen compounds were synthesized with varying substituents. pI_{50} was measured for each analog. Values for the octanol–water partition logarithm, π, and Hammett's constant, σ, were computed for each of the substituent sets.

These data were regressed to develop a relationship between pI_{50} and the biochemical parameters π and σ. The model developed was

$$pI_{50} = -0.72 - 0.14\Sigma\pi - 1.35(\Sigma\sigma)^2. \qquad (14)$$

Group-contribution estimation techniques were developed for π and σ. The contributions were obtained by regressing the substituent constants tabulated by Hansch and Leo (1979). Overall the estimation techniques have considerable error and cannot be used for quantitative estimation. However, they provide correct overall trends and thus inform the designer of promising substitutions.

Equation (14) shows that to perform an interactive design we would use a two-dimensional σ vs. π physical property design space. The contours represent solutions of Eq. (14) with pI_{50} equal to -1.0, -0.5, and 0.0. The symbols correspond to the values:

Symbol	pI_{50}
Square	-1.0
Triangle	-0.5
Circle	0.0

The direction of maximum activity is thus near σ equal to zero and π large and negative. Figure 8 shows a pair of metasubstituents in our target

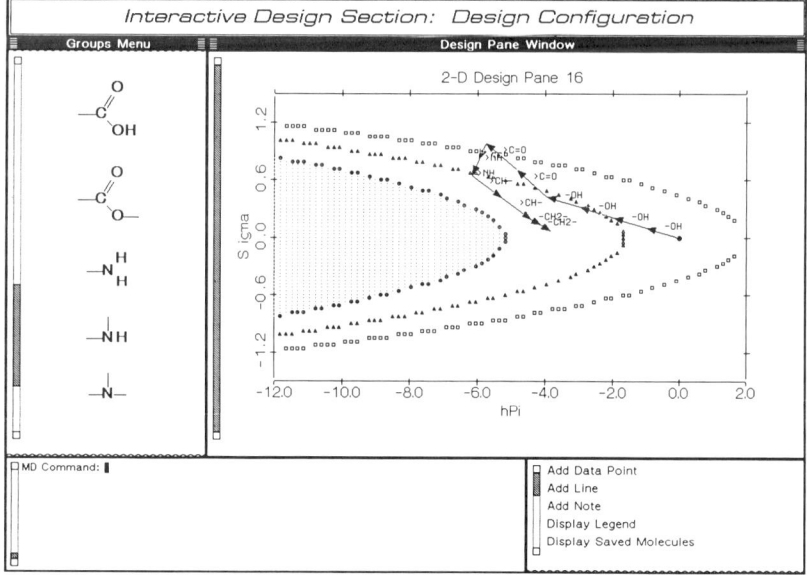

FIG. 8. Two high-activity metasubstituents during the interactive drug design.

TABLE XIX

COMPARISON OF EXPERIMENTAL AND MODEL ACTIVITIES

Substituents	π	σ_M	pI_{50} Model	pI_{50} Experimental
1. 3 — NHCO(CHOH)$_2$H 5 — NHCO(CHOH)$_2$H	−3.812	0.064	−0.19	3.0
2. 3 — NHCOCH$_2$CH$_3$ 5 — NHCOCH$_2$CH$_3$	−0.964	0.269	−0.68	2.5
3. 3 — NHCOCH$_3$ 5 — NHCOCH$_3$	−1.552	0.391	−0.71	1.9
4. 3 — NHCOCH$_3$ 5 — OH	−1.763	0.301	−0.60	1.7
5. 3 — NHCOCOOCH$_2$CH$_3$ 5 — NHCOCOOCH$_2$CH$_3$	−1.890	0.838	−1.40	1.7
6. 3 — NHCOCH$_2$CH$_2$CH$_3$ 5 — NHCOCH$_2$CH$_2$CH$_3$	−0.376	0.147	−0.70	1.3
7. 3 — NHCO(CHOH)$_2$H	−1.906	0.032	−0.45	1.3
8. 3 — NHCOCH$_3$ 5 — NH$_2$	−1.653	0.157	−0.49	1.0
9. 3 — NHCOCH$_2$CH$_3$	−0.482	0.134	−0.68	0.7
10. 3 — NHCOCH$_3$	−0.776	0.196	−0.66	0.7

direction. These metasubstituents correspond to entry 1 of Table XIX. Cramer found these substituents to have one thousand times more activity than the original unsubstituted compound.

IV. The *Molecule-Designer* Software System

A. GENERAL DESCRIPTION

The *Molecule-Designer* is the software system constructed to implement the interactive and automatic procedures for the design of molecules discussed in Sections II and III. It consists of approximately 20,000 lines of LISP code with an additional 17,000-line databank. It is implemented in Common LISP on a LISP Machine. The system is divided into eight sections each corresponding to a section of the overall methodology. The eight sections of the system are (1) login section configuration, (2) database section, (3) interactive design section, (4) molecule evaluation section, (5) problem formulation section, (6) target transformation section, (7) automatic design section, and (8) group-contribution section.

B. Interactive-Design-Relevant Sections

The following sections are specifically relevant to interactive design.

1. Problem Formulation Section

The problem formulation section provides an interface with which the designer can enter physical property constraints. The system displays the properties stored in its database (currently about 40 properties are present) and provides a simple constraint editor for entering and modifying physical property constraints.

2. Group-Contribution Section

Physical property estimation procedures are at the heart of the design procedures. The group-contribution section provides facilities for entering group contribution and equation oriented correlations. Models in the form of LISP code are entered for both types of estimation techniques. Additionally, groups and their contributions are specified for group contribution techniques.

The section is divided into two configurations: (1) an editing configuration that provides facilities for entering new groups to the system's database and (2) a model entry configuration that provides facilities for entering contributions for these groups, models for group-contribution techniques, and models for equation oriented estimation techniques. Figure 9 shows an example screen of the editing configuration with a new group being constructed.

3. Target Transformation Section

Once physical property constraints are entered in the problem formulation section, it is necessary to instruct the system how to estimate the properties contained in these constraints. This involves the creation of estimation procedures for each physical property used in the design constraints. The target transformation section provides facilities for collecting estimation techniques into estimation procedures. The chosen

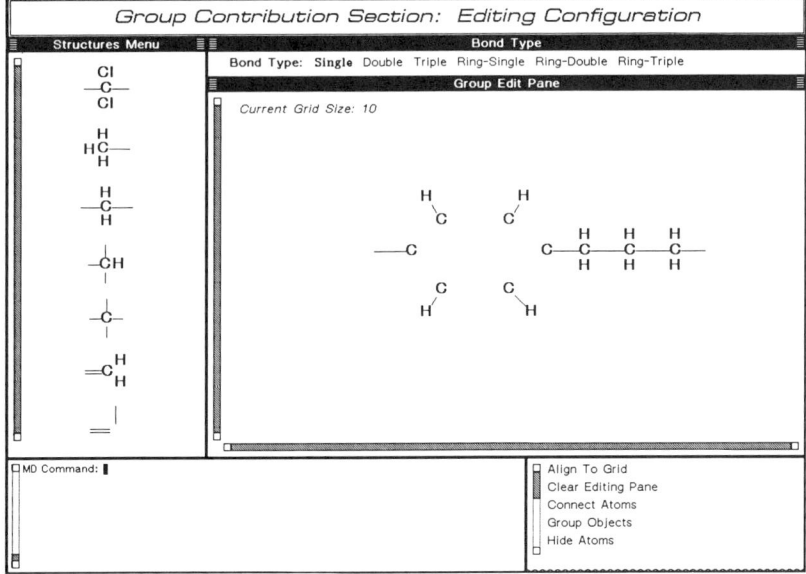

FIG. 9. Group contribution editing, configuration screen.

estimation techniques are used by the system to form an evaluation function to determine the feasibility of each point in a design space.

Figure 10 shows an example screen of the target transformation section. The constraints being transformed are shown in the window on the left. The estimation techniques that the designer can choose are shown in the window on the right. Currently the system has over 80 estimation techniques stored in its database.

4. Interactive Design Section

The interactive design section displays design spaces and group vectors. The configuration provides facilities to assist in selecting groups and in manipulating design constraints. Group vectors can be sorted by angle, magnitude, or closeness to another vector. The groups available can be restricted to those satisfying constraints on global valence, bond or atom types, and designer preferences. Each constraint can be interactively respecified enabling a designer to investigate the sensitivity of the feasible region. Example screens were previously shown when discussing the case studies in Sections III.B–D.

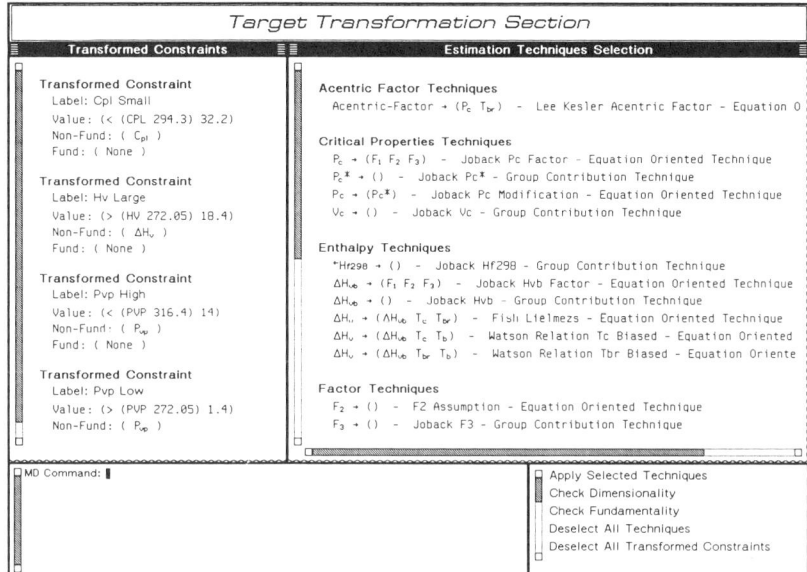

FIG. 10. Target transformation section, configuration screen.

5. Molecule Evaluation Section

The evaluation section provides facilities for estimating a molecules' physical properties. Estimating physical properties is especially useful when formulating the design target and evaluating designed molecules. During problem formulation we may need to estimate the properties of compounds currently in use. In final evaluation we would like to have the property profile of the designed compounds available for inspection.

One of the major objectives of the evaluation section is to provide estimation techniques of the highest accuracy. These estimation techniques may not be appropriate for use in the design procedures. They would thus serve as an additional check to verify the efficacy of any molecules designed.

V. Concluding Remarks

The identification of molecules with desired physical properties' values is becoming at an increasing frequency, a deliberate task with a dominant feedforward design philosophy and a decreasing number of feedback adjustments, coming from experimental results. This attitude will be reinforced as new theoretical and empirical physical property estimation techniques will provide improved estimates.

Using functional groups as the essential building blocks, one may compose any molecular structure and simultaneously employ group contribution techniques for the estimation of the molecule's properties. Clearly, a common alphabet for (1) the description of molecules and (2) the evaluation of their properties, is the most efficient but not necessarily the only alternative. The best representation for the design of molecules will be determined by the representation offering the best property estimation techniques, provided that it is at a higher-than-atomic level. Functional groups and group contribution techniques offer, at the present, the only available, consistent framework.

Once the representation of a molecular structure has been resolved, selecting the molecule(s), which satisfies a set of constraints, is simply a task of searching through a large space of discrete solutions. One could bring forth any technique for solving such problem, e.g., specific branch-and-bound algorithm to solve the underlying MINLP problem. Instead of *one, implicitly located solution*, we have opted for a search strategy that *explicitly locates all feasible solutions*. Such a preference has been motivated by the fact that during the initial design phase one is interested in all options rather than just one. Several additional considerations will be taken into account before a final choice is made; considerations that are not normally included during the initial phase of the design. To tame the combinatorial complexity of the design problem, we have used a hierarchical strategy with successive refinement of molecular representation. So, the clearly infeasible molecules can be easily screened out fairly early on. As more detail is added to the molecular representations, if the number of retained molecules is still very high, then, clearly, the initial design specs are fairly loose and should be tightened. Also, as the number of alternative molecules satisfying the set of property constraints decreases as a result of systematic search, more advanced techniques can be employed for the estimation of physical properties, and increased experimentation offers high returns.

As the limitations of the additive group-contribution techniques become more apparent, new representational models will be required to solve the product design problem. These models must maintain some simplicity in the *structure–property* relationships, which can be inverted in an efficient and explicit manner to yield the structure(s) of the feasible molecule(s). Such models have yet to be invented, but it is important to keep in mind the needs of the product design, as new theories and techniques are being written for the estimation of physical properties.

References

Alternburg, von K., Die Abhängigkeit der Siedetemperatur isomerer Kohlenwasserstoffe von der Form der Moleküle. *Brennst.-Chem.* **47**(11), 331–336 (1966).
American Society of Heating, Refrigerating and Air-Conditioning Engineers (ASHRAE), "Handbook of Fundamentals." ASHRAE, New York, 1972.
Barton, A. F. M., "CRC Handbook of Solubility Parameters and Other Cohesion Parameters." CRC Press, Boca Raton, FL, 1983.
Berg, L., Selecting the agent for distillation processes. *Chem. Eng. Prog.* **65**(9), 52–57 (1969).
Brignole, E. A., Bottini, S., and Gani, R., A strategy for the design and selection of solvents for separation processes. *Fluid Phase Equilib.* **29**, 125–132 (1986).
Constantinou, L., Gani, R., Fredenslund, Aa., Klein, J. A., and Wu, D. J., Computer-aided product design, problem formulation and application. *Proc. PSE'94*, Kyongju, Korea (1994).
Cramer, R. D., BC(DEF) Parameters. 1. The intrinsic dimensionality of intermolecular interactions in the liquid state. *J. Am. Chem. Soc.* **102**(6), 1837–1849 (1980a).
Cramer, R. D., BC(DEF) Parameters. 2. An empirical structure-based for the prediction of some physical properties. *J. Am. Chem. Soc.* **102**(6), 1849–1859 (1980b).
Cramer, R. D., A QSAR success story. *CHEMTECH*, December, pp. 744–747 (1980c).
Cramer, R. D., Snader, K. M., Willis, C. R., Chakrin, L. W., Thomas, J., and Sutton, B. M., Application of quantitative structure-activity relationships in the development of the antiallergic pyranenamines. *J. Med. Chem.* **22**(6), 714–725 (1979).
Derringer, G. C., and Markham, R. L., A computer-based methodology for matching polymer structures with required properties. *J. Appl. Polym. Sci.* **30**, 4609–4617 (1985).
Dillinger, T., "VLSI Engineering." Prentice-Hall, Englewood Cliffs, NJ, 1988.
Dossat, R. J., "Principles of Refrigeration." Wiley, New York, 1981.
Douglas, J. M., "Conceptual Design of Chemical Processes." McGraw-Hill, New York, 1988.
Fogiel, M., "Modern Microelectronics." Research and Education Association, New York, 1972.
Francis, A. W., Solvent selectivity for hydrocarbons. *Ind. Eng. Chem.* **36**(8), 764–771 (1944).
Franke, R., "Theoretical Drug Design Methods." Elsevier, Amsterdam, 1984.
Fredenslund, A., Gmehling, J., and Rasmussen, P., "Vapor-Liquid Equilibria using UNIFAC." Elsevier, Amsterdam, 1977.
Gani, R., and Brignole, E. A., Molecular design of solvents for liquid extraction based on UNIFAC. *Fluid Phase Equilib.* **13**, 331–340 (1983).
Gani, R., and Fredenslund, Aa., Computer-aided molecular and mixture design with specific property constraints. *Fluid Phase Equilib.* **82**, (1993).
Gani, R., Nielsen, B., and Fredenslund, Aa., A group contribution approach to computer-aided molecular design. *AIChE J.* **37**, 1318 (1991).
Godfrey, N. B., Solvent selection via miscibility number. *CHEMTECH*, June, pp. 359–363 (1972).
Goosey, M. T., Permeability of coatings and encapsulants for electronic and optoelectronic devices, *in* "Polymer Permeability" (J. Comyn, ed.), pp. 309–339. Elsevier, London, 1985.
Gordon, M., and Scantlebury, G. R., Non-random polycondensation, statistical theory of the substitution effect. *Trans. Faraday Soc.* **60**, 604–621 (1964).
Gray, N. A. B., "Computer-Assisted Structure Elucidation." Wiley, New York, 1986.
Hammett, L. P., Some relations between reaction rates and equilibrium constants. *Chem. Rev.* **17**(1), 125–136 (1935).

Hansch, C., and Leo, A., "Substituent Constants for Correlation Analysis in Chemistry and Biology." Wiley, New York, 1979.

Hansch, C., Muir, R. M., Fujita, T., Maloney, P. P., Geiger, F., and Streich, M., The correlation of biological activity of plant growth regulators and chloromycetin derivatives with Hammett constants and partition coefficients. *J. Am. Chem. Soc.* **85**, 2817–2824 (1963).

Hayes-Roth, F., Waterman, D. A., and Lenat, D. B., "Building Expert Systems." Addison-Wesley, Reading, MA, 1983.

Horvath, A. L., "Molecular Design–Chemical Structure Generation from the Properties of Pure Organic Compounds." Elsevier, Amsterdam, 1992.

Hosoya, H., and Murakami, M., Topological index as applied to p-electronic systems. II. Topological bond order. *Bull. Chem. Soc. J.* **48**(12), 3512–3517 (1975).

Joback, K. G., A unified approach to physical property estimation using multivariate statistical techniques. Master's Thesis, Massachusetts Institute of Technology, Cambridge, MA (1984).

Joback, K. G., and Reid, R. C. Estimation of pure-component properties from group-contributions. *Chem. Eng. Commun.* **57**, 233–243 (1987).

Joback, K. G., and Stephanopoulos, G. Designing molecules possessing desired physical property values. "Foundations of Computer-Aided Process Design." Elsevier, Amsterdam, 1990.

Kier, L. B., and Hall, L. H., "Molecular Connectivity in Structure-Activity Analysis." Wiley, New York, 1986.

Klincewicz, K. M., Prediction of critical temperatures, pressures, and volumes of organic compounds from molecular structure. Master's Thesis, Massachusetts Institute of Technology, Cambridge, MA, (1982).

Langley, B. C., "Refrigeration and Air Conditioning." Prentice-Hall, Englewood Cliffs, NJ, 1986.

Lo, T. C., Baird, M. H. I., and Hanson, C., "Handbook of Solvent Extraction." Wiley, New York, 1983.

Lyman, W. J., Reehl, W. F., and Rosenblatt, D. H., "Handbook of Chemical Property Estimation Methods." McGraw-Hill, New York, 1982.

Macchietto, S., Odele, O., and Omatsone, O., Design of optimal solvents for liquid–liquid extraction and gas absorption processes. *Trans. Inst. Chem. Eng.* **68**, 429 (1990).

Martin, Y. C., "Quantitative Drug Design: A Critical Introduction." Dekker, Inc., New York, 1978.

Mih, W. C., Catalysts for epoxy molding compounds in microelectronic encapsulation. *ACS Symp. Ser.* **242**, 273–283 (1984).

Moore, R. E., "Methods and Applications of Interval Analysis." Society for Industrial and Applied Mathematics, Philadelphia, 1979.

Nielsen, B., Gani, R., and Fredenslund, Aa., A group contribution approach to computer aided molecular design. *AIChE J.* (1995) (in press).

Odele, O., and Macchietto, S., Computer-aided molecular design, a novel method for optimal solvent selection. *Fluid Phase Equilib.* **82**, 47 (1993).

Perry, R. H., and Chilton, C. H., "Chemical Engineers' Handbook." McGraw-Hill, New York, 1973.

Randić, M., On characterization of molecular branching. *J. Am. Chem. Soc.* **97**(23), 6609–6615 (1975).

Reid, R. C., Prausnitz, J. M., and Poling, B. E., "The Properties of Gases and Liquid." McGraw-Hill, New York, 1987.

Rouvray, D. H., Predicting chemistry from topology. *Sci. Am.* **255**(3), 40–47 (1986).

Salame, M., Prediction of gas barrier properties of high polymers. *Polym. Eng. Sci.* **26**(33), 1543–1546 (1986).
Stephanopoulos, G., and Townsend, D. W., Synthesis in process development. *Chem. Eng. Res. Dev.* **64**, 160–174 (1986).
Taft, R. W., Separation of polar, steric, and resonance effects in reactivity. *In* "Steric Effects in Organic Chemistry" (M. S. Newman, ed.). Wiley, New York, 1956.
Tortorello, A., and Kinsella, M. A., Solubility parameter concept in the design of polymers for high performance coatings. I. *J. Coat. Technol.* **55**(696), 99–38 (1983a).
Torterello, A., and Kinsella, M. A., Solubility parameter concept in the design of polymers for high performance coatings, II. *J. Coat. Technol.* **55**(697), 29–38 (1983b).
van Krevelen, D. W., "Properties of Polymers, Correlations with Chemical Structure." Elsevier, Amsterdam (1976).
Venkatasubramanian, V., Chan, K., and Carathers, J. M., Computer-aided molecular design using genetic algorithms. *Comput. Chem. Eng.* **18**, 883 (1994).
Verloop, A., The use of linear free energy parameters and other experimental constants in structure-activity studies. *In* "Drug Design" (E. J. Ariëns, ed.) Academic Press, New York, 1972.
Watson, K. M., Thermodynamics of the liquid state. *Ind. Eng. Chem.* **35**, 398–405 (1943).
Wiener, H., Correlation of heats of isomerization, and differences in heats of vaporization of isomers, among the paraffin hydrocarbons. *J. Am. Chem. Soc.* **69**(11), 2636–2638 (1947).

The index for volume 21 appears in volume 22.

ISBN 0-12-008521-6